T0310524

FILTERING, CONTROL AND FAULT DETECTION WITH RANDOMLY OCCURRING INCOMPLETE INFORMATION

FILTERING, CONTROL AND FAULT DETECTION WITH RANDOMLY OCCURRING INCOMPLETE INFORMATION

Hongli Dong
Northeast Petroleum University, China

Zidong Wang
Brunel University, UK

Huijun Gao
Harbin Institute of Technology, China

WILEY

Library of Congress Cataloging-in-Publication Data

Dong, Hongli, 1977–
    Filtering, control and fault detection with randomly occurring incomplete information / Hongli Dong, Zidong Wang, Huijun Gao.
        pages cm
    Includes bibliographical references and index.
    ISBN 978-1-118-64791-2 (cloth)
    1. Automatic control. 2. Electric filters, Digital. 3. Fault tolerance (Engineering) I. Wang, Zidong, 1966–
II. Gao, Huijun. III. Title.
    TJ213.D655 2013
    003′.75–dc23

                                                                                        2013007460

A catalogue record for this book is available from the British Library

ISBN 9781118647912

Typeset in 10/12pt Times by Aptara Inc., New Delhi, India

Printed and bound in Singapore by Markono Print Media Pte Ltd

1   2013

The time is boring without random occurrence

The research is monotonous without incomplete information

The life is tedious without fault detection

The living is tough without noise filtering

The power is nothing without control

This book is dedicated to the Dream Dynasty, consisting of a group of simple yet happy people who are falling in love with both the random incompleteness and the incomplete randomness by detecting the faults, filtering the noises, and controlling the powers . . .

Contents

Preface

In the context of systems and control, incomplete information refers to a dynamical system in which knowledge about the system states is limited due to the difficulties in modeling complexity in a quantitative way. The well-known types of incomplete information include parameter uncertainties and norm-bounded nonlinearities. Recently, in response to the development of network technologies, the phenomenon of randomly occurring incomplete information has become more and more prevalent. Such a phenomenon typically appears in a networked environment. Examples include, but are not limited to, randomly varying nonlinearities (RVNs), randomly occurring mixed time-delays (ROMDs), randomly occurring multiple time-varying communication delays (ROMTCDs), and randomly occurring quantization errors (ROQEs). Randomly occurring incomplete information, if not properly handled, would seriously deteriorate the performance of a control system.

In this book, we investigate the filtering, control, and fault detection problems for several classes of nonlinear systems with randomly occurring incomplete information. Some new concepts are proposed which include RVNs, ROMDs, ROMTCDs, and ROQEs. The incomplete information under consideration mainly includes missing measurements, time delays, sensor and actuator saturations, quantization effects, and time-varying nonlinearities. The content of this book can be divided into three parts. In the first part, we focus on the filtering, control, and fault detection problems for several classes of nonlinear stochastic discrete-time systems with missing measurements, sensor and actuator saturations, RVNs, ROMDs, and ROQEs. Some sufficient conditions are derived for the existence of the desired filters, controllers, and fault detection filters by developing new techniques for the considered nonlinear stochastic systems. In the second part, the theories and techniques developed in the previous part are extended to deal with distributed filtering issues over sensor networks, and some distributed filters are designed for nonlinear time-varying systems and Markovian jump nonlinear time-delay systems. Finally, we apply a new stochastic H_∞ filtering approach to study the mobile robot localization problem, which shows the promising application potential of our main results.

The book is organized as follows. Chapter 1 introduces some recent advances on the analysis and synthesis problems with randomly occurring incomplete information. The developments of the filtering, control, and fault detection problems are systematically reviewed, and the research problems to be addressed in each individual chapter are also outlined. Chapter 2 is concerned with the finite-horizon filtering and control problems for nonlinear time-varying stochastic systems where sensor and actuator saturations, variance-constrained and missing measurements are considered. In Chapters 3 and 4, the H_∞ filtering and control problems are addressed for several classes of nonlinear discrete systems where ROMTCDs and multiple packet dropouts are taken into account. Chapter 5 investigates the robust H_∞ filtering and fault detection problems for nonlinear Markovian jump systems with sensor saturation and RVNs. In Chapter 6, the fault detection problem is considered for two classes of discrete-time systems with randomly occurring nonlinearities, ROMDs, successive packet dropouts and measurement quantizations. Chapters 7, 8, and 9 discuss the distributed H_∞ filtering problem over sensor networks. In Chapter 10, a new stochastic H_∞ filtering approach is proposed to deal with the localization problem of the mobile robots modeled by a class of discrete nonlinear time-varying systems subject to missing measurements and quantization effects. Chapter 11 summarizes the results of the book and discusses some future work to be investigated further.

This book is a research monograph whose intended audience is graduate and postgraduate students and researchers.

Acknowledgments

We would like to express our deep appreciation to those who have been directly involved in various aspects of the research leading to this book. Special thanks go to Professor Daniel W. C. Ho from City University of Hong Kong, Professor James Lam from the University of Hong Kong, Professor Xiaohui Liu from Brunel University in the UK, Professor Steven X. Ding from the University of Duisburg–Essen in Germany, and Professor Ligang Wu from Harbin Institute of Technology of China for their valuable suggestions, constructive comments, and support. We also extend our thanks to the many colleagues who have offered support and encouragement throughout this research effort. In particular, we would like to acknowledge the contributions from Bo Shen, Liang Hu, Jun Hu, Yurong Liu, Jinling Liang, Guoliang Wei, Xiao He, Lifeng Ma, Derui Ding, Yao Wang, Xiu Kan, Sunjie Zhang, and Nianyin Zeng. Finally, we are especially grateful to our families for their encouragement and never-ending support when it was most required.

The writing of this book was supported in part by the National 973 Project under Grant 2009CB320600, the National Natural Science Foundation of China under Grants 61273156, 61134009, 61004067, and 61104125, the Engineering and Physical Sciences Research Council (EPSRC) of the UK, the Royal Society of the UK, and the Alexander von Humboldt Foundation of Germany. The support of these organizations is gratefully acknowledged.

List of Abbreviations

CCL	cone complementarity linearization
DFD	distributed filter design
DKF	distributed Kalman filtering
FHFD	finite-horizon H_∞ filter design
HCMDL	H_∞ control with multiple data losses
HFDL	H_∞ filtering with data loss
HinfFC	H_∞ fuzzy control
HinfF	H_∞ filtering
LMI	linear matrix inequality
MJS	Markovian jump system
NCS	networked control system
OFDFD	optimized fault detection filter design
RFD	robust filter design
RHFD	robust H_∞ filter design
RLMI	recursive linear matrix inequality
RMM	randomly missing measurement
ROMD	randomly occurring mixed time-delay
ROMTCD	randomly occurring multiple time-varying communication delay
ROPD	randomly occurring packet dropout
ROQE	randomly occurring quantization error
ROSS	randomly occurring sensor saturation
RVN	randomly varying nonlinearity
RDE	Riccati difference equation
SAS	sensor and actuator saturation
SPD	successive packet dropout
TP	transition probability
T–S	Takagi–Sugeno

List of Notations

$\mathbb{R}^n$	the n-dimensional Euclidean space
$\mathbb{R}^{n \times m}$	the set of all $n \times m$ real matrices
$\mathbb{R}^+$	the set of all nonnegative real numbers
$\mathbb{I}^+$	the set of all nonnegative integers
$\mathbb{Z}^-$	the set of all negative integers
OL	the class of all continuous nondecreasing convex functions $\phi : \mathbb{R}^+ \to \mathbb{R}^+$ such that $\phi(0) = 0$ and $\phi(r) > 0$ for $r > 0$
$\|A\|$	the norm of matrix A defined by $\|A\| = \sqrt{\mathrm{tr}(A^T A)}$
A^T	the transpose of the matrix A
$A^\dagger \in \mathbb{R}^{n \times m}$	the Moore–Penrose pseudo inverse of $A \in \mathbb{R}^{m \times n}$
I	the identity matrix of compatible dimension
0	the zero matrix of compatible dimension
$\mathrm{Prob}(\cdot)$	the occurrence probability of the event "$\cdot$"
$\mathbb{E}\{x\}$	the expectation of the stochastic variable x
$\mathbb{E}\{x|y\}$	the expectation of the stochastic variable x conditional on y
$(\Omega, \mathcal{F}, \mathrm{Prob})$	the complete probability space
$\lambda_{\min}(A)$	the smallest eigenvalue of a square matrix A
$\lambda_{\max}(A)$	the largest eigenvalue of a square matrix A
$*$	the ellipsis for terms induced by symmetry, in symmetric block matrices
$\mathrm{diag}\{\cdots\}$	the block-diagonal matrix
$l_2[0, \infty)$	the space of square summable sequences
$\|\cdot\|_2$	the usual l_2 norm
$\mathrm{tr}(A)$	the trace of a matrix A
$\min \mathrm{tr}(A)$	the minimization of $\mathrm{tr}(A)$
$\mathrm{Var}\{x_i\}$	the variance of x_i
$\otimes$	the Kronecker product
$\mathbf{1}_n$	$\mathbf{1}_n = [1, 1, \ldots, 1]^T \in \mathbb{R}^n$
e_i	$e_i = [\underbrace{0, \ldots, 0}_{i-1}, 1, \underbrace{0, \ldots, 0}_{n-i}]^T$
$X > Y$	the $X - Y$ is positive definite, where X and Y are real symmetric matrices
$X \geq Y$	the $X - Y$ is positive semi-definite, where X and Y are real symmetric matrices

1

Introduction

In the past decade, networked control systems (NCSs) have attracted much attention owing to their successful applications in a wide range of areas for the advantage of decreasing the hard-wiring, the installation cost, and implementation difficulties. Nevertheless, network-related challenging problems inevitably arise due to the physical equipment constraints, the complexity, and uncertainty of the external environment in the process of modeling or information transmission, which would drastically degrade the system performance. Such network-induced problems include, but are not limited to, missing measurements, communication delays, sensor and actuator saturations, signal quantization, and randomly varying nonlinearities. These phenomena may occur in a probabilistic way that is customarily referred to as randomly occurring incomplete information.

For several decades, nonlinear analysis and stochastic analysis have arguably been two of the most active research areas in systems and control. This is simply because (1) nonlinear control problems are of interest to engineers, physicists, and mathematicians as most physical systems are inherently nonlinear in nature, and (2) stochastic modeling has come to play an important role in many branches of science and industry as many real-world system and natural processes may be subject to stochastic disturbances. There has been a rich literature on the general nonlinear stochastic control problems. A great number of techniques have been developed on filtering, control, and fault detection problems for nonlinear stochastic systems in order to meet the needs of practical engineering. Recently, with the development of NCSs, the analysis and synthesis problems for nonlinear stochastic systems with the aforementioned network-induced phenomena have become interesting and imperative, yet challenging, topics. Therefore, the aim of this book is to deal with the filtering, control, and fault detection problems for nonlinear stochastic systems with randomly occurring incomplete information.

The focus of this chapter is to provide a timely review on the recent advances of the analysis and synthesis issues for complex systems with randomly occurring incomplete information. Most commonly used methods for modeling randomly occurring incomplete information are summarized. Based on the models established, various filtering, control, and fault detection problems with randomly occurring incomplete information are discussed in great detail. Subsequently, some challenging issues for future research are pointed out. Finally, we give the outline of this book.

Filtering, Control and Fault Detection with Randomly Occurring Incomplete Information, First Edition.
Hongli Dong, Zidong Wang, and Huijun Gao.
© 2013 John Wiley & Sons, Ltd. Published 2013 by John Wiley & Sons, Ltd.

1.1 Background, Motivations, and Research Problems

1.1.1 Randomly Occurring Incomplete Information

Accompanied by the rapid development of communication and computer technology, NCSs have become more and more popular for their successful applications in modern complicated industry processes, e.g., aircraft and space shuttle, nuclear power stations, high-performance automobiles, etc. However, the insertion of network makes the analysis and synthesis problems much more complex due to the randomly occurring incomplete information that is mainly caused by the limited bandwidth of the digital communication channel. The randomly occurring incomplete information under consideration mainly includes randomly missing measurements (RMMs), randomly occurring communication delays, sensor and actuator saturations (SASs), randomly occurring quantization and randomly varying nonlinearities (RVNs).

Missing Measurements

In practical systems within a networked environment, measurement signals are usually subject to missing probabilistic information (data dropouts or packet losses). This may be caused for a variety of reasons, such as the high maneuverability of the tracked target, a fault in the measurement, intermittent sensor failures, network congestion, accidental loss of some collected data, or some of the data may be jammed or coming from a very noisy environment, and so on. Such a missing measurement phenomenon that typically occurs in NCSs has attracted considerable attention during the past few years; see Refs [1–24] and the references cited therein. Various approaches have been presented in the literature to model the packet dropout phenomenon. For example, the data packet dropout phenomenon has been described as a binary switching sequence that is specified by a conditional probability distribution taking on values of 0 or 1 [25, 26]. A discrete-time linear system with Markovian jumping parameters was employed by Shu *et al.* [27] and Xiong and Lam [28] to construct the random packet dropout model. A model that comprises former measurement information of the process output was introduced by Sahebsara *et al.* [29] to account for the successive packet dropout phenomenon. A model of multiple missing measurements was proposed by Wei *et al.* [18] using a diagonal matrix to describe the different missing probability for individual sensors.

Communication Delays

Owing to the fact that time delays commonly reside in practical systems and constitute a main source for system performance degradation or even instability, the past decade has witnessed significant progress on analysis and synthesis for systems with various types of delays, and a large amount of literature has appeared on the general topic of time-delay systems. For example, the stability of NCSs under a network-induced delay was studied by Zhao *et al.* [30] using a hybrid system technique. The optimal stochastic control method was proposed by Nilsson [31] to control the communication delays in NCSs. A networked controller was designed in the frequency domain using robust control theory by Gokas [32] in which the network delays were considered as an uncertainty. However, most of the relevant literature mentioned above has focused on the *constant time-delays*. Delays resulting from network transmissions are inherently *random* and *time varying* [33–41]. This is particularly true when signals are transmitted over the internet and, therefore, existing control methods for constant time-delay cannot be utilized directly [42]. Recently, some researchers have started to model the network-induced time delays in multi-form probabilistic ways and, accordingly, some initial results have been reported. For example, the random communication delays have been modeled as Markov chains and the resulting closed-loop systems have been represented as Markovian jump linear systems with two jumping parameters [43, 44]. Two kinds of random delays, which happen in the channels from the controller to the plant and from the sensor to the controller, were simultaneously considered by Yang *et al.* [45]. The random delays were modeled by Yang *et al.* [45] as a linear function of the stochastic variable satisfying a Bernoulli random binary distribution. Different from Yang *et al.* [45], the problem of stability analysis and stabilization control design was studied by Yue *et al.* [46] for Takagi–Sugeno (T–S) fuzzy systems with probabilistic interval delay, and the Bernoulli distributed sequence was utilized to describe the probability distribution of the time-varying delay taking values in an interval. It should be mentioned that, among others, the binary representation of the random delays has been fairly popular because of its practicality and simplicity in describing communication delays.

However, most research attention has been centered on the *single* random delay having a *fixed* value if it occurs. This would lead to conservative results or even degradation of the system performance since, at a certain time, the NCSs could give rise to multiple time-varying delays but with different occurrence probabilities. Therefore, a more advanced methodology is needed to handle time-varying network-induced time delays in a closed-loop control system.

Signal Quantization

As is well known, quantization always exists in computer-based control systems employing finite-precision arithmetic. Moreover, the performance of NCSs will be inevitably subject to the effect of quantization error owing to the limited network bandwidth caused possibly by strong signal attenuation and perturbation in the operational environment. Hence, the quantization problem of NCSs has long been studied and many important results have been reported; see Refs [47–64] and references cited therein. For example, in Brockett and Liberzon [65], the time-varying quantization strategy was first proposed where the number of quantization levels is fixed and finite while at the same time the quantization resolution can be manipulated over time. The problem of input-to-state stable with respect to the quantization error for nonlinear continuous-time systems has been studied by Liberzon [66]. In this framework, the effect of quantization is treated as an additional disturbance whose effect is overcome by a Lyapunov redesign of the control law. A switching control strategy with dwell time was proposed by Ishii and Francis [67] to use as a quantizer for single-input systems. The quantizer employed in this framework is in fact an extension of the static logarithmic quantizer in [68] to the continuous case. So far, there have been mainly two different types of quantized communication models adopted in the literature: uniform quantization [62–64] and logarithmic quantization [56–59, 61]. It has been proved that, compared with a uniform quantizer, logarithmic quantization is more preferable since fewer bits need to be communicated. A sector bound scheme to handle the logarithmic quantization effects in feedback control systems was proposed by Fu and Xie [69], and such an elegant scheme was then extensively employed later on; for example, see Refs [58, 70, 71] and references cited therein. However, we note that the methods in most of the references cited above could not be directly applied to NCSs, because in NCSs the effects of network-included delay and packet dropout should also be considered.

Sensor and Actuator Saturations

In practical control systems, sensors and actuators cannot provide unlimited amplitude signal due primarily to the physical, safety, or technological constraints. In fact, actuator/sensor saturation is probably the most common nonlinearity encountered in practical control systems, which can degrade the system performance or even cause instability if such a nonlinearity is ignored in the controller/filter design. Because of their theoretical significance and practical importance, considerable attention has been focused on the filtering and control problems for systems with *actuator saturation* [72–82]. As for *sensor saturation*, the associated results have been relatively few due probably to the technical difficulty [83–88]. Nevertheless, in the scattered literature regarding sensor saturation, it has been implicitly assumed that the occurrence of sensor saturations is deterministic; that is, the sensor always undergoes saturation. Such an assumption, however, does have its limitations, especially in a sensor network. The sensor saturations may occur in a probabilistic way and are randomly changeable in terms of their types and/or levels due to the random occurrence of networked-induced phenomena such as random sensor failures, sensor aging, or sudden environment changes. To reflect the reality in networked sensors, it would make practical sense to consider the randomly occurring sensor saturations (ROSSs) where the occurrence probability can be estimated via statistical tests. Also, it should be mentioned that very few results have dealt with the systems with simultaneous presence of actuator and sensor saturations [89], although such a presence is quite typical in engineering practice.

Randomly Varying Nonlinearities

It is well known that nonlinearities exist universally in practice, and it is quite common to describe them as additive nonlinear disturbances that are caused by environmental circumstances. In a networked system such as the internet-based three-tank system for leakage fault diagnosis, such nonlinear disturbances may occur in a probabilistic way due to the random occurrence of a networked-induced phenomenon. For example, in a particular moment, the transmission channel for a large amount of packets may encounter severe network-induced congestions due to the bandwidth limitations, and the resulting phenomenon could be reflected by certain randomly occurring nonlinearities where the occurrence probability can be estimated via statistical tests. As discussed in Refs [90–93], in the NCSs that are prevalent nowadays, the nonlinear disturbances themselves may experience random abrupt changes due to random changes and failures arising from networked-induced phenomena, which give rise to the so-called RVNs. In other words, the type and intensity of the so-called RVNs could be changeable in a probabilistic way.

1.1.2 The Analysis and Synthesis of Nonlinear Stochastic Systems

For several decades now, stochastic systems have received considerable research attention in which stochastic differential equations are the most useful stochastic models with broad applications in aircraft, chemical, or process control systems and distributed networks. Generally speaking, stochastic systems can be categorized into two types, namely internal stochastic systems and external stochastic systems [94].

As a class of internal stochastic systems with finite operation modes, Markovian jump systems (MJSs) have received particular research interest in the past two decades because of their practical applications in a variety of areas, such as power systems, control systems of a solar thermal central receiver, NCSs, manufacturing systems, and financial markets. So far, the existing results for MJSs have covered a wide range of research problems, including those for stability analysis [95–97], filter design [98–104], and controller design [105, 106]. Nevertheless, compared with the fruitful results for MJSs for filtering and control problems, MJS use for the corresponding fault detection problem has received much less attention [107, 108], due primarily to the difficulty in accommodating the multiple fault detection performances. In the literature concerning MJSs, most results have been reported by supposing that the transition probabilities (TPs) in the jumping process are completely accessible. However, this is not always true for many practical systems. For example, in NCSs, it would be extremely difficult to obtain precisely all the TPs via time-consuming yet expensive statistical tests. In other words, some of TPs are very likely to be incomplete (i.e., uncertain or even unknown). So far, some initial efforts have been made to address the incomplete probability issue for MJSs. For example, the problems of uncertain TPs and partially unknown TPs have been addressed by Xiong and coworkers [95, 98] and Zhang and coworkers [100, 109], respectively. Furthermore, the concept of deficient statistics for modes transitions has been put forward [110] to reflect different levels of the limitations in acquiring accurate TPs. Unfortunately, the filtering/control/fault detection problem for discrete-time MJSs with RVNs has yet to be fully investigated.

For external stochastic systems, stochasticity is always caused by an external stochastic noise signal, and can be modeled by stochastic differential equations with stochastic processes [94, 111]. Furthermore, recognizing that nonlinearities exist universally in practice and both nonlinearity and stochasticity are commonly encountered in engineering practice, the robust H_∞ filtering, H_∞ control, and fault detection problems for nonlinear stochastic systems have stirred a great deal of research interest. For the fault detection problems, we refer the readers to [82, 112–114] and references cited therein. With respect to the H_∞ control and filtering problems, we mention the following representative work. The stochastic H_∞ filtering problem for time-delay systems subject to sensor nonlinearities has been dealt with by Wang and coworkers [115, 116]. The robust stability and controller design problems for NCSs with uncertain parameters have been studied by Zhang *et al.* [44] and Jiang and Han [117], respectively. The stability issue was addressed by Wang *et al.* [118] for a class of T–S fuzzy dynamical systems with time delays and uncertain parameters. In Zhang *et al.* [119], the robust H_∞ filtering problem for affine nonlinear stochastic systems with state and external disturbance-dependent noise was studied, where the filter can be designed by solving second-order nonlinear Hamilton–Jacobi inequalities. So far, in comparison with the fruitful literature available for continuous-time systems, the corresponding H_∞ filtering results for discrete-time systems has been relatively sparse. Also, to the best of our knowledge, the analysis and design problems for *nonlinear discrete-time stochastic systems with randomly occurring incomplete information* have not been properly investigated yet, and constitutes the main motivation for this book.

1.1.3   Distributed Filtering over Sensor Networks

In the past decade, sensor networks have attracted increasing attention from many researchers in different disciplines owing to the extensive applications of sensor networks in many areas, including in surveillance, environment monitoring, information collection, industrial automation, and wireless networks [120–127]. A sensor network typically consists of a large number of sensor nodes and also a few control nodes, all of which are distributed over a spatial region. The distributed filtering or estimation, as an important issue for sensor networks, has been an area of active research for many years. Different from the traditional filtering for a single sensor [111, 103, 128, 129], the information available for the filter algorithm on an individual node of a sensor network is not only from its own measurement, but also from its neighboring sensors' measurements according to the given topology. As such, the objective of filtering based on a sensor network can be achieved in a distributed yet collaborative way. It is noted that one of the main challenges for distributed filtering lies in how to handle the complicated coupling issues between one sensor and its neighboring sensors.

In recent years, the distributed filtering problem for sensor networks has received considerable research interest and a lot of research results have been available in the literature; for example, see Refs [122–124, 126, 130–142]. The distributed diffusion filtering strategy was established by Cattivelli and Sayed [140, 122] for the design of distributed Kalman filters and smoothers, where the information is diffused across the network through a sequence of Kalman iterations and data aggregation. A distributed Kalman filtering (DKF) algorithm was introduced by Olfati-Saber and Shamma [142], through which a crucial part of the solution is used to estimate the average of n signals in a distributed way. Furthermore, three novel DKF algorithms were introduced by Olfati-Saber [141], with the first one being a modification of the previous DKF algorithm [142]. Olfati-Saber also rigorously derived and analyzed a continuous-time DKF algorithm [141] and the corresponding extension to the discrete-time setting [124], which included an optimality and stability analysis.

It should be pointed out that, so far, most reported distributed filter algorithms for sensor networks have been mainly based on the traditional Kalman filtering theory that requires exact information about the plant model. In the presence of unavoidable parameter drifts and external disturbances, a desired distributed filtering scheme should be made as robust as possible. However, the robust performance of the available distributed filters has not yet been thoroughly studied, and this would inevitably restrict the application potential in practical engineering. Therefore, it is of great significance to introduce the H_∞ performance requirement with the hope to enhance the disturbance rejection attenuation level of designed distributed filters. Note that some initial efforts have been made to address the robustness issue. Very recently, a new distributed H_∞-consensus performance was proposed by Shen *et al.* [143] to quantify the consensus degree over a finite-horizon and the distributed filtering problem has been addressed for a class of linear time-varying systems in the sensor network, and the filter parameters were designed recursively by resorting to the difference linear matrix inequalities (LMIs). Ugrinovskii [144] included an H_∞-type performance measure of disagreement between adjacent nodes of the network and a robust filtering approach was proposed to design the distributed filters for uncertain plants. On the other hand, since nonlinearities are ubiquitous in practice, it is necessary to consider the distributed filtering problem for target plants described by nonlinear systems.

Unfortunately, up to now, the distributed nonlinear H_∞ filtering problem for sensor networks has gained very little research attention despite its practical importance, and it remains as a challenging research topic.

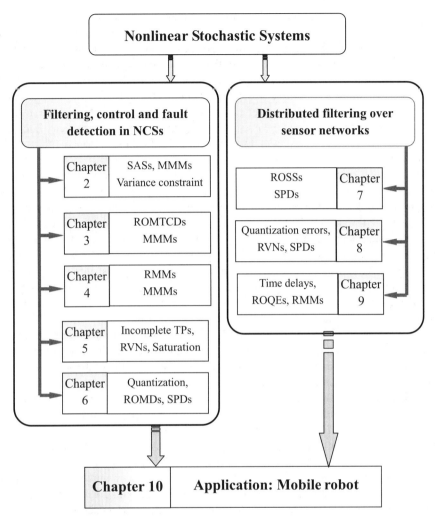

Figure 1.1 Organizational structure of the book (see List of Abbreviations for the meanings of the abbreviations)

1.2   Outline

The organization of this book is shown in Figure 1.1 and the outline of the book is as follows:

1. Chapter 1 has introduced the research background, motivations, and research problems (mainly involving incomplete information, nonlinear stochastic systems, and sensor networks), and concludes by presenting the outline of the book.
2. Chapter 2 addresses the robust H_∞ finite-horizon filtering and output feedback control problems for a class of uncertain discrete stochastic nonlinear time-varying systems with

sensor and actuator saturations, error variance constraints, and multiple missing measurements. In the system under investigation, all the system parameters are allowed to be time varying, and the stochastic nonlinearities are described by statistical means which can cover several classes of well-studied nonlinearities. First, we develop a new robust H_∞ filtering technique for the nonlinear discrete time-varying stochastic systems with norm-bounded uncertainties, multiple missing measurements, and error variance constraints. Sufficient conditions are derived for a finite-horizon filter to satisfy both the estimation error variance constraints and the prescribed H_∞ performance requirement. Such a technique relies on the forward solution to a set of recursive linear matrix inequalities (RLMIs) and, therefore, is suitable for online computation. Second, the corresponding robust H_∞ finite-horizon output feedback control problem is investigated for nonlinear system with both sensor and actuator saturations. An RLMI approach is employed to design the desired output feedback controller achieving the prescribed H_∞ disturbance rejection level.

3. In Chapter 3, the robust H_∞ filtering and control problems are studied for two classes of uncertain nonlinear systems with both multiple stochastic time-varying communication delays and multiple packet dropouts. A sequence of random variables, all of which are mutually independent but obey a Bernoulli distribution, are first introduced to account for the randomly occurring communication delays. The packet dropout phenomenon occurs in a random way and the occurrence probability for each sensor is governed by an individual random variable satisfying a certain probabilistic distribution on the interval [0, 1]. First, the robust H_∞ filtering problem is investigated for the discrete-time system with parameter uncertainties, state-dependent stochastic disturbances, and sector-bounded nonlinearities. Intensive stochastic analysis is carried out to obtain sufficient conditions for ensuring the exponential stability, as well as prescribed H_∞ performance. Furthermore, the phenomena of multiple probabilistic delays and multiple missing measurements are extended, in a parallel way, to fuzzy systems, and a set of parallel results is derived.

4. In Chapter 4, the H_∞ filtering and control problems are investigated for systems with repeated scalar nonlinearities and missing measurements. The nonlinear system is described by a discrete-time state equation involving a repeated scalar nonlinearity which typically appears in recurrent neural networks. The communication links, existing between the plant and filter, are assumed to be imperfect and a stochastic variable satisfying the Bernoulli random binary distribution is utilized to model the phenomenon of the missing measurements. The stable full- and reduced-order filters are designed such that the filtering process is stochastically stable and the filtering error satisfies the H_∞ performance constraint. Moreover, the multiple missing measurements are included to model the randomly intermittent behaviors of the individual sensors, where the missing probability for each sensor/actuator is governed by a random variable satisfying a certain probabilistic distribution on the interval [0, 1]. By employing the cone complementarity linearization procedure, the observer-based H_∞ control problem is also studied for systems with repeated scalar nonlinearities and multiple packet losses, and a set of parallel results is derived.

5. Chapter 5 addresses the filtering and fault detection problems for discrete-time MJSs with incomplete knowledge of TPs, RVNs, and sensor saturations. The issue of RVNs is first addressed to reflect the limited capacity of the communication networks resulting from the noisy environment. Two kinds of TP matrices for the Markovian process are considered: those with polytopic uncertainties and those with partially unknown entries.

Sufficient conditions are established for the existence of the desired filter satisfying the H_∞ performance constraint in terms of a set of RLMIs. The other research focus of this chapter is to investigate the fault detection problem for discrete-time MJSs with incomplete knowledge of TPs, RVNs, and sensor saturations. Two energy norm indices are used for the fault detection problem: one to account for the restraint of disturbance and the other to account for sensitivity of faults. The characterization of the gains of the desired fault detection filters is derived in terms of the solution to a convex optimization problem that can be easily solved by using the semi-definite program method.

6. Chapter 6 is concerned with the quantized fault detection problem for two classes of discrete-time nonlinear systems with stochastic mixed time-delays and successive packet dropouts. The mixed time-delays comprise both the multiple discrete time-delays and the infinite distributed delays that occur in a random way. The fault detection problem is first considered for a class of discrete-time systems with randomly occurring nonlinearities, mixed stochastic time-delays, and measurement quantizations. A sequence of stochastic variables is introduced to govern the random occurrences of the nonlinearities, discrete time-delays, and distributed time-delays, where all the stochastic variables are mutually independent but obey the Bernoulli distribution. In addition, by using similar analysis techniques, the network-based robust fault detection problem is also investigated for a class of uncertain discrete-time T–S fuzzy systems with stochastic mixed time-delays and successive packet dropouts.

7. Chapter 7 is concerned with the distributed H_∞ filtering problem for a class of nonlinear systems with ROSSs and successive packet dropouts over sensor networks. The issue of ROSSs is brought up to account for the random nature of sensor saturations in a networked environment of sensors and, accordingly, a novel sensor model is proposed to describe both the ROSSs and successive packet dropouts within a unified framework. Two sets of Bernoulli-distributed white sequences are introduced to govern the random occurrences of the sensor saturations and successive packet dropouts. Through available output measurements from not only the individual sensor but also its neighboring sensors, a sufficient condition is established for the desired distributed filter to ensure that the filtering dynamics is exponentially mean-square stable and the prescribed H_∞ performance constraint is satisfied. The solution of the distributed filter gains is characterized by solving an auxiliary convex optimization problem.

8. Chapter 8 is concerned with the distributed finite-horizon filtering problem for a class of time-varying systems over lossy sensor networks. The time-varying system (target plant) is subject to RVNs caused by environmental circumstances. The lossy sensor network suffers from quantization errors and successive packet dropouts that are described in a unified framework. Two mutually independent sets of Bernoulli-distributed white sequences are introduced to govern the random occurrences of the RVNs and successive packet dropouts. Through available output measurements from both the individual sensor and its neighboring sensors according to the given topology, a sufficient condition is established for the desired distributed finite-horizon filter to ensure that the prescribed average filtering performance constraint is satisfied. The solution of the distributed filter gains is characterized by solving a set of RLMIs.

9. Chapter 9 is concerned with the distributed H_∞ filtering problem for a class of discrete-time Markovian jump nonlinear time-delay systems with deficient statistics of modes transitions. The system measurements are collected through a lossy sensor network

subject to randomly occurring quantization errors (ROQEs) and randomly occurring packet dropouts (ROPDs). The system model (dynamical plant) includes the mode-dependent Lipschitz-like nonlinearities. The description of deficient statistics of modes transitions is comprehensive, accounting for known, unknown, and uncertain TPs. Two sets of Bernoulli-distributed white sequences are introduced to govern the phenomena of ROQEs and ROPDs in the lossy sensor network. We aim to design the distributed H_∞ filters through available system measurements from both the individual sensor and its neighboring sensors according to a given topology. The stability analysis is first carried out to obtain sufficient conditions for ensuring stochastic stability, as well as the prescribed average H_∞ performance constraint for the dynamics of the estimation errors, and then a filter design scheme is outlined by explicitly characterizing the filter gains in terms of some matrix inequalities.

10. In Chapter 10, a new stochastic H_∞ filtering approach is proposed to deal with the localization problem of the mobile robots modeled by a class of discrete nonlinear time-varying systems subject to missing measurements and quantization effects. The missing measurements are modeled via a diagonal matrix consisting of a series of mutually independent random variables satisfying certain probabilistic distributions on the interval [0, 1]. The measured output is quantized by a logarithmic quantizer. Attention is focused on the design of a stochastic H_∞ filter such that the H_∞ estimation performance is guaranteed over a given finite horizon in the simultaneous presence of plant nonlinearities (in the robot kinematic model and the distance measurements), probabilistic missing measurements, quantization effects, linearization error, and external non-Gaussian disturbances. A *necessary and sufficient* condition is first established for the existence of the desired time-varying filters in virtue of the solvability of certain coupled recursive Riccati difference equations (RDEs). Both theoretical analysis and simulation results are provided to demonstrate the effectiveness of the proposed localization approach.

11. In Chapter 11, we sum up the results of the book and discuss some related topics for future research work.

2

Variance-Constrained Finite-Horizon Filtering and Control with Saturations

This chapter addresses the robust H_∞ finite-horizon filtering and output feedback control problems for a class of uncertain discrete stochastic nonlinear time-varying systems with sensor and actuator saturations, error variance constraints, and multiple missing measurements. In the system under investigation, all the system parameters are allowed to be time varying, and the stochastic nonlinearities are described by statistical means which can cover several classes of well-studied nonlinearities. First, we develop a new robust H_∞ filtering technique for the nonlinear discrete time-varying stochastic systems with norm-bounded uncertainties, multiple missing measurements, and error variance constraints. The missing measurement phenomenon occurs in a random way, and the missing probability for each sensor is governed by an individual random variable satisfying a certain probabilistic distribution on the interval [0, 1]. Sufficient conditions are derived for a finite-horizon filter to satisfy both the estimation error variance constraints and the prescribed H_∞ performance requirement. Such a technique relies on the forward solution to a set of RLMIs and, therefore, is suitable for online computation. Second, the corresponding robust H_∞ finite-horizon output feedback control problem is investigated for such types of stochastic nonlinearities with both sensor and actuator saturations. The parameter uncertainties are assumed to be of the polytopic type. Sufficient conditions are first established for the robust H_∞ performance through intensive stochastic analysis, and then an RLMI approach is employed to design the desired output feedback controller achieving the prescribed H_∞ disturbance rejection level. Finally, some illustrative examples are exploited to show the effectiveness and applicability of the proposed filter and controller design schemes.

Filtering, Control and Fault Detection with Randomly Occurring Incomplete Information, First Edition.
Hongli Dong, Zidong Wang, and Huijun Gao.
© 2013 John Wiley & Sons, Ltd. Published 2013 by John Wiley & Sons, Ltd.

2.1   Problem Formulation for Finite-Horizon Filter Design

Consider the following discrete uncertain nonlinear time-varying stochastic system defined on $k \in [0, N]$:

$$
\begin{cases}
x_{k+1} = (A_k + \Delta A_k)x_k + f_k + D_{1k}w_k \\
y_{ck} = B_k x_k + g_k + D_{2k}w_k \\
z_k = C_k x_k
\end{cases}
, \qquad (2.1)
$$

where $x_k \in \mathbb{R}^n$ represents the state vector, $y_{ck} \in \mathbb{R}^r$ is the process output, $z_k \in \mathbb{R}^m$ is the signal to be estimated, $w_k \in \mathbb{R}^p$ is a zero-mean Gaussian white-noise sequence with covariance $W > 0$, and A_k, B_k, C_k, D_{1k}, and D_{2k} are known, real, time-varying matrices with appropriate dimensions. The parameter uncertainty ΔA_k is a real-valued time-varying matrix of the form

$$
\Delta A_k = H_k F_k E_k, \qquad (2.2)
$$

where H_k and E_k are known time-varying matrices with appropriate dimensions, and F_k is an unknown time-varying matrix satisfying $F_k F_k^{\mathrm{T}} \leq I$.

The functions $f_k = f(x_k, k)$ and $g_k = g(x_k, k)$ are stochastic nonlinear functions which are described by their statistical characteristics as follows:

$$
\mathbb{E}\left\{ \begin{bmatrix} f_k \\ g_k \end{bmatrix} \middle| x_k \right\} = 0, \qquad (2.3)
$$

$$
\mathbb{E}\left\{ \begin{bmatrix} f_k \\ g_k \end{bmatrix} \begin{bmatrix} f_j^{\mathrm{T}} & g_j^{\mathrm{T}} \end{bmatrix} \middle| x_k \right\} = 0, \quad k \neq j, \qquad (2.4)
$$

and

$$
\begin{aligned}
\mathbb{E}\left\{ \begin{bmatrix} f_k \\ g_k \end{bmatrix} \begin{bmatrix} f_k^{\mathrm{T}} & g_k^{\mathrm{T}} \end{bmatrix} \middle| x_k \right\} &= \sum_{i=1}^{q} \pi_i \pi_i^{\mathrm{T}} \mathbb{E}\left\{ x_k^{\mathrm{T}} \Gamma_i x_k \right\} \\
&:= \sum_{i=1}^{q} \begin{bmatrix} \pi_{1i} \\ \pi_{2i} \end{bmatrix} \begin{bmatrix} \pi_{1i} \\ \pi_{2i} \end{bmatrix}^{\mathrm{T}} \mathbb{E}\left\{ x_k^{\mathrm{T}} \Gamma_i x_k \right\} \\
&:= \sum_{i=1}^{q} \begin{bmatrix} \theta_{11}^i & \theta_{12}^i \\ (\theta_{12}^i)^{\mathrm{T}} & \theta_{22}^i \end{bmatrix} \mathbb{E}\left\{ x_k^{\mathrm{T}} \Gamma_i x_k \right\},
\end{aligned} \qquad (2.5)
$$

where $\pi_{1i}, \pi_{2i}, \theta_{jl}^i$, and Γ_i ($j, l = 1, 2$; $i = 1, 2, \ldots, q$) are known matrices.

Remark 2.1   *As pointed out by Yaz and coworkers [145, 146], the nonlinearity description in (2.3)–(2.5) encompasses many well-studied nonlinearities in stochastic systems such as (1) linear system with state- and control-dependent multiplicative noise; (2) nonlinear systems with random vectors dependent on the norms of states and control input; and (3) nonlinear systems with a random sequence dependent on the sign of a nonlinear function of states and control inputs.*

In this chapter, the measurement with sensor data missing is paid special attention, where the multiple missing measurements are described by

$$y_k = \Xi B_k x_k + g_k + D_{2k} w_k$$
$$= \sum_{i=1}^{r} \alpha_i B_{ki} x_k + g_k + D_{2k} w_k, \tag{2.6}$$

where $y_k \in \mathbb{R}^r$ is the actual measurement signal of the plant (2.1) and $\Xi := \text{diag}\{\alpha_1, \ldots, \alpha_r\}$, with α_i $(i = 1, \ldots, r)$ being r unrelated random variables which are also unrelated to w_k. It is assumed that α_i has the probabilistic density function $q_i(s)$ $(i = 1, \ldots, r)$ on the interval $[0, 1]$ with mathematical expectation μ_i and variance σ_i^2. B_{ki} is defined by

$$B_{ki} := \text{diag}\{\underbrace{0, \ldots, 0}_{i-1}, 1, \underbrace{0, \ldots, 0}_{r-i}\} B_k.$$

Note that α_i could satisfy any discrete probabilistic distributions on the interval $[0, 1]$, which include the widely used Bernoulli distribution as a special case. In the following, we denote $\bar{\Xi} = \mathbb{E}\{\Xi\}$.

In this section, we consider the following time-varying filter for system (2.1):

$$\begin{cases} \hat{x}_{k+1} = A_{fk} \hat{x}_k + B_{fk} y_k \\ \hat{z}_k = C_{fk} \hat{x}_k \end{cases}, \tag{2.7}$$

where $\hat{x}_k \in \mathbb{R}^n$ represents the state estimate, $\hat{z}_k \in \mathbb{R}^m$ is the estimated output, and A_{fk}, B_{fk}, and C_{fk} are appropriately dimensioned filter parameters to be determined.

Setting $\bar{x}_k = [\, x_k^T \quad \hat{x}_k^T \,]^T$, we subsequently obtain an augmented system as follows:

$$\begin{cases} \bar{x}_{k+1} = (\bar{A}_k + \check{A}_k) \bar{x}_k + \bar{G}_k h_k + \bar{D}_k w_k \\ \bar{z}_k = \bar{C}_k \bar{x}_k \end{cases}, \tag{2.8}$$

where

$$h_k = [\, f_k^T \quad g_k^T \,]^T, \quad \bar{z}_k = z_k - \hat{z}_k, \quad \bar{C}_k = [\, C_k \quad -C_{fk} \,],$$

$$\bar{A}_k = \begin{bmatrix} A_k + \Delta A_k & 0 \\ B_{fk} \bar{\Xi} B_k & A_{fk} \end{bmatrix}, \qquad \bar{G}_k = \begin{bmatrix} I & 0 \\ 0 & B_{fk} \end{bmatrix},$$

$$\check{A}_k = \begin{bmatrix} 0 & 0 \\ B_{fk}(\Xi - \bar{\Xi}) B_k & 0 \end{bmatrix}, \qquad \bar{D}_k = \begin{bmatrix} D_{1k} \\ B_{fk} D_{2k} \end{bmatrix}. \tag{2.9}$$

The state covariance matrix of the augmented system (2.8) can be defined as

$$\bar{X}_k := \mathbb{E}\{\bar{x}_k \bar{x}_k^T\} = \mathbb{E}\left\{ \begin{bmatrix} x_k \\ \hat{x}_k \end{bmatrix} \begin{bmatrix} x_k \\ \hat{x}_k \end{bmatrix}^T \right\} \tag{2.10}$$

Our aim in this chapter is to design a finite-horizon filter in the form of (2.7) such that the following two requirements are satisfied simultaneously:

(R1) For given scalar $\gamma > 0$, matrix $S > 0$, and the initial state $\bar{x}_0$, the H_∞ performance index

$$J := \mathbb{E}\left\{ \|\bar{z}_k\|_{[0,N-1]}^2 - \gamma^2 \|\omega_k\|_{[0,N-1]}^2 \right\} - \gamma^2 (x_0 - \hat{x}_0)^\mathrm{T} S(x_0 - \hat{x}_0) < 0 \quad (2.11)$$

is achieved for all admissible parameter uncertainties and all stochastic nonlinearities.
(R2) For a sequence of specified definite matrices $\{\Psi_k\}_{0 < k \leqslant N}$, at each sampling instant k, the estimation error covariance satisfies

$$\mathbb{E}\{(x_k - \hat{x}_k)(x_k - \hat{x}_k)^\mathrm{T}\} \leqslant \Psi_k, \quad \forall k. \quad (2.12)$$

Remark 2.2 *In the desired performance requirement (R2), the estimation error variance at each sampling time point is required to be not more than an individual upper bound. Note that the specified error variance constraint may not be minimal but should meet engineering requirements, which gives rise to a practically acceptable "window" with the hope to keep the estimated states within such a "window." On the other hand, since the variance constraint is relaxed from the minimum to the acceptable one, there would exist much freedom that can be used to attempt to directly achieve other desired performance requirements, such as the robustness and $\mathcal{H}_\infty$ disturbance rejection attenuation level as discussed in this chapter.*

The finite-horizon filter problem in the presence of missing measurements addressed above is referred to as the robust finite-horizon $\mathcal{H}_\infty$ filter problem for uncertain nonlinear discrete time-varying stochastic systems with variance constraint and multiple missing measurements.

2.2   Analysis of H_∞ and Covariance Performances

2.2.1 H_∞ Performance

We start by analyzing the H_∞ performance; that is, presenting sufficient conditions under which the H_∞ performance index is achieved for a given filter.

Theorem 2.2.1 *Consider the system (2.1) and suppose that the filter parameters $A_{\mathrm{f}k}$, $B_{\mathrm{f}k}$, and $C_{\mathrm{f}k}$ in (2.7) are given. For a positive scalar $\gamma > 0$ and a positive-definite matrix $S > 0$, the $\mathcal{H}_\infty$ performance requirement defined in (2.11) is achieved for all nonzero ω_k if, with the initial condition $Q_0 \leqslant \gamma^2 [I \quad -I]^\mathrm{T} S[I \quad -I]$, there exists a sequence of positive-definite matrices $\{Q_k\}_{1 \leqslant k \leqslant N+1}$ satisfying the following recursive matrix inequalities:*

$$\Lambda := \begin{bmatrix} \Lambda_1 & * \\ 0 & \bar{D}_k^\mathrm{T} Q_{k+1} \bar{D}_k - \gamma^2 I \end{bmatrix} < 0, \quad (2.13)$$

where

$$\Lambda_1 = \bar{A}_k^{\mathrm{T}} Q_{k+1} \bar{A}_k + \sum_{i=1}^{q} \hat{\Gamma}_i \cdot \mathrm{tr}[\bar{G}_k^{\mathrm{T}} Q_{k+1} \bar{G}_k \hat{\theta}_i] + \bar{C}_k^{\mathrm{T}} \bar{C}_k - Q_k + \sum_{i=1}^{r} \sigma_i^2 \bar{B}_{ki}^{\mathrm{T}} Q_{k+1} \bar{B}_{ki},$$

$$\hat{\theta}_i = \begin{bmatrix} \theta_{11}^i & \theta_{12}^i \\ (\theta_{12}^i)^{\mathrm{T}} & \theta_{22}^i \end{bmatrix}, \quad \hat{\Gamma}_i = \begin{bmatrix} \Gamma_i & 0 \\ 0 & 0 \end{bmatrix}, \quad \bar{B}_{ki} = \begin{bmatrix} 0 & 0 \\ B_{\mathrm{f}k} B_{ki} & 0 \end{bmatrix}.$$

Proof.   Define

$$J_k := \bar{x}_{k+1}^{\mathrm{T}} Q_{k+1} \bar{x}_{k+1} - \bar{x}_k^{\mathrm{T}} Q_k \bar{x}_k. \tag{2.14}$$

Noticing (2.3) and the filter structure in (2.7), we have

$$\mathbb{E}\left\{ \begin{bmatrix} f_k \\ g_k \end{bmatrix} \middle| \bar{x}_k \right\} = \mathbb{E}\{h_k | \bar{x}_k\} = 0. \tag{2.15}$$

Substituting (2.8) into (2.14), we have

$$\mathbb{E}\{J_k\} = \mathbb{E}\left\{ \bar{x}_k^{\mathrm{T}} \bar{A}_k^{\mathrm{T}} Q_{k+1} \bar{A}_k \bar{x}_k + \bar{x}_k^{\mathrm{T}} \check{A}_k^{\mathrm{T}} Q_{k+1} \check{A}_k \bar{x}_k + w_k^{\mathrm{T}} \bar{D}_k^{\mathrm{T}} Q_{k+1} \bar{D}_k w_k \right. \\ \left. + h_k^{\mathrm{T}} \bar{G}_k^{\mathrm{T}} Q_{k+1} \bar{G}_k h_k - \bar{x}_k^{\mathrm{T}} Q_k \bar{x}_k \right\}. \tag{2.16}$$

Taking (2.5) into consideration, we have

$$\mathbb{E}\left\{ h_k^{\mathrm{T}} \bar{G}_k^{\mathrm{T}} Q_{k+1} \bar{G}_k h_k \right\} = \mathbb{E}\left\{ \mathrm{tr}\left[\bar{G}_k^{\mathrm{T}} Q_{k+1} \bar{G}_k h_k h_k^{\mathrm{T}} \right] \right\} \\ = \mathbb{E}\left\{ \bar{x}_k^{\mathrm{T}} \sum_{i=1}^{q} \hat{\Gamma}_i \cdot \mathrm{tr}[\bar{G}_k^{\mathrm{T}} Q_{k+1} \bar{G}_k \hat{\theta}_i] \bar{x}_k \right\}. \tag{2.17}$$

Adding the zero term $\bar{z}_k^{\mathrm{T}} \bar{z}_k - \gamma^2 \omega_k^{\mathrm{T}} \omega_k - \bar{z}_k^{\mathrm{T}} \bar{z}_k + \gamma^2 \omega_k^{\mathrm{T}} \omega_k$ to $\mathbb{E}\{J_k\}$ results in

$$\mathbb{E}\{J_k\} = \mathbb{E}\left\{ \begin{bmatrix} \bar{x}_k^{\mathrm{T}} & \omega_k^{\mathrm{T}} \end{bmatrix} \Lambda \begin{bmatrix} \bar{x}_k \\ \omega_k \end{bmatrix} - \bar{z}_k^{\mathrm{T}} \bar{z}_k + \gamma^2 \omega_k^{\mathrm{T}} \omega_k \right\}. \tag{2.18}$$

Summing up (2.18) on both sides from 0 to $N - 1$ with respect to k, we obtain

$$
\begin{aligned}
\sum_{k=0}^{N-1} \mathbb{E}\{J_k\} &= \mathbb{E}\left\{\bar{x}_N^{\mathrm{T}} Q_N \bar{x}_N\right\} - \bar{x}_0^{\mathrm{T}} Q_0 \bar{x}_0 \\
&= \mathbb{E}\left\{\sum_{k=0}^{N-1} [\,\bar{x}_k^{\mathrm{T}} \quad \omega_k^{\mathrm{T}}\,]\Lambda \begin{bmatrix} \bar{x}_k \\ \omega_k \end{bmatrix}\right\} \\
&\quad -\mathbb{E}\left\{\sum_{k=0}^{N-1} (\bar{z}_k^{\mathrm{T}} \bar{z}_k - \gamma^2 \omega_k^{\mathrm{T}} \omega_k)\right\}.
\end{aligned}
\tag{2.19}
$$

Hence, the H_∞ performance index defined in (2.11) is given by

$$
\begin{aligned}
J &= \mathbb{E}\left\{\sum_{k=0}^{N-1} [\,\bar{x}_k^{\mathrm{T}} \quad \omega_k^{\mathrm{T}}\,]\Lambda \begin{bmatrix} \bar{x}_k \\ \omega_k \end{bmatrix}\right\} - \mathbb{E}\{\bar{x}_N^{\mathrm{T}} Q_N \bar{x}_N\} \\
&\quad + \bar{x}_0^{\mathrm{T}}(Q_0 - \gamma^2[\,I \quad -I\,]^{\mathrm{T}} S[\,I \quad -I\,])\bar{x}_0.
\end{aligned}
\tag{2.20}
$$

Noting that $\Lambda < 0$, $Q_N > 0$, and the initial condition $Q_0 \leqslant \gamma^2[\,I \quad -I\,]^{\mathrm{T}} S[\,I \quad -I\,]$, we know $J < 0$ and the proof is now complete. $\qquad\square$

2.2.2 Variance Analysis

Let us now deal with the error variance analysis issue for the nonlinear stochastic time-varying systems addressed.

Theorem 2.2.2   *Consider the system (2.1) and let the filter parameters $A_{\mathrm{f}k}$, $B_{\mathrm{f}k}$, and $C_{\mathrm{f}k}$ in (2.7) be given. We have $P_k \geqslant \bar{X}_k$ ($\forall k \in \{1, 2, \ldots, N + 1\}$) if, with initial condition $P_0 = \bar{X}_0$, there exists a sequence of positive-definite matrices $\{P_k\}_{1 \leqslant k \leqslant N+1}$ satisfying the following matrix inequality:*

$$
P_{k+1} \geqslant \Phi(P_k)
\tag{2.21}
$$

where

$$
\Phi(P_k) := \bar{A}_k P_k \bar{A}_k^{\mathrm{T}} + \sum_{i=1}^{r} \sigma_i^2 \bar{B}_{ki} P_k \bar{B}_{ki}^{\mathrm{T}} + \sum_{i=1}^{q} \bar{G}_k \hat{\theta}_i \bar{G}_k^{\mathrm{T}} \cdot \mathrm{tr}[\hat{\Gamma}_i P_k] + \bar{D}_k W_k \bar{D}_k^{\mathrm{T}}.
\tag{2.22}
$$

Proof.   As we know from (2.12), the Lyapunov-type equation that governs the evolution of state covariance $\bar{X}_k$ is given by

$$
\begin{aligned}
\bar{X}_{k+1} &= \mathbb{E}\{\bar{x}_{k+1} \bar{x}_{k+1}^{\mathrm{T}}\} \\
&= \mathbb{E}\{(\bar{A}_k \bar{x}_k + \check{A}_k \bar{x}_k + \bar{G}_k h_k + \bar{D}_k w_k) \\
&\quad \times (\bar{A}_k \bar{x}_k + \check{A}_k \bar{x}_k + \bar{G}_k h_k + \bar{D}_k w_k)^{\mathrm{T}}\}.
\end{aligned}
\tag{2.23}
$$

Since

$$\mathbb{E}\{\bar{G}_k h_k h_k^{\mathrm{T}} \bar{G}_k^{\mathrm{T}}\} = \bar{G}_k \sum_{i=1}^{q} \hat{\theta}_i \mathbb{E}(x_k^{\mathrm{T}} \Gamma_i x_k) \bar{G}_k^{\mathrm{T}}$$

$$= \sum_{i=1}^{q} \bar{G}_k \hat{\theta}_i \bar{G}_k^{\mathrm{T}} \cdot \mathrm{tr}[\hat{\Gamma}_i \bar{X}_k],$$

we obtain

$$\bar{X}_{k+1} = \bar{A}_k \bar{X}_k \bar{A}_k^{\mathrm{T}} + \sum_{i=1}^{r} \sigma_i^2 \bar{B}_{ki} \bar{X}_k \bar{B}_{ki}^{\mathrm{T}} + \bar{D}_k W_k \bar{D}_k^{\mathrm{T}} + \sum_{i=1}^{q} \bar{G}_k \hat{\theta}_i \bar{G}_k^{\mathrm{T}} \cdot \mathrm{tr}[\hat{\Gamma}_i \bar{X}_k]$$

$$= \Phi(\bar{X}_k).$$

We now complete the proof by induction. Obviously, $P_0 \geqslant \bar{X}_0$. Letting $P_k \geqslant \bar{X}_k$, we arrive at

$$P_{k+1} \geqslant \Phi(P_k) \geqslant \Phi(\bar{X}_k) = \bar{X}_{k+1}, \tag{2.24}$$

and therefore the proof is finished.   □

Furthermore, in light of Theorem 2.2.2, we have the following corollary.

Corollary 2.2.3   *The inequality holds*

$$\mathbb{E}\{(x_k - \hat{x}_k)(x_k - \hat{x}_k)^{\mathrm{T}}\} = [\, I \quad -I\,]\bar{X}_k[\, I \quad -I\,]^{\mathrm{T}}$$

$$\leq [\, I \quad -I\,]P_k[\, I \quad -I\,]^{\mathrm{T}}, \forall k.$$

To conclude the above analysis, we present a theorem which intends to take both the H_∞ performance index and the covariance constraint into consideration in a unified framework via the RLMI method.

Theorem 2.2.4   *Consider the system (2.1) and let the filter parameters $A_{\mathrm{f}k}$, $B_{\mathrm{f}k}$, and $C_{\mathrm{f}k}$ in (2.7) be given. For a positive scalar $\gamma > 0$ and a positive-definite matrix $S > 0$, if there exist families of positive-definite matrices $\{Q_k\}_{1 \leqslant k \leqslant N+1}$, $\{P_k\}_{1 \leqslant k \leqslant N+1}$, and $\{\eta_{ik}\}_{0 \leqslant k \leqslant N}$ ($i = 1, 2, \cdots, q$) satisfying the recursive matrix inequalities*

$$\begin{bmatrix} -\eta_{ik} & * \\ *\bar{G}_k \pi_i & -Q_{k+1}^{-1} \end{bmatrix} < 0, \tag{2.25}$$

$$
\begin{bmatrix}
\hat{\Lambda}_1 & * & * & * & * \\
0 & -\gamma^2 I & * & * & * \\
\bar{A}_k & 0 & -\bar{Q}_{k+1}^{-1} & * & * \\
\bar{B}_k & 0 & 0 & -\bar{Q}_{k+1}^{-1} & * \\
0 & \bar{D}_k & 0 & 0 & -\bar{Q}_{k+1}^{-1}
\end{bmatrix} < 0,
\tag{2.26}
$$

and

$$
\begin{bmatrix}
-P_{k+1} & * & * & * & * \\
P_k \bar{A}_k^{\mathrm{T}} & -P_k & * & * & * \\
\Theta_{31} & 0 & \Theta_{33} & * & * \\
\bar{D}_k^{\mathrm{T}} & 0 & 0 & -W_k^{-1} & * \\
\hat{B}_{kp}^{\mathrm{T}} & 0 & 0 & 0 & -\bar{P}_k
\end{bmatrix} \leq 0
\tag{2.27}
$$

with the initial condition

$$
\begin{cases}
Q_0 \leq \gamma^2 [\, I \quad -I \,]^{\mathrm{T}} S [\, I \quad -I \,] \\
P_0 = \bar{X}_0
\end{cases},
\tag{2.28}
$$

where

$$
\hat{\Lambda}_1 = -Q_k + \sum_{i=1}^{q} \hat{\Gamma}_i \eta_{ik} + \bar{C}_k^{\mathrm{T}} \bar{C}_k, \quad \bar{Q}_{k+1} = \mathrm{diag}\{\underbrace{Q_{k+1}, \ldots, Q_{k+1}}_{r}\},
$$

$$
\bar{B}_k = [\sigma_1 \bar{B}_{k1}^{\mathrm{T}}, \ldots, \sigma_r \bar{B}_{kr}^{\mathrm{T}}]^{\mathrm{T}}, \quad \hat{B}_{kp} = [\sigma_1 \bar{B}_{k1} P_k, \ldots, \sigma_r \bar{B}_{kr} P_k],
$$

$$
\Theta_{31} = [\bar{G}_k \pi_1, \bar{G}_k \pi_2, \ldots, \bar{G}_k \pi_q]^{\mathrm{T}}, \quad \Theta_{33} = \mathrm{diag}\{-\rho_1 I, -\rho_2 I, \ldots, -\rho_q I\},
$$

$$
\rho_i = (\mathrm{tr}[\hat{\Gamma}_i P_k])^{-1}, \quad i = 1, 2 \ldots, q, \quad \bar{P}_k = \mathrm{diag}\{\underbrace{P_{k,}, \ldots, P_k}_{r}\},
$$

then, for the filtering error system (2.8), we have $J < 0$ and $\mathbb{E}\{(x_k - \hat{x}_k)(x_k - \hat{x}_k)^{\mathrm{T}}\} \leqslant [\, I \quad -I \,] P_k [\, I \quad -I \,]^{\mathrm{T}}, \forall k \in \{0, 1, \ldots, N + 1\}$.

Proof. Based on the previous analysis of the H_∞ performance and state estimation covariance, we just need to show that, under initial conditions (2.28), the inequalities (2.25) and (2.26) imply (2.13), and the inequality (2.27) is equivalent to (2.21).

From the Schur complement lemma, (2.25) is equivalent to

$$
\pi_i^{\mathrm{T}} \bar{G}_k^{\mathrm{T}} Q_{k+1} \bar{G}_k \pi_i < \eta_{ik} \quad (i = 1, 2, \ldots, q),
\tag{2.29}
$$

which, by the property of matrix trace, can be rewritten as

$$
\mathrm{tr}\left[\bar{G}_k^{\mathrm{T}} Q_{k+1} \bar{G}_k \hat{\theta}_i\right] < \eta_{ik},
$$

and (2.26) is equivalent to

$$\hat{\Lambda} = \begin{bmatrix} \hat{\Lambda}_{11} & * \\ 0 & \bar{D}_k^T Q_{k+1} \bar{D}_k - \gamma^2 I \end{bmatrix} < 0, \tag{2.30}$$

where

$$\hat{\Lambda}_{11} = \bar{A}_k^T Q_{k+1} \bar{A}_k - Q_k + \bar{C}_k^T \bar{C}_k + \sum_{i=1}^{q} \hat{\Gamma}_i \eta_{ik} + \sum_{i=1}^{r} \sigma_i^2 \bar{B}_{ki}^T Q_{k+1} \bar{B}_{ki}.$$

Hence, it is easy to see that (2.13) can be obtained by (2.25) and (2.26) under the same initial condition.

In the same way, we can easily obtain that (2.27) is equivalent to (2.21). Thus, according to Theorem 2.2.1, Theorem 2.2.2 and Corollary 2.2.3, the H_∞ index defined in (2.20) satisfies $J < 0$ and, at the same time, the system error covariance achieves $\mathbb{E}\{(x_k - \hat{x}_k)(x_k - \hat{x}_k)^T\} \leqslant [I \quad -I]P_k[I \quad -I]^T, \forall k \in \{0, 1, \ldots, N+1\}$. The proof is complete. □

Up to now, the analysis problem has been dealt with for the $\mathcal{H}_\infty$ filtering problem for a class of uncertain nonlinear discrete time-varying stochastic systems with error variance constraints and multiple missing measurements. In the next section, we proceed to solve the filter design problem using the RLMI approach developed.

2.3   Robust Finite-Horizon Filter Design

In this section, an algorithm is proposed to cope with the addressed filter design problem for an uncertain discrete time-varying nonlinear stochastic system (2.1). It is shown that the filter matrices can be obtained by solving a certain set of RLMIs. In other words, at each sampling instant k ($k > 0$), a set of LMIs will be solved to obtain the desired filter matrices and, at the same time, certain key parameters are obtained which are needed in solving the LMIs for the $(k + 1)$th instant.

Theorem 2.3.1   *For a given disturbance attenuation level $\gamma > 0$, a positive-definite matrix $S > 0$, and a sequence of prespecified variance upper bounds $\{\Psi_k\}_{0 \leqslant k \leqslant N+1}$, if there exist families of positive-definite matrices $\{\hat{M}_k\}_{1 \leqslant k \leqslant N+1}$, $\{\hat{N}_k\}_{1 \leqslant k \leqslant N+1}$, $\{P_{1k}\}_{1 \leqslant k \leqslant N+1}$, and $\{P_{2k}\}_{1 \leqslant k \leqslant N+1}$, positive scalars $\{\varepsilon_{1k}\}_{0 \leqslant k \leqslant N}$, $\{\varepsilon_{2k}\}_{0 \leqslant k \leqslant N}$, and $\{\eta_{ik}\}_{0 \leqslant k \leqslant N}$ ($i = 1, 2, \ldots, q$), and families of real-valued matrices $\{P_{3k}\}_{1 \leqslant k \leqslant N+1}$, $\{A_{fk}\}_{0 \leqslant k \leqslant N}$, $\{B_{fk}\}_{0 \leqslant k \leqslant N}$, and $\{C_{fk}\}_{0 \leqslant k \leqslant N}$, under initial conditions*

$$\begin{cases} \begin{bmatrix} M_0 - \gamma^2 S & \gamma^2 S \\ \gamma^2 S & N_0 - \gamma^2 S \end{bmatrix} \leqslant 0 \\ \mathbb{E}\{(x_0 - \hat{x}_0)(x_0 - \hat{x}_0)^T\} = P_{10} + P_{20} - P_{30} - P_{30}^T \leq \Psi_0 \end{cases}, \tag{2.31}$$

such that the recursive LMIs

$$\begin{bmatrix} -\eta_{ik} & * & * \\ \pi_{1i} & -\hat{M}_{k+1} & * \\ B_{fk}\pi_{2i} & 0 & -\hat{N}_{k+1} \end{bmatrix} < 0, \tag{2.32}$$

$$\begin{bmatrix} \Upsilon_{11} & * \\ \Upsilon_{21} & \Upsilon_{22} \end{bmatrix} < 0, \tag{2.33}$$

$$\begin{bmatrix} \hat{\Upsilon}_{11} & * \\ \hat{\Upsilon}_{21} & \hat{\Upsilon}_{22} \end{bmatrix} < 0, \tag{2.34}$$

and

$$P_{1k+1} + P_{2k+1} - P_{3k+1} - P_{3k+1}^{\mathrm{T}} - \Psi_{k+1} \leqslant 0 \tag{2.35}$$

are satisfied with the parameters updated by

$$M_{k+1} = \hat{M}_{k+1}^{-1} \quad and \quad N_{k+1} - \hat{N}_{k+1}^{-1}, \tag{2.36}$$

where

$$\Upsilon_{11} = \begin{bmatrix} \bar{\Pi}_{11k} & * \\ \bar{\Pi}_{21k} & \bar{\Pi}_{22k} \end{bmatrix}, \quad \bar{\Pi}_{11k} = \mathrm{diag}\{\Phi_1, -N_k, -\gamma^2 I\},$$

$$\bar{\Pi}_{21k} = \begin{bmatrix} A_k & 0 & 0 \\ B_{fk}\bar{\Xi}B_k & A_{fk} & 0 \end{bmatrix}, \quad \bar{\Pi}_{22k} = \mathrm{diag}\{-\hat{M}_{k+1}, -\hat{N}_{k+1}\},$$

$$\Upsilon_{21} = \begin{bmatrix} \bar{L} & 0 & 0 & 0 & 0 \\ C_k & -C_{fk} & 0 & 0 & 0 \\ 0 & 0 & D_{1k} & 0 & 0 \\ 0 & 0 & B_{fk}D_{2k} & 0 & 0 \\ 0 & 0 & 0 & H_k^{\mathrm{T}} & 0 \end{bmatrix},$$

$$\Upsilon_{22} = \mathrm{diag}\{-\bar{M}_{k+1}, -I, -\hat{M}_{k+1}, -\hat{N}_{k+1}, -\varepsilon_{1k}I\},$$

$$\hat{\Upsilon}_{11} = \begin{bmatrix} \Phi_2 & * & * & * \\ -P_{3k+1} & -P_{2k+1} & * & * \\ P_{1k}A_k^{\mathrm{T}} & \tilde{\Omega}_{1k} & -P_{1k} & * \\ P_{3k}A_k^{\mathrm{T}} & \tilde{\Omega}_{2k} & -P_{3k} & -P_{2k} \end{bmatrix},$$

$$\hat{\Upsilon}_{21} = \begin{bmatrix} \hat{L}_{1k} & \hat{L}_{2k} & 0 & 0 \\ D_{1k}^{\mathrm{T}} & D_{2k}^{\mathrm{T}}B_{fk}^{\mathrm{T}} & 0 & 0 \\ 0 & \check{L} & 0 & 0 \\ 0 & 0 & E_k P_{1k} & E_k P_{3k}^{\mathrm{T}} \end{bmatrix},$$

$$\tilde{\Omega}_{1k} = P_{1k}B_k^{\mathrm{T}}\bar{\Xi}B_{fk}^{\mathrm{T}} + P_{3k}^{\mathrm{T}}A_{fk}^{\mathrm{T}}, \quad \tilde{\Omega}_{2k} = P_{3k}B_k^{\mathrm{T}}\bar{\Xi}B_{fk}^{\mathrm{T}} + P_{2k}A_{fk}^{\mathrm{T}},$$

$$\hat{\Upsilon}_{22} = \text{diag}\{\hat{\Theta}_{33}, -W_k^{-1}, -\hat{P}_k, -\varepsilon_{2k}I\}, \quad \Phi_1 = -M_k + \sum_{i=1}^{q} \Gamma_i \eta_{ik} + \varepsilon_{1k} E_k^{\mathrm{T}} E_k,$$

$$\Phi_2 = -P_{1k+1} + \varepsilon_{2k} H_k H_k^{\mathrm{T}}, \quad \bar{L} = [\sigma_1 \hat{\Pi}_{k1}^{\mathrm{T}}, \ldots, \sigma_r \hat{\Pi}_{kr}^{\mathrm{T}}]^{\mathrm{T}},$$

$$\bar{M}_{k+1} = \text{diag}\underbrace{\{-\bar{\Pi}_{22k}, \ldots, -\bar{\Pi}_{22k}\}}_{r}, \quad \hat{\Pi}_{ki} = \begin{bmatrix} 0 & 0 \\ j B_{fk} B_{ki} & 0 \end{bmatrix}, \quad i = 1, 2, \ldots, r,$$

$$\hat{L}_{1k} = [\pi_{11}, \pi_{12}, \ldots, \pi_{1q}]^{\mathrm{T}}, \quad \hat{L}_{2k} = \begin{bmatrix} B_{fk}\pi_{21}, B_{fk}\pi_{22}, \ldots, B_{fk}\pi_{2q} \end{bmatrix}^{\mathrm{T}},$$

$$\hat{\Theta}_{33} = \text{diag}\{-\hat{\rho}_1 I, -\hat{\rho}_2 I, \ldots, -\hat{\rho}_q I\},$$

$$\hat{\rho}_i = \left(\text{tr} \begin{bmatrix} \Gamma_i P_{1k} & \Gamma_i P_{3k}^{\mathrm{T}} \\ 0 & 0 \end{bmatrix} \right)^{-1}, \quad i = 1, 2, \ldots, q,$$

$$\hat{P}_k = \text{diag}\underbrace{\left\{ \begin{bmatrix} P_{1k} & * \\ P_{3k} & P_{2k} \end{bmatrix}, \ldots, \begin{bmatrix} P_{1k} & * \\ P_{3k} & P_{2k} \end{bmatrix} \right\}}_{r},$$

$$\check{L} = [\check{L}_1, \ldots, \check{L}_r]^{\mathrm{T}}, \quad \check{L}_j = [\sigma_j B_{fk} B_{kj} P_{1k} \quad \sigma_j B_{fk} B_{kj} P_{3k}^{\mathrm{T}}], \quad j = 1, 2, \ldots, r,$$

then the addressed robust H_∞ finite-horizon filter design problem is solved for the stochastic nonlinear system (2.1).

Proof. The proof is based on Theorem 2.2.4. We suppose that the variables P_k and Q_k can be decomposed as follows:

$$P_k = \begin{bmatrix} P_{1k} & * \\ P_{3k} & P_{2k} \end{bmatrix}, \quad Q_k = \begin{bmatrix} M_k & 0 \\ 0 & N_k \end{bmatrix}, \quad Q_k^{-1} = \begin{bmatrix} \hat{M}_k & 0 \\ 0 & \hat{N}_k \end{bmatrix}.$$

It is easy to see that (2.32) and (2.25) are equivalent.

In order to eliminate the parameter uncertainty ΔA_k in (2.26), we rewrite it in the following form:

$$\begin{bmatrix} \hat{\Lambda}_1 & * & * & * & * \\ 0 & -\gamma^2 I & * & * & * \\ \tilde{A}_k & 0 & -Q_{k+1}^{-1} & * & * \\ \bar{B}_k & 0 & 0 & -\bar{Q}_{k+1}^{-1} & * \\ 0 & \bar{D}_k & 0 & 0 & -Q_{k+1}^{-1} \end{bmatrix} + \bar{H}_k F_k \bar{E}_k + \bar{E}_k^{\mathrm{T}} F_k^{\mathrm{T}} \bar{H}_k^{\mathrm{T}} < 0, \quad (2.37)$$

where

$$\tilde{A}_k = \begin{bmatrix} A_k & 0 \\ B_{fk} \bar{\Xi} B_k & A_{fk} \end{bmatrix}, \quad \bar{H}_k = [0 \quad 0 \quad \hat{H}_k^{\mathrm{T}} \quad 0]^{\mathrm{T}}, \quad \hat{H}_k = [H_k^{\mathrm{T}} \quad 0]^{\mathrm{T}},$$

$$\bar{E}_k = [\hat{E}_k \quad 0 \quad 0 \quad 0], \quad \hat{E}_k = [E_k \quad 0].$$

Then, from Schur complement lemma and S-procedure, it follows that (2.26) is equivalent to (2.33). Similarly, we can see that (2.27) is also equivalent to (2.34). Therefore, according to Theorem 2.2.4, we have $J < 0$ and $\mathbb{E}\{(x_k - \hat{x}_k)(x_k - \hat{x}_k)^{\mathrm{T}}\} \leqslant [\, I \quad -I\,]P_k[\, I \quad -I\,]^{\mathrm{T}}, \forall k \in \{0, 1, \ldots, N + 1\}$. From (2.35), it is obvious that $\mathbb{E}\{(x_k - \hat{x}_k)(x_k - \hat{x}_k)^{\mathrm{T}}\} \leqslant [\, I \quad -I\,]P_k[\, I \quad -I\,]^{\mathrm{T}} \leq \Psi_k, \forall k \in \{0, 1, \ldots, N\}$. It can now be concluded that the requirements (R1) and (R2) are simultaneously satisfied. The proof is complete. □

By means of Theorem 2.3.1, we can summarize the robust filter design (RFD) algorithm as follows:

Algorithm RFD

    Step 1. Given the H_∞ performance index γ, the positive-definite matrix S, and the error initial condition $x_0 - \hat{x}_0$, select the initial values for matrices $\{P_{10}, P_{20}, P_{30}, M_0, N_0\}$ which satisfy the condition (2.31) and set $k = 0$.

    Step 2. Obtain the values of matrices $\{\hat{M}_{k+1}, \hat{N}_{k+1}, P_{1k+1}, P_{2k+1}, P_{3k+1}\}$ and the desired filter parameters $\{A_{fk}, B_{fk}, C_{fk}\}$ for the sampling instant k by solving the LMIs (2.32)–(2.35).

    Step 3. Set $k = k + 1$ and obtain $\{M_{k+1}, N_{k+1}\}$ by the parameter update formula (2.36).

    Step 4. If $k < N$, then go to Step 2, else go to Step 5.

    Step 5. Stop.

Remark 2.3   *From an engineering viewpoint, the recursive Kalman filter is efficient because only the estimated state from the previous time step and the current measurement are needed to compute the estimate for the current state. In fact, the main aim of this chapter is to modify the traditional Kalman filtering approach to handle a class of nonlinearities and missing measurements with variance constraints. For the techniques used, we propose to replace the traditional recursive Riccati equations by the RLMIs for computational convenience. On the other hand, it would be interesting to deal with the corresponding robust steady-state filtering problem when the system parameters become time invariant. This is one of our future research topics.*

Remark 2.4   *In Theorem 2.3.1, the robust $\mathcal{H}_\infty$ finite-horizon filter is designed by solving a series of RLMIs where both the current measurement and the previous state estimation are employed to estimate the current state. Such a recursive filtering process is particularly useful for real-time implementation such as online tracking of highly maneuvering targets. On the other hand, we point out that our main results can be extended to the case of dynamic output feedback control for the same class of nonlinear stochastic time-varying systems, and the results will be given in Section 2.4.*

2.4   Robust H_∞ Finite-Horizon Control with Sensor and Actuator Saturations

In this section, for the stochastic nonlinearities as described in (2.3)–(2.5), the corresponding robust H_∞ finite-horizon output feedback control problem is investigated with both sensor and actuator saturations.

2.4.1   Problem Formulation

Consider the following uncertain discrete stochastic nonlinear time-varying system with both the sensor and actuator saturations:

$$
\begin{cases}
x(k+1) = A^{(\varepsilon)}(k)x(k) + B^{(\varepsilon)}(k)\sigma_u(u(k)) + f(x(k),k) + D_1^{(\varepsilon)}(k)w(k), \\
y_s(k) = \sigma_y(y(k)) + g(x(k),k) + D_2^{(\varepsilon)}(k)w(k), \\
y(k) = C(k)x(k),
\end{cases}
\tag{2.38}
$$

where $x(k) \in \mathbb{R}^n$ is the state vector, $y_s(k) \in \mathbb{R}^r$ is the output, $u(k) \in \mathbb{R}^m$ is the control input, and $w(k) \in \mathbb{R}^p$ is the disturbance input which belongs to $l_2\,[0,\infty)$. All the system matrices in (2.38) are appropriately dimensioned, of which $C(k)$ is a known time-varying matrix, and $A^{(\varepsilon)}(k)$, $B^{(\varepsilon)}(k)$, $D_1^{(\varepsilon)}(k)$, and $D_2^{(\varepsilon)}(k)$ are unknown time-varying matrices which contain polytopic uncertainties (e.g., see Refs [97, 147]) given as follows:

$$
\Xi^{(\varepsilon)} := (A^{(\varepsilon)}(k), B^{(\varepsilon)}(k), D_1^{(\varepsilon)}(k), D_2^{(\varepsilon)}(k)) \in \mathfrak{R}
\tag{2.39}
$$

where $\mathfrak{R}$ is a given convex bounded polyhedral domain described by ν vertices

$$
\mathfrak{R} := \left\{ \Xi^{(\varepsilon)} \,\middle|\, \Xi^{(\varepsilon)} = \sum_{i=1}^{\nu} \varepsilon_i \Xi^{(i)}, \sum_{i=1}^{\nu} \varepsilon_i = 1, \varepsilon_i \geq 0, i = 1, 2, \ldots, \nu \right\}
\tag{2.40}
$$

and $\Xi^{(i)} := (A^{(i)}(k), B^{(i)}(k), D_1^{(i)}(k), D_2^{(i)}(k))$ are known matrices for $i = 1, 2, \ldots, \nu$.

The saturation function $\sigma(\cdot) : \mathbb{R}^r \mapsto \mathbb{R}^r$ is defined as

$$
\sigma(v) = [\sigma_1^{\mathrm{T}}(v_1) \quad \sigma_2^{\mathrm{T}}(v_2) \quad \cdots \quad \sigma_r^{\mathrm{T}}(v_r)]^{\mathrm{T}},
\tag{2.41}
$$

with $\sigma_i(v_i) = \mathrm{sign}(v_i)\min\{v_{i,\max}, |v_i|\}$, where $v_{i,\max}$ is the ith element of the vector $v_{\max}$, the saturation level.

Definition 2.4.1   *[148] A nonlinearity* $\Psi : \mathbb{R}^m \mapsto \mathbb{R}^m$ *is said to satisfy a sector condition if*

$$
(\Psi(v) - H_1 v)^{\mathrm{T}}(\Psi(v) - H_2 v) \leq 0, \quad \forall v \in \mathbb{R}^r
\tag{2.42}
$$

for some real matrices $H_1, H_2 \in \mathbb{R}^{r \times r}$, *where* $H = H_2 - H_1$ *is a positive-definite symmetric matrix. In this case, we say that* Ψ *belongs to the sector* $[H_1, H_2]$.

As in Refs [87, 88, 148], assuming that there exist diagonal matrices K_1, K_2 and R_1, R_2 such that $0 \leq K_1 < I \leq K_2$ and $0 \leq R_1 < I \leq R_2$, then the saturation functions $\sigma_y(y(k))$ and $\sigma_u(u(k))$ in (2.38) can be decomposed into linear and nonlinear parts as

$$
\sigma_y(y(k)) = K_1 C(k)x(k) + \Psi_y(y(k)),
\tag{2.43}
$$

$$
\sigma_u(u(k)) = R_1 u(k) + \Psi_u(u(k)),
\tag{2.44}
$$

where $\Psi_y(y(k))$ and $\Psi_u(u(k))$ are two nonlinear vector-valued functions satisfying two sector conditions, respectively, with $H_1 = 0$, $H_2 = K$ and with $H_1 = 0$, $H_2 = R$, which can be described as follows:

$$\Psi_y^{\mathrm{T}}(y(k))(\Psi_y(y(k)) - KC(k)x(k)) \leq 0, \tag{2.45}$$

$$\Psi_u^{\mathrm{T}}(u(k))(\Psi_u(u(k)) - Ru(k)) \leq 0, \tag{2.46}$$

where $K = K_2 - K_1$ and $R = R_2 - R_1$.

In this section, we consider the following time-varying full-order dynamic output feedback controller for the system (2.38):

$$\begin{cases} x_c(k+1) = A_c(k)x_c(k) + B_c(k)y_s(k) \\ u(k) = C_c(k)x_c(k) \end{cases}, \tag{2.47}$$

where $x_c(k) \in \mathbb{R}^{n_c}$ is the controller state and $A_c(k)$, $B_c(k)$, and $C_c(k)$ are controller parameters of appropriate dimensions to be designed.

Under the output feedback controller (2.47), the closed-loop system becomes

$$\begin{cases} \bar{x}(k+1) = \bar{A}^{(\varepsilon)}(k)\bar{x}(k) + G^{(c)}(k)\bar{\Psi}(k) + H(k)h(x(k), k) + \bar{D}^{(\varepsilon)}(k)w(k) \\ y_s(k) = \bar{C}(k)\bar{x}(k) + \bar{H}\bar{\Psi}(k) + \bar{H}h(x(k), k) + D_2^{(\varepsilon)}(k)w(k) \end{cases}, \tag{2.48}$$

where

$$\bar{x}(k) = \begin{bmatrix} x(k) \\ x_c(k) \end{bmatrix}, \quad \bar{\Psi}(k) = \begin{bmatrix} \Psi_u(u(k)) \\ \Psi_y(y(k)) \end{bmatrix}, \quad \bar{C}(k) = [K_1 C(k) \quad 0],$$

$$\bar{A}^{(\varepsilon)}(k) = \begin{bmatrix} A^{(\varepsilon)}(k) & B^{(\varepsilon)}(k)R_1 C_c(k) \\ B_c(k)K_1 C(k) & A_c(k) \end{bmatrix}, \quad G^{(\varepsilon)}(k) = \begin{bmatrix} B^{(\varepsilon)}(k) & 0 \\ 0 & B_c(k) \end{bmatrix},$$

$$\bar{D}^{(\varepsilon)}(k) = \begin{bmatrix} D_1^{(\varepsilon)}(k) \\ B_c(k)D_2^{(\varepsilon)}(k) \end{bmatrix}, \quad h(x(k), k) = \begin{bmatrix} f(x(k), k) \\ g(x(k), k) \end{bmatrix}, \quad \bar{H} = [0 \quad I],$$

$$H(k) = \begin{bmatrix} I & 0 \\ 0 & B_c(k) \end{bmatrix}. \tag{2.49}$$

Our aim in this chapter is to design a finite-horizon dynamic output feedback controller of the form (2.47) such that, for the given disturbance attenuation level $\gamma > 0$, the positive-definite matrix S and the initial state $x(0)$, the saturated output $y_s(k)$ satisfies the following H_∞ performance constraint:

$$J := \mathbb{E}\{\|y_s(k)\|_{[0,N-1]}^2 - \gamma^2 \|w(k)\|_{[0,N-1]}^2\} - \gamma^2 x^{\mathrm{T}}(0)Sx(0) < 0 \tag{2.50}$$

The finite-horizon control problem in the presence of actuator and sensor saturations addressed above is referred to as the robust finite-horizon $\mathcal{H}_\infty$ control problem for the uncertain nonlinear discrete time-varying stochastic system (2.38).

2.4.2   Main Results

Before proceeding further, we introduce the following lemma which will be needed for the derivation of our main results.

Lemma 2.4.2   [148] Let $Y_0(\eta), Y_1(\eta), \ldots, Y_p(\eta)$ be quadratic functions of $\eta \in \mathbb{R}^n$, $Y_i(\eta) = \eta^T T_i \eta$, $i = 0, 1, \ldots, p$, with $T_i = T_i^T$. Then, the implication $Y_1(\eta) \leq 0, \ldots, Y_p(\eta) \leq 0 \Rightarrow Y_0(\eta) \leq 0$ holds if there exist $\tau_1, \ldots, \tau_p > 0$ such that

$$T_0 - \sum_{i=1}^{p} \tau_i T_i \leq 0. \qquad (2.51)$$

We are now in a position to provide the analysis results in the following theorem.

Theorem 2.4.3   *Let the disturbance attenuation level $\gamma > 0$, families of scalars $\{\tau_1(k)\}_{0 \leq k \leq N} > 0$, $\{\tau_2(k)\}_{0 \leq k \leq N} > 0$, a positive-definite matrix $S > 0$, and the controller feedback gain matrices $\{A_c(k)\}_{0 \leq k \leq N}$, $\{B_c(k)\}_{0 \leq k \leq N}$, $\{C_c(k)\}_{0 \leq k \leq N}$ be given. For the system (2.38) subject to the sensor and actuator saturation (2.43) and (2.44), the $\mathcal{H}_\infty$ performance index requirement defined in (2.50) is achieved for all nonzero $w(k)$ if, with the initial condition $P(0) \leq \gamma^2 \hat{S}$, there exist a family of positive-definite matrices $\{P(k)\}_{0 \leq k \leq N+1}$ satisfying the following recursive matrix inequalities*

$$\Omega^{(\varepsilon)} = \begin{bmatrix} \Omega_{11}^{(\varepsilon)}(k) & * & * \\ \Omega_{21}^{(\varepsilon)}(k) & \Omega_{22}^{(\varepsilon)}(k) & * \\ \Omega_{31}^{(\varepsilon)}(k) & \Omega_{32}^{(\varepsilon)}(k) & \Omega_{33}^{(\varepsilon)}(k) \end{bmatrix} \leq 0 \qquad (2.52)$$

for all $0 \leq k \leq N$, where

$$\Omega_{11}^{(\varepsilon)}(k) = \bar{A}^{(\varepsilon)T}(k)P(k+1)\bar{A}^{(\varepsilon)}(k) - P(k) + \bar{C}^T(k)\bar{C}(k)$$
$$+ \sum_{i=1}^{q} \hat{\Gamma}_i \cdot (\mathrm{tr}[H^T(k)P(k+1)H(k)\hat{\Theta}_i] + \mathrm{tr}[\Theta_{22}^i]),$$

$$\Omega_{21}^{(\varepsilon)}(k) = G^{(\varepsilon)T}(k)P(k+1)\bar{A}^{(\varepsilon)}(k) + \bar{H}^T\bar{C}(k) - \tfrac{1}{2}\tau_1(k)\hat{C}(k) - \tfrac{1}{2}\tau_2(k)\hat{C}_c(k),$$

$$\Omega_{22}^{(\varepsilon)}(k) = G^{(\varepsilon)T}(k)P(k+1)G^{(\varepsilon)}(k) + \bar{H}^T\bar{H} - \tfrac{1}{2}\tau_1(k)\hat{H} - \tfrac{1}{2}\tau_2(k)\check{H},$$

$$\Omega_{31}^{(\varepsilon)}(k) = \bar{D}^{(\varepsilon)T}(k)P(k+1)\bar{A}^{(\varepsilon)}(k) + D_2^{(\varepsilon)T}(k)\bar{C}(k), \qquad (2.53)$$

$$\Omega_{32}^{(\varepsilon)}(k) = \bar{D}^{(\varepsilon)T}(k)P(k+1)G^{(\varepsilon)}(k) + D_2^{(\varepsilon)T}(k)\bar{H},$$

$$\Omega_{33}^{(\varepsilon)}(k) = \bar{D}^{(\varepsilon)T}(k)P(k+1)\bar{D}^{(\varepsilon)}(k) + D_2^{(\varepsilon)T}(k)D_2^{(\varepsilon)}(k) - \gamma^2 I, \quad \hat{\Gamma}_i = \mathrm{diag}\{\Gamma_i, 0\},$$

$$\hat{C}(k) = \begin{bmatrix} 0 \\ -\tilde{C}(k) \end{bmatrix}, \quad \tilde{C}(k) = [KC(k) \quad 0], \quad \hat{C}_c(k) = \begin{bmatrix} -\bar{C}_c(k) \\ 0 \end{bmatrix},$$

$$\bar{C}_c(k) = \begin{bmatrix} 0 & RC_c(k) \end{bmatrix}, \quad \check{H} = \begin{bmatrix} 2I & 0 \\ 0 & 0 \end{bmatrix}, \quad \hat{H} = \begin{bmatrix} 0 & 0 \\ 0 & 2I \end{bmatrix}, \quad \hat{S} = \begin{bmatrix} S & 0 \\ 0 & 0 \end{bmatrix}.$$

Proof.   Define

$$J(k) = \bar{x}^{\mathrm{T}}(k+1)P(k+1)\bar{x}(k+1) - \bar{x}^{\mathrm{T}}(k)P(k)\bar{x}(k). \tag{2.54}$$

Taking (2.5) into consideration, we have

$$
\begin{aligned}
&\mathbb{E}\left\{h^{\mathrm{T}}(x(k), k)H^{\mathrm{T}}(k)P(k+1)H(k)h(x(k), k)\right\} \\
&= \mathbb{E}\{\mathrm{tr}[H^{\mathrm{T}}(k)P(k+1)H(k)h(x(k), k)h^{\mathrm{T}}(x(k), k)]\} \\
&= \mathbb{E}\left\{\mathrm{tr}\left[H^{\mathrm{T}}(k)P(k+1)H(k) \cdot \sum_{i=1}^{q}\hat{\Theta}_i x^{\mathrm{T}}(k)\Gamma_i x(k)\right]\right\} \\
&= \mathbb{E}\left\{\bar{x}^{\mathrm{T}}(k)\sum_{i=1}^{q}\hat{\Gamma}_i \cdot \mathrm{tr}[H^{\mathrm{T}}(k)P(k+1)H(k)\hat{\Theta}_i]\bar{x}(k)\right\}
\end{aligned}
\tag{2.55}
$$

and then obtain from (2.48) that

$$
\begin{aligned}
\mathbb{E}\{J(k)\} = \mathbb{E}\Bigg\{&\bar{x}^{\mathrm{T}}(k)\Bigg(\bar{A}^{(\varepsilon)T}(k)P(k+1)\bar{A}^{(\varepsilon)}(k) \\
&+ \sum_{i=1}^{q}\hat{\Gamma}_i \cdot \mathrm{tr}[H^{\mathrm{T}}(k)P(k+1)H(k)\hat{\Theta}_i] - P(k)\Bigg)\bar{x}(k) \\
&+ 2\bar{x}^{\mathrm{T}}(k)\bar{A}^{(\varepsilon)T}(k)P(k+1)G^{(\varepsilon)}(k)\bar{\Psi}(k) \\
&+ \bar{\Psi}^{\mathrm{T}}(k)G^{(\varepsilon)T}(k)P(k+1)G^{(\varepsilon)}(k)\bar{\Psi}(k) \\
&+ 2w^{\mathrm{T}}(k)\bar{D}^{(\varepsilon)T}(k)P(k+1)\bar{A}^{(\varepsilon)}(k)\bar{x}(k) \\
&+ 2w^{\mathrm{T}}(k)\bar{D}^{(\varepsilon)T}(k)P(k+1)G^{(\varepsilon)}(k)\bar{\Psi}(k) \\
&+ w^{\mathrm{T}}(k)\bar{D}^{(\varepsilon)T}(k)P(k+1)\bar{D}^{(\varepsilon)}(k)w(k)\Bigg\}.
\end{aligned}
\tag{2.56}
$$

Adding the zero term $y_{\mathrm{s}}^{\mathrm{T}}(k)y_{\mathrm{s}}(k) - \gamma^2\omega^{\mathrm{T}}(k)\omega(k) - y_{\mathrm{s}}^{\mathrm{T}}(k)y_{\mathrm{s}}(k) + \gamma^2\omega^{\mathrm{T}}(k)\omega(k)$ to $\mathbb{E}\{J(k)\}$ results in

$$
\begin{aligned}
&\mathbb{E}\{J(k)\} \\
&= \mathbb{E}\left\{\left[\bar{x}^{\mathrm{T}}(k) \quad \bar{\Psi}^{\mathrm{T}}(k) \quad \omega^{\mathrm{T}}(k)\right]\Lambda_k \begin{bmatrix}\bar{x}(k) \\ \bar{\Psi}(k) \\ \omega(k)\end{bmatrix} - y_{\mathrm{s}}^{\mathrm{T}}(k)y_{\mathrm{s}}(k) + \gamma^2\omega^{\mathrm{T}}(k)\omega(k)\right\} \\
&= \mathbb{E}\{\eta^{\mathrm{T}}(k)\Lambda_k\eta(k) - y_{\mathrm{s}}^{\mathrm{T}}(k)y_{\mathrm{s}}(k) + \gamma^2\omega^{\mathrm{T}}(k)\omega(k)\},
\end{aligned}
\tag{2.57}
$$

where

$$
\Lambda_k = \begin{bmatrix} \Omega_{11}^{(\varepsilon)}(k) & * & * \\ G_{Pk} & G^{(\varepsilon)T}(k)P(k+1)G^{(\varepsilon)}(k) + \bar{H}^T \bar{H} & * \\ \Omega_{31}^{(\varepsilon)}(k) & \Omega_{32}^{(\varepsilon)}(k) & \Omega_{33}^{(\varepsilon)}(k) \end{bmatrix},
$$

(2.58)

$$
\eta(k) = [\bar{x}^T(k) \quad \bar{\Psi}^T(k) \quad \omega^T(k)]^T,
$$

$$
G_{Pk} = G^{(\varepsilon)T}(k)P(k+1)\bar{A}^{(\varepsilon)}(k) + \bar{H}^T \bar{C}(k).
$$

Summing up (2.57) on both sides from 0 to $N-1$ with respect to k, we obtain

$$
\sum_{k=0}^{N-1} \mathbb{E}\{J(k)\}
$$

$$
= \mathbb{E}\left\{\bar{x}^T(N)P(N)\bar{x}(N)\right\} - \bar{x}^T(0)P(0)\bar{x}(0)
$$

(2.59)

$$
= \mathbb{E}\left\{\sum_{k=0}^{N-1} \eta^T(k)\Lambda_k \eta(k)\right\} - \mathbb{E}\left\{\sum_{k=0}^{N-1}(y_s^T(k)y_s(k) - \gamma^2 \omega^T(k)\omega(k))\right\}.
$$

Hence, the H_∞ performance index defined in (2.50) is given by

$$
J = \mathbb{E}\left\{\sum_{k=0}^{N-1} \eta^T(k)\Lambda_k \eta(k)\right\} - \mathbb{E}\{\bar{x}^T(N)P(N)\bar{x}(N)\} + \bar{x}^T(0)(P(0) - \gamma^2 \hat{S})\bar{x}(0).
$$

(2.60)

Noting that $P(N) > 0$ and the initial condition $P(0) \le \gamma^2 \hat{S}$, we have $J < 0$ when the following inequality holds:

$$
\eta^T(k)\Lambda_k \eta(k) \le 0.
$$

(2.61)

Noticing the sensor saturation constraint in (2.45), we have

$$
\Psi_y^T(y(k))(\Psi_y(y(k)) - \tilde{C}(k)\bar{x}(k)) \le 0,
$$

(2.62)

which can be written by means of $\eta(k)$ as follows:

$$
\eta^T(k)\Phi_{yk}\eta(k) \le 0,
$$

(2.63)

where

$$
\Phi_{yk} = \frac{1}{2}\begin{bmatrix} 0 & * & * \\ \hat{C}(k) & \hat{H} & * \\ 0 & 0 & 0 \end{bmatrix}.
$$

(2.64)

In the same way, we have from the actuator saturation constraint in (2.46) that

$$
\Psi_u^T(u(k))\left(\Psi_u(u(k)) - \bar{C}_c(k)\bar{x}(k)\right) \le 0
$$

(2.65)

or

$$\eta^{\mathrm{T}}(k)\Phi_{uk}\eta(k) \le 0, \tag{2.66}$$

where

$$\Phi_{uk} = \frac{1}{2}\begin{bmatrix} 0 & * & * \\ \hat{C}_{\mathrm{c}}(k) & \check{H} & * \\ 0 & 0 & 0 \end{bmatrix}. \tag{2.67}$$

Therefore, what we need to do is to find a condition under which (2.61) holds subject to the constraints (2.63) and (2.66). By using Lemma 2.4.2, such a sufficient condition under which (2.63) and (2.66) imply (2.61) is that there exist positive scalars τ_1 and τ_2 such that

$$\Lambda_k - \tau_1(k)\Phi_{yk} - \tau_2(k)\Phi_{uk} \le 0, \tag{2.68}$$

which is equivalent to (2.52). The proof is now complete.   □

Up to now, the analysis problem has been dealt with for the H_∞ output feedback control problem for a class of stochastic nonlinear discrete time-varying systems with sensor and actuator saturation constraints. In the following, we proceed to solve the controller design problem by developing an RLMI approach.

Theorem 2.4.4 *Let a disturbance attenuation level $\gamma > 0$ and a positive-definite matrix $S > 0$ be given. The robust H_∞ controller (2.47) can be designed for system (2.38) with sensor and actuator saturation constraints if there exist families of positive-definite matrices $\{M(k)\}_{0 \le k \le N+1}$, $\{N(k)\}_{0 \le k \le N+1}$, families of positive scalars $\{\lambda_i(k)\}_{0 \le k \le N} > 0$ ($i = 1, 2, \ldots, q$), $\{\tau_1(k)\}_{0 \le k \le N} > 0$, $\{\tau_2(k)\}_{0 \le k \le N} > 0$, and families of real-valued matrices $\{A_{\mathrm{c}}(k)\}_{0 \le k \le N}$, $\{B_{\mathrm{c}}(k)\}_{0 \le k \le N}$, and $\{C_{\mathrm{c}}(k)\}_{0 \le k \le N}$ satisfying the initial condition*

$$\begin{bmatrix} P_1(0) - \gamma^2 S & 0 \\ 0 & P_2(0) \end{bmatrix} \le 0 \tag{2.69}$$

and the recursive LMIs

$$\begin{bmatrix} -\lambda_i(k) & * & * \\ \pi_{1i} & -M_{k+1} & * \\ B_{\mathrm{c}}(k)\pi_{2i} & 0 & -N_{k+1} \end{bmatrix} < 0 \tag{2.70}$$

and

$$\begin{bmatrix} \Upsilon_{11}(k) & * & * \\ \Upsilon_{21}^{(i)}(k) & \Upsilon_{22}^{(i)}(k) & * \\ \Upsilon_{31}(k) & \Upsilon_{32}^{(i)}(k) & \Upsilon_{33}(k) \end{bmatrix} \le 0, \quad i = 1, 2, \ldots, \nu. \tag{2.71}$$

are satisfied with the parameter updated by

$$\begin{aligned} P_1(k+1) &= M^{-1}(k+1) \\ P_2(k+1) &= N^{-1}(k+1), \end{aligned} \tag{2.72}$$

where

$$\Upsilon_{11}(k) = \begin{bmatrix} -P_1(k) + \sum_{i=1}^{q} \Gamma_i(\lambda_i(k) + \mathrm{tr}[\Theta_{22}^i]) & * & * \\ 0 & -P_2(k) & * \\ 0 & \frac{1}{2}RC_c(k) & -\bar{\tau}_2(k)I \end{bmatrix},$$

$$\Upsilon_{21}^{(i)}(k) = \begin{bmatrix} \frac{1}{2}\tau_1(k)KC(k) & 0 & 0 \\ D_2^{(i)T}(k)K_1C(k) & 0 & 0 \\ A^{(i)}(k) & B^{(i)}(k)R_1C_c(k) & B^{(i)}(k) \end{bmatrix},$$

$$\Upsilon_{22}^{(i)}(k) = \begin{bmatrix} -\tau_1(k)I & * & * \\ D_2^{(i)T}(k) & D_2^{(i)T}(k)D_2^{(i)}(k) - \gamma^2 I & * \\ 0 & D_1^{(i)}(k) & -M(k+1) \end{bmatrix},$$

$$\Upsilon_{31}(k) = \begin{bmatrix} B_c(k)K_1C(k) & A_c(k) & 0 \\ K_1C(k) & 0 & 0 \end{bmatrix}, \quad \Upsilon_{32}^{(i)}(k) = \begin{bmatrix} B_c(k) & B_c(k)D_2^{(i)}(k) & 0 \\ I & 0 & 0 \end{bmatrix},$$

$$\Upsilon_{33}(k) = \begin{bmatrix} -N(k+1) & * \\ 0 & -I \end{bmatrix}, \quad \tau_2 = \bar{\tau}_2^{-1}. \tag{2.73}$$

Here, $A^{(i)}(k)$, $B^{(i)}(k)$, $D_1^{(i)}(k)$, and $D_2^{(i)}(k)$ are the matrices at the ith vertex of the polytope.

Proof. Since the set of system matrices

$$\Xi^{(\varepsilon)} := (A^{(\varepsilon)}(k), B^{(\varepsilon)}(k), D_1^{(\varepsilon)}(k), D_2^{(\varepsilon)}(k))$$

belongs to the convex polyhedral set $\mathfrak{R}$, there always exist scalars $\varepsilon_i \geq 0$ ($i = 1, 2, \ldots, \nu$) such that $\Xi^{(\varepsilon)} = \sum_{i=1}^{\nu} \varepsilon_i \Xi^{(i)}$ with $\sum_{i=1}^{\nu} \varepsilon_i = 1$, where

$$\Xi^{(i)} = (A^{(i)}(k), B^{(i)}(k), D_1^{(i)}(k), D_2^{(i)}(k)) \quad (i = 1, 2, \ldots, \nu)$$

are ν vertexes of the polytope. By considering (2.11) together with (2.53), one can easily see that (2.52) holds if and only if

$$\begin{bmatrix} \Omega_{11}^{(i)}(k) & * & * \\ \Omega_{21}^{(i)}(k) & \Omega_{22}^{(i)}(k) & * \\ \Omega_{31}^{(i)}(k) & \Omega_{32}^{(i)}(k) & \Omega_{33}^{(i)}(k) \end{bmatrix} \leq 0. \tag{2.74}$$

Subsequently, we choose the variables $P(k)$ and $P^{-1}(k)$ that can be decomposed as follows:

$$P(k) = \mathrm{diag}\{P_1(k), P_2(k)\}, \quad P^{-1}(k) = \mathrm{diag}\{M(k), N(k)\}. \tag{2.75}$$

By the Schur complement lemma [149], (2.70) is equivalent to

$$\pi_i^T H^T(k) P(k+1) H(k) \pi_i < \lambda_i(k) \quad (i = 1, 2, \ldots, q), \tag{2.76}$$

which, by the property of matrix trace, can be rewritten as

$$\text{tr}[H^T(k) P(k+1) H(k) \hat{\Theta}_i] < \lambda_i(k). \tag{2.77}$$

Noticing (2.77) and using the Schur complement lemma, (2.74) can be rewritten as

$$\begin{bmatrix} \Sigma_{11} & * & * & * & * \\ \Sigma_{21} & \Sigma_{22} & * & * & * \\ D_2^{(i)T}(k)\bar{C}(k) & D_2^{(i)T}(k)\bar{H} & \Sigma_{33} & * & * \\ \bar{A}^{(i)}(k) & G^{(i)}(k) & \bar{D}^{(i)}(k) & -P^{-1}(k+1) & * \\ \bar{C}(k) & \bar{H} & 0 & 0 & -I \end{bmatrix} \leq 0,$$

$$\Sigma_{11} = -P(k) + \sum_{i=1}^{q} \hat{\Gamma}_i(\lambda_i(k) + \text{tr}[\Theta_{22}^i]), \quad \Sigma_{21} = -\frac{1}{2}\tau_1(k)\hat{C}(k) - \frac{1}{2}\tau_2(k)\hat{C}_c(k),$$

$$\Sigma_{22} = -\frac{1}{2}\tau_1(k)\hat{H} - \frac{1}{2}\tau_2(k)\check{H}, \quad \Sigma_{33} = D_2^{(i)T}(k)D_2^{(i)}(k) - \gamma^2 I. \tag{2.78}$$

It follows that (2.78) is guaranteed by (2.71) after some straightforward algebraic manipulations, and the proof of this theorem is then complete. $\qquad\square$

2.5 Illustrative Examples

In this section we give some simulation examples to demonstrate the theory presented in this chapter.

2.5.1 Example 1

In this example, we consider the finite-horizon case of variance-constrained H_∞ filtering for a class of nonlinear time-varying systems with multiple missing measurements.

First of all, let us discuss the practical application of the theory developed to the target-tracking problem through networked transmission, which is an important branch of signal processing. Let the maneuvering target be accelerating with random bursts of gas from its reaction control system thrusters and the state vector consist of the position and velocity. Obviously, owing to the high maneuverability of the tracked target, it is neither possible nor necessary to track the target in a precise way. Instead, as discussed in Remark 2.2, an acceptable compromise is to keep the target within a given "window" as frequently as possible, and such a requirement can be expressed as an upper bound on the estimation error covariance. For online tracking, the system parameters would have to be time varying that contain some uncertainties. Also, because of the bandwidth limit of the signal transmission channel, there are inevitably probabilistic measurements missing and probabilistic nonlinearities. In such a case, there is an urgent need to investigate the robust H_∞ filtering problem for uncertain nonlinear discrete time-varying stochastic systems with error variance constraints and multiple missing measurements.

Motivated by the background discussed above, we consider the following discretized maneuvering target-tracking system that is uncertain, time-varying with stochastic nonlinearities:

$$
\begin{cases}
x_{k+1} = \left(\begin{bmatrix} 0 & -0.3 \\ 0.2 + 0.2\sin(3k) & -0.3 \end{bmatrix} + \begin{bmatrix} 0.1 \\ 0.3 \end{bmatrix} \delta_k \begin{bmatrix} 0.2 \\ 0 \end{bmatrix}^{\mathrm{T}} \right) x_k + f_k + \begin{bmatrix} \sin(3k) \\ -0.03 \end{bmatrix} w_k, \\[2ex]
y_k = \Xi \begin{bmatrix} -2 + 0.3\sin(5k) & 0.5 \\ 0 & 1 \end{bmatrix} x_k + g_k + \begin{bmatrix} 0.1\sin(3k) \\ 0.2 \end{bmatrix} w_k, \\[2ex]
z_k = [\,0.5 \quad 0.5\sin(3k)\,] x_k,
\end{cases} \tag{2.79}
$$

with the state initial value $x_0 = [\,0.26 \quad -0.2\,]^{\mathrm{T}}$, $\hat{x}_0 = [\,0.2 \quad -0.16\,]^{\mathrm{T}}$, and $S = \mathrm{diag}\{2, 2\}$, where the uncertain parameter δ_k satisfies $|\delta_k| \le 2$.

Let ω_k have the unity covariance and the nonlinear functions f_k and g_k be given as follows:

$$
f_k = \begin{bmatrix} 0.1 \\ 0.3 \end{bmatrix} \times \left(0.2 x_{1k} \xi_{1k} + 0.3 x_{2k} \xi_{2k} \right),
$$

$$
g_k = \begin{bmatrix} 0.1 \\ 0.1 \end{bmatrix} \times \left(0.2 x_{1k} \xi_{1k} + 0.3 x_{2k} \xi_{2k} \right),
$$

where x_{ik} $(i = 1, 2)$ is the ith element of x_k and ξ_{ik} $(i = 1, 2)$ are zero-mean, uncorrelated Gaussian white-noise processes with unity variances that are also uncorrelated with ω_k. It can be easily checked that the above class of stochastic nonlinearities satisfies

$$
\mathbb{E} \left\{ \begin{bmatrix} f_k \\ g_k \end{bmatrix} \middle| x_k \right\} = 0,
$$

$$
\mathbb{E} \left\{ \begin{bmatrix} f_k \\ g_k \end{bmatrix} [\, f_k^{\mathrm{T}} \quad g_k^{\mathrm{T}} \,] \middle| x_k \right\} = \begin{bmatrix} 0.1 \\ 0.3 \\ 0.1 \\ 0.1 \end{bmatrix} \begin{bmatrix} 0.1 \\ 0.3 \\ 0.1 \\ 0.1 \end{bmatrix}^{\mathrm{T}} \mathbb{E} \left\{ x_k^{\mathrm{T}} \begin{bmatrix} 0.04 & 0 \\ 0 & 0.09 \end{bmatrix} x_k \right\}.
$$

Assume that the probabilistic density functions of α_1 and α_2 in $[0, 1]$ are described by

$$
q_1(s_1) = \begin{cases} 0 & s_1 = 0 \\ 0.1 & s_1 = 0.5 \\ 0.9 & s_1 = 1 \end{cases}, \quad q_2(s_2) = \begin{cases} 0 & s_2 = 0 \\ 0.2 & s_2 = 0.5 \\ 0.8 & s_2 = 1 \end{cases}, \tag{2.80}
$$

from which the expectations and variances can be easily calculated as $\mu_1 = 0.95$, $\mu_2 = 0.9$, $\sigma_1 = 0.15$, and $\sigma_2 = 0.2$.

Set the disturbance attenuation level as $\gamma = 0.8$ and $\{\Psi_k\}_{1 \le k \le N} = \mathrm{diag}\{0.3, 0.3\}$, and choose the parameters' initial values satisfying (2.31). According to *Algorithm RFD* (see

Table 2.1 Recursive process

k	0	1	2	$\cdots$
P_{1k}	$\begin{bmatrix} 0.0095 & 0.0099 \\ 0.0099 & 0.0194 \end{bmatrix}$	$\begin{bmatrix} 0.0182 & 0.0056 \\ 0.0056 & 0.0306 \end{bmatrix}$	$\begin{bmatrix} 0.0061 & 0.0032 \\ 0.0032 & 0.0107 \end{bmatrix}$	$\cdots$
P_{2k}	$\begin{bmatrix} 0.0005 & 0.0000 \\ 0.0000 & 0.0005 \end{bmatrix}$	$\begin{bmatrix} 0.0155 & -0.0001 \\ -0.0001 & 0.0153 \end{bmatrix}$	$\begin{bmatrix} 0.0037 & -0.0000 \\ -0.0000 & 0.0037 \end{bmatrix}$	$\cdots$
P_{3k}	$\begin{bmatrix} -0.0000 & 0.0000 \\ 0.0000 & -0.0000 \end{bmatrix}$	$\begin{bmatrix} 0.0005 & -0.0001 \\ -0.0001 & 0.0003 \end{bmatrix}$	$\begin{bmatrix} 0.0001 & -0.0000 \\ -0.0000 & 0.0000 \end{bmatrix}$	$\cdots$
A_{fk}	$\begin{bmatrix} 0.2636 & 0.2636 \\ 0.2617 & 0.2617 \end{bmatrix}$	$\begin{bmatrix} -0.1906 & 0.6516 \\ -0.1686 & 0.6284 \end{bmatrix}$	$\begin{bmatrix} -0.27116 & 0.3895 \\ -0.2849 & 0.4025 \end{bmatrix}$	$\cdots$
B_{fk}	$\begin{bmatrix} -0.0980 & 0.2668 \\ 0.0972 & 0.2961 \end{bmatrix}$	$\begin{bmatrix} 0.3462 & 0.4978 \\ 0.3317 & 0.4740 \end{bmatrix}$	$\begin{bmatrix} 0.6405 & 0.8414 \\ 0.5276 & 0.7679 \end{bmatrix}$	$\cdots$
C_{fk}	$[\,0.4545\quad 0.4545\,]$	$[\,0.2165\quad -0.9380\,]$	$[\,0.3982\quad -0.32261\,]$	$\cdots$

Section 2.3), the time-varying LMIs in Theorem 2.3.1 can be solved recursively subject to given initial conditions and prespecified performance indices. Table 2.1 lists the matrix variables P_{1k}, P_{2k}, and P_{3k} and the desired parameters of filter A_{fk}, B_{fk}, and C_{fk} from time $k = 0$ to $k = 2$.

The simulation results are shown in Figure 2.1–Figure 2.5. Figure 2.1 gives the error variance upper bound and actual error variance with $e_{1k} = x_{1k} - \hat{x}_{1k}$ and $e_{2k} = x_{2k} - \hat{x}_{2k}$. Figure 2.2

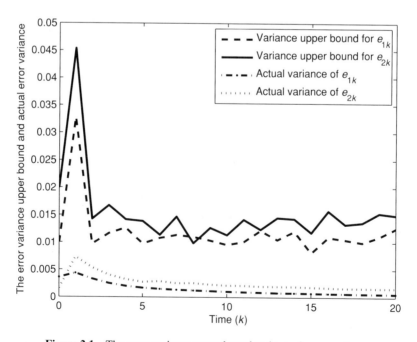

Figure 2.1 The error variance upper bound and actual error variance

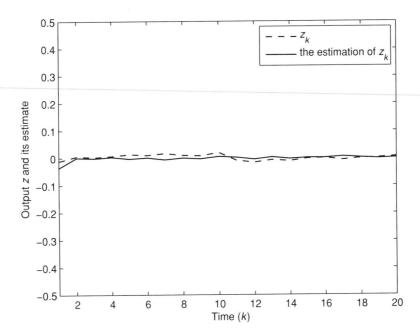

Figure 2.2   Output z_k and its estimate

plots the output z_k and its estimation $\hat{z}_k$, whereas the estimation error $\tilde{z}_k$ is shown in Figure 2.3. The actual states x_{1k}, x_{2k} and their estimates $\hat{x}_{1k}, \hat{x}_{2k}$ are depicted in Figure 2.4 and Figure 2.5, respectively. All the simulation results confirm that the desired finite-horizon performance is well achieved and the proposed RFD algorithm is indeed effective.

2.5.2   Example 2

In this example, we consider robust H_∞ finite-horizon control for a class of stochastic nonlinear time-varying systems subject to sensor and actuator saturations.

Consider the following discrete time-varying stochastic nonlinear systems with sensor and actuator saturations:

$$
\begin{cases}
x(k+1) = \begin{bmatrix} -0.6 & 0.2 \\ 1.1\sin(5k) & 0.5 \end{bmatrix} x(k) + \begin{bmatrix} -2 \\ 3\sin(5k) \end{bmatrix} \sigma_u(u(k)) + f(x(k), k) \\[2mm]
\qquad + \begin{bmatrix} 0.1\sin(3k) \\ -0.3 \end{bmatrix} w(k), \\[2mm]
y_s(k) = \sigma_y(y(k)) + g(x(k), k) + 0.2w(k), \\[1mm]
y(k) = [\,-0.4 \quad 0.5\sin(5k)\,]x(k),
\end{cases}
\tag{2.81}
$$

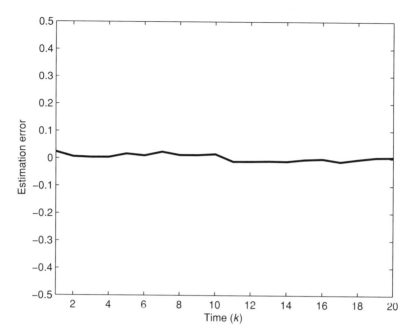

Figure 2.3   Estimation error

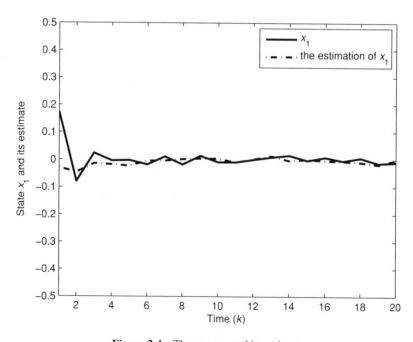

Figure 2.4   The state x_1 and its estimate

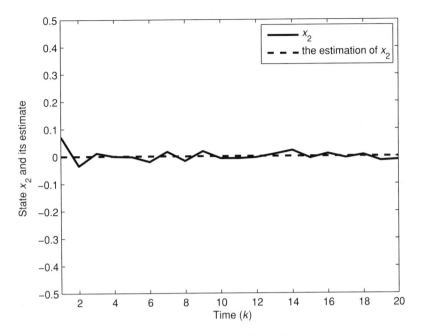

Figure 2.5 The state x_2 and its estimate

where $\sigma_u(u(k))$ and $\sigma_y(y(k))$ are saturation functions described as follows:

$$\begin{cases} \sigma_u(u(k)) = u(k), & \text{if } -V_{ui,\max} \leq u(k) \leq V_{ui,\max}; \\ \sigma_u(u(k)) = V_{ui,\max}, & \text{if } u(k) > V_{ui,\max}; \\ \sigma_u(u(k)) = -V_{ui,\max}, & \text{if } u(k) < -V_{ui,\max}; \end{cases} \tag{2.82}$$

$$\begin{cases} \sigma_y(y(k)) = y(k), & \text{if } -V_{yj,\max} \leq y(k) \leq V_{yj,\max}; \\ \sigma_y(y(k)) = V_{yj,\max}, & \text{if } y(k) > V_{yj,\max}; \\ \sigma_y(y(k)) = -V_{yj,\max}, & \text{if } y(k) < -V_{yj,\max}. \end{cases} \tag{2.83}$$

In this example, we have $i = j = 1$. Take the saturation values as $V_{u1,\max} = 0.02$ and $V_{y1,\max} = 0.04$ with the state initial value $\bar{x}(0) = [\,0.26 \quad -0.2 \quad 0 \quad 0\,]^{\mathrm{T}}$ and $S = \text{diag}\{1, 1\}$. The exogenous disturbance input is selected as $w(k) = 0.5\sin(4k)$, and other parameters are chosen as $K = 0.3$, $K_1 = 0.7$, $R = 0.4$, and $R_1 = 0.6$.

The nonlinear functions $f(x(k), k)$ and $g(x(k), k)$ are

$$f(x(k), k) = \begin{bmatrix} 0.3 \\ 0.4 \end{bmatrix} \times (0.5x_1(k)\xi_1(k) + 0.4x_2(k)\xi_2(k)), \tag{2.84}$$

$$g(x(k), k) = 0.8 \times (0.5x_1(k)\xi_1(k) + 0.4x_2(k)\xi_2(k)), \tag{2.85}$$

where $x_i(k)$ $(i = 1, 2)$ is the ith element of $x(k)$, and $\xi_i(k)$ $(i = 1, 2)$ are zero-mean, uncorrelated Gaussian white-noise processes with unity variances that are also uncorrelated with $\omega(k)$.

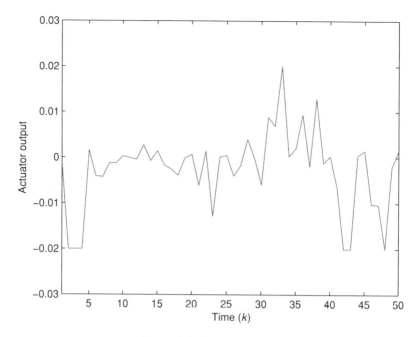

Figure 2.6   Actuator output

Let $\gamma = 0.5$ and choose the parameters' initial values satisfying (2.69). The uncertain parameter ε is unknown but assumed to belong to the known range $[-0.2, 0.2]$. According to the controller design algorithm, the RLMIs in Theorem 2.4.4 can be solved recursively subject to given initial conditions and prespecified performance indices.

The simulation results are shown in Figure 2.6–Figure 2.10, where Figure 2.6 plots the actuator output and Figure 2.7 depicts the sensor output. Note that both the actuator and sensor outputs are saturated. Figure 2.8 shows the state simulation results of the closed-loop system (2.8), and the system output $y_s(k)$ is depicted in Figure 2.9. When the saturation levels are changed to $V_{u1,\max} = 0.01$ and $V_{y1,\max} = 0.03$, the state simulation for system (2.8) is given in Figure 2.10, from which we can observe that (1) the saturations do influence the control performances, and (2) the saturation range would have a serious impact on the feasibility of the RLMIs.

2.6   Summary

In this chapter, the problem of robust H_∞ finite-horizon filtering and output feedback control problems have been discussed for a class of uncertain stochastic nonlinear discrete time-varying systems with sensors and actuators subject to saturation, error variance constraints, and multiple missing measurements. The stochastic nonlinearities under consideration here have been widely used in engineering applications. A new robust H_∞ filtering technique has been first considered for the nonlinear discrete time-varying stochastic

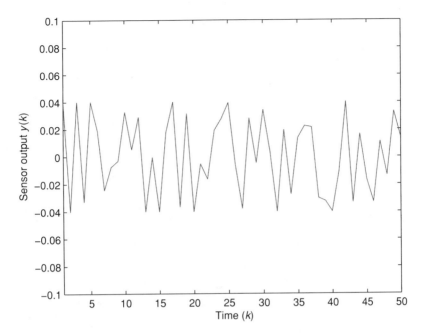

Figure 2.7   Sensor output

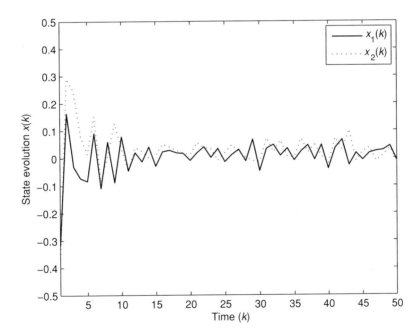

Figure 2.8   The state evolution $x(k)$

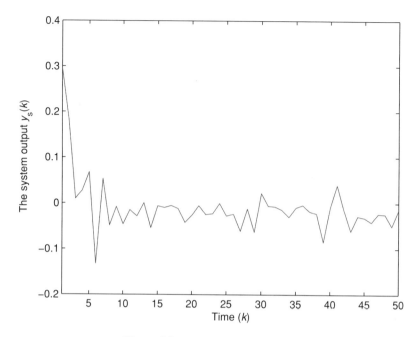

Figure 2.9   The system output $y_s(k)$

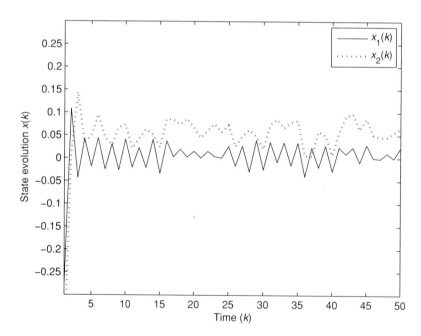

Figure 2.10   The state evolution $x(k)$ when the saturation gets severe

systems with norm-bounded uncertainties, multiple missing measurements, and error variance constraints. Sufficient conditions for the finite-horizon filter to satisfy state estimation error variance constraints and prescribed H_∞ performance have been given in terms of the feasibility of a series of RLMIs. Moreover, by using similar analysis techniques, some parallel results have also been derived for the corresponding robust H_∞ finite-horizon output feedback control problem with both sensor and actuator saturations. Finally, the results of this chapter have been demonstrated by some simulation examples.

3

Filtering and Control with Stochastic Delays and Missing Measurements

In this chapter, the robust H_∞ filtering and control problems are studied for two classes of uncertain nonlinear systems with both multiple stochastic time-varying communication delays and multiple packet dropouts. A sequence of random variables, all of which are mutually independent but obey the Bernoulli distribution, are introduced to account for the randomly occurring communication delays. The packet dropout phenomenon occurs in a random way and the occurrence probability for each sensor is governed by an individual random variable satisfying a certain probabilistic distribution on the interval [0, 1]. First, the robust H_∞ filtering problem is investigated for the discrete-time system with parameter uncertainties, state-dependent stochastic disturbances, and sector-bounded nonlinearities. We aim to design a linear full-order filter such that the estimation error converges to zero exponentially in the mean square while the disturbance rejection attenuation is constrained to a give level by means of the H_∞ performance index. Intensive stochastic analysis is carried out to obtain sufficient conditions for ensuring the exponential stability, as well as prescribed H_∞ performance. Furthermore, the phenomena of multiple probabilistic delays and multiple missing measurements are extended, in a parallel way, to fuzzy systems. Attention is focused on the analysis and design of H_∞ fuzzy output feedback controllers such that the closed-loop T–S fuzzy control system is exponentially stable in the mean square. The disturbance rejection attenuation is constrained to a given level by means of the H_∞ performance index. Intensive analysis is carried out to obtain sufficient conditions for the existence of admissible output feedback controllers ensuring the exponential stability, as well as the prescribed H_∞ performance. The cone complementarity linearization (CCL) procedure is employed to cast the controller design problem into a sequential minimization one solved by the semi-definite program method. Simulation results are utilized to demonstrate the effectiveness of the proposed design technique in this chapter.

Filtering, Control and Fault Detection with Randomly Occurring Incomplete Information, First Edition.
Hongli Dong, Zidong Wang, and Huijun Gao.
© 2013 John Wiley & Sons, Ltd. Published 2013 by John Wiley & Sons, Ltd.

3.1   Problem Formulation for Robust Filter Design

To start with, let us denote the following for presentation clarity:

$$\tilde{x}(k) := \sum_{i=1}^{q} \alpha_i(k)x(k - \tau_i(k)),$$ (3.1)

where $\tau_i(k)$ $(i = 1, 2, \ldots, q)$ are the random communication delays to be discussed in detail.

Consider the following discrete-time uncertain nonlinear networked system with multiple stochastic communication delays:

$$\begin{cases} x(k+1) = (A + \Delta A)x(k) + A_d\tilde{x}(k) + Ff(x(k)) \\ \qquad + g(x(k), \tilde{x}(k), k)w(k) + D_1v(k) \\ \tilde{y}(k) = Cx(k) + D_2v(k) \\ z(k) = Lx(k) \\ x(j) = \varphi(j), \ j = -d_M, -d_M + 1, \ldots, 0 \end{cases}$$ (3.2)

where $x(k) \in \mathbb{R}^n$ represents the state vector, $\tilde{x}(k) \in \mathbb{R}^n$ is defined in (3.1), $\tilde{y}(k) \in \mathbb{R}^r$ is the process output, $z(k) \in \mathbb{R}^m$ is the signal to be estimated, and $v(k) \in \mathbb{R}^q$ is the exogenous disturbance signal belonging to $l_2[0, \infty)$. $\varphi(j)$ $(j = -d_M, -d_M + 1, \ldots, 0)$ are the initial conditions. $w(k)$ is a scalar Wiener process (Brownian motion) satisfying

$$\mathbb{E}\{w(k)\} = 0, \quad \mathbb{E}\{w^2(k)\} = 1, \quad \mathbb{E}\{w(k)w(j)\} = 0 (k \neq j),$$

and $g : \mathbb{R}^n \times \mathbb{R}^n \times \mathbb{N} \to \mathbb{R}^n$ is the continuous function quantifying the noise intensity which satisfies

$$g^{\mathrm{T}}(x(k), \tilde{x}(k), k)g(x(k), \tilde{x}(k), k) \leq \rho_1 x^{\mathrm{T}}(k)x(k) + \rho_2 \tilde{x}^{\mathrm{T}}(k)\tilde{x}(k),$$

where $\rho_1 > 0$ and $\rho_2 > 0$ are known constant scalars. The parameter uncertainties ΔA are a real-valued matrix of the form

$$\Delta A = HF(k)E,$$ (3.3)

where H and E are known real constant matrices with appropriate dimensions, and $F(k)$ is the unknown time-varying matrix function satisfying $F^{\mathrm{T}}(k)F(k) \leq I$.

The vector-valued nonlinear function f is assumed to satisfy the following sector-bounded conditions with $f(0) = 0$:

$$[f(x) - f(y) - R_1(x - y)]^{\mathrm{T}}[f(x) - f(y) - R_2(x - y)] \leq 0,$$ (3.4)

where $R_1, R_2 \in \mathbb{R}^{n \times n}$, and $R_1 - R_2$ is a positive-definite matrix.

Remark 3.1 *It is customary that the nonlinear function f is said to belong to $[R_1, R_2]$ (see Ref. [150]). The nonlinear description in (3.4) is quite general and includes the usual Lipschitz conditions as a special case. Note that both the control analysis and model reduction problems for systems with sector nonlinearities have been intensively studied; for example, see Refs [151, 152].*

The random variables $\alpha_i(k) \in \mathbb{R}$ $(i = 1, 2, \ldots, q)$ in (3.1) are mutually uncorrelated Bernoulli-distributed white sequences obeying the following probability distribution law:

$$\text{Prob}\{\alpha_i(k) = 1\} = \mathbb{E}\{\alpha_i(k)\} = \bar{\alpha}_i, \quad \text{Prob}\{\alpha_i(k) = 0\} = 1 - \bar{\alpha}_i. \tag{3.5}$$

The following assumption is needed on the random communication time delays considered.

Assumption 3.1.1 *The communication delays $\tau_i(k)$ $(i = 1, 2, \ldots, q)$ are time varying and satisfy $d_m \le \tau_i(k) \le d_M$, where d_m and d_M are constant positive scalars representing the lower and upper bounds on the communication delays, respectively.*

Remark 3.2 *The way the communication delays are described in (3.1) is believed to be new because, different from most existing literature, such a description exhibits the following two features: (1) the communication delays are allowed to occur in any fashion, either discretely, successively, or even distributely; and (2) each possible delay could occur independently and randomly according to an individual probability distribution which can be specified a priori through a statistical test.*

In this chapter, the packet dropout (missing measurement) phenomenon constitutes another focus of our present research. The multiple packet dropouts are described by

$$
\begin{aligned}
y(k) &= \Xi C x(k) + D_2 v(k) \\
&= \sum_{j=1}^{r} \beta_j C_j x(k) + D_2 v(k),
\end{aligned} \tag{3.6}
$$

where $y(k) \in \mathbb{R}^r$ is the *actual* measurement signal of (3.2), $\Xi := \text{diag}\{\beta_1, \ldots, \beta_r\}$ with β_j $(j = 1, \ldots, r)$ being r unrelated random variables which are also unrelated with $\alpha_i(k)$ and $w(k)$. It is assumed that β_j has the probabilistic density function $q_j(s)$ $(j = 1, \ldots, r)$ on the interval $[0, 1]$ with mathematical expectation μ_j and variance σ_j^2. C_j is defined by

$$C_j := \text{diag}\{\underbrace{0, \ldots, 0}_{j-1}, 1, \underbrace{0, \ldots, 0}_{r-j}\}C.$$

Note that β_j could satisfy any discrete probabilistic distributions on the interval $[0, 1]$, which include the widely used Bernoulli distribution as a special case. In the following, we denote $\bar{\Xi} = \mathbb{E}\{\Xi\}$.

According to the above analysis, we have the following system to be investigated:

$$
\begin{cases}
x(k+1) = (A + \Delta A)x(k) + A_d \tilde{x}(k) + F f(x(k)) \\
\qquad\qquad + g(x(k), \tilde{x}(k), k)w(k) + D_1 v(k) \\
y(k) = \Xi C x(k) + D_2 v(k) = \displaystyle\sum_{j=1}^{r} \beta_j C_j x(k) + D_2 v(k) \ . \\
z(k) = L x(k) \\
x(j) = \varphi(j), \ j = -d_M, -d_M + 1, \ldots, 0
\end{cases}
\tag{3.7}
$$

In this chapter, we are interested in obtaining $\hat{z}(k)$, the estimate of the signal $z(k)$, from the *actual* measured output $y(k)$. The full-order filter to be considered is given as follows:

$$
\begin{cases}
\hat{x}(k+1) = A_f \hat{x}(k) + B_f y(k) \\
\qquad\qquad \hat{z}(k) = C_f \hat{x}(k)
\end{cases}
,
\tag{3.8}
$$

where $\hat{x}(k) \in \mathbb{R}^n$ represents the state estimate, $\hat{z}(k) \in \mathbb{R}^m$ is the estimated output, and A_f, B_f, and C_f are appropriately dimensioned filter matrices to be determined.

Augmenting the model of (3.7) to include the states of the filter (3.8), the filtering error system is given by

$$
\begin{cases}
\bar{x}(k+1) = (\bar{A} + \tilde{A})\bar{x}(k) + \displaystyle\sum_{i=1}^{q}(\bar{A}_{di} + \tilde{A}_{di})\bar{x}(k - \tau_i(k)) \\
\qquad\qquad + \bar{D}w(k) + \bar{F} f(x(k)) + \bar{D}_1 v(k) \\
\bar{z}(k) = \bar{L}\bar{x}(k)
\end{cases}
,
\tag{3.9}
$$

where

$$
\bar{x}(k) = [x^T(k) \quad \hat{x}^T(k)t]^T, \quad \bar{z}(k) = z(k) - \hat{z}(k),
$$

$$
\bar{L} = [L \quad -C_f], \quad \bar{F} = [F^T \quad 0]^T,
$$

$$
\bar{A} = \begin{bmatrix} A + \Delta A & 0 \\ B_f \bar{\Xi} C & A_f \end{bmatrix}, \quad
\tilde{A} = \begin{bmatrix} 0 & 0 \\ B_f(\Xi - \bar{\Xi})C & 0 \end{bmatrix},
$$

$$
\bar{A}_{di} = \begin{bmatrix} \bar{\alpha}_i A_d & 0 \\ 0 & 0 \end{bmatrix}, \quad
\tilde{A}_{di} = \begin{bmatrix} \tilde{\alpha}_i(k) A_d & 0 \\ 0 & 0 \end{bmatrix},
$$

$$
\bar{D} = \begin{bmatrix} g(x(k), \tilde{x}(k), k) \\ 0 \end{bmatrix}, \quad
\bar{D}_1 = \begin{bmatrix} D_1 \\ B_f D_2 \end{bmatrix},
$$

with $\tilde{\alpha}_i(k) = \alpha_i(k) - \bar{\alpha}_i$. It is clear that $\mathbb{E}\{\tilde{\alpha}_i(k)\} = 0$ and $\mathbb{E}\{\tilde{\alpha}_i^2(k)\} = \bar{\alpha}_i(1 - \bar{\alpha}_i)$.

Owing to the existence of the stochastic variable $\alpha_i(k)$ and β_j, the definition for the exponential stability in the mean square is needed for the forthcoming issue of stochastic analysis.

Definition 3.1.1 *[25] The filtering error system (3.9) is said to be exponentially stable in the mean square if, in the case of $v(k) = 0$, for any initial conditions, there exist constants $\delta > 0$ and $0 < \kappa < 1$ such that*

$$\mathbb{E}\left\{\|\bar{x}(k)\|^2\right\} \le \delta\kappa^k \sup_{-d_M \le i \le 0} \mathbb{E}\left\{\|\varphi(i)\|^2\right\}, \quad \forall k \ge 0.$$

Our aim in this chapter is to develop techniques to deal with the robust H_∞ filtering problem for uncertain discrete nonlinear systems with multiple communication delays and packet dropouts. More specifically, given a disturbance attenuation level $\gamma > 0$, we like to design the filter of the form (3.8) for the system (3.7) such that, for all admissible parameter uncertainties, nonlinearities, multiple communication delays, and packet dropouts, the following two requirements are satisfied simultaneously:

(R1) The filter error system (3.9) is exponentially stable in the mean square.
(R2) Under zero initial condition, the filtering error $\bar{z}(k)$ satisfies

$$\sum_{k=0}^{\infty} \mathbb{E}\left\{\|\bar{z}(k)\|^2\right\} \le \gamma^2 \sum_{k=0}^{\infty} \mathbb{E}\{\|v(k)\|^2\} \tag{3.10}$$

for all nonzero $v(k)$, where $\gamma > 0$ is a prescribed scalar.

3.2 Robust H_∞ Filtering Performance Analysis

Before proceeding further, we introduce the following lemmas which will be needed for the derivation of our main results.

Lemma 3.2.1 (Schur complement.) Given constant matrices $\mathcal{S}_1, \mathcal{S}_2$, and $\mathcal{S}_3$, where $\mathcal{S}_1 = \mathcal{S}_1^T$ and $0 < \mathcal{S}_2 = \mathcal{S}_2^T$, then $\mathcal{S}_1 + \mathcal{S}_3^T \mathcal{S}_2^{-1} \mathcal{S}_3 < 0$ if and only if

$$\begin{bmatrix} \mathcal{S}_1 & \mathcal{S}_3^T \\ \mathcal{S}_3 & -\mathcal{S}_2 \end{bmatrix} < 0 \quad \text{or} \quad \begin{bmatrix} -\mathcal{S}_2 & \mathcal{S}_3 \\ \mathcal{S}_3^T & \mathcal{S}_1 \end{bmatrix} < 0. \tag{3.11}$$

Lemma 3.2.2 (S-procedure.) Let $L = L^T$ and H and E be real matrices of appropriate dimensions with F satisfying $FF^T \le I$, then $L + HFE + E^T F^T H^T < 0$ if and only if there exists a positive scalar $\varepsilon > 0$ such that $L + \varepsilon^{-1} HH^T + \varepsilon E^T E < 0$, or equivalently:

$$\begin{bmatrix} L & H & \varepsilon E^T \\ H^T & -\varepsilon I & 0 \\ \varepsilon E & 0 & -\varepsilon I \end{bmatrix} < 0. \tag{3.12}$$

Let us first consider the robust exponential stability analysis problem for the filter error system (3.9) with $v(k) = 0$.

Theorem 3.2.3 *Let the filter parameters A_f, B_f, and C_f be given and the admissible conditions hold. Then, the filtering error system (3.9) with $v(k) = 0$ is robustly exponentially stable in the* *mean square if there exist matrices $P > 0$, $Q_j > 0$ ($j = 1, 2, \ldots, q$) and positive constant scalars λ_1, λ_2 satisfying*

$$\Omega = \begin{bmatrix} \Omega_{11} & * & * \\ \hat{Z}^T P \bar{A} & \Omega_{22} & * \\ \Omega_{31} & \bar{F}^T P \hat{Z} & \Omega_{33} \end{bmatrix} < 0, \tag{3.13}$$

$$P \leq \lambda_1 I, \tag{3.14}$$

where

$$\Omega_{11} = \lambda_1 A_\rho - P + \sum_{j=1}^{q}(d_M - d_m + 1)Q_j + \bar{A}^T P \bar{A} + \sum_{j=1}^{r} \sigma_j^2 \bar{C}_j^T P \bar{C}_j - \lambda_2 G^T \tilde{R}_1 G,$$

$$\Omega_{22} = \text{diag}\{-Q_1 + \tilde{A}_1, -Q_2 + \tilde{A}_2, \ldots, -Q_q + \tilde{A}_q\} + \hat{Z}^T P \hat{Z} + \lambda_1 \rho_2 \hat{Z}_a^T \hat{Z}_a,$$

$$\Omega_{31} = \bar{F}^T P \bar{A} - \lambda_2 \tilde{R}_2^T G, \quad \Omega_{33} = \bar{F}^T P \bar{F} - \lambda_2 I,$$

$$\tilde{A}_i = \bar{\alpha}_i(1 - \bar{\alpha}_i)\hat{A}_d^T P \hat{A}_d + \lambda_1 \rho_2 W_i, \ i = 1, 2, \ldots, q, \quad \tilde{R}_1 = (R_1^T R_2 + R_2^T R_1)/2,$$

$$\tilde{R}_2 = -(R_1^T + R_2^T)/2, \quad G = [I \quad 0], \quad \hat{Z} = [\bar{A}_{d1} \quad \bar{A}_{d2} \quad \cdots \quad \bar{A}_{dq}],$$

$$\hat{Z}_a = [T_1 \quad T_2 \quad \cdots \quad T_q], \quad \bar{C}_j = \begin{bmatrix} 0 & 0 \\ B_f C_j & 0 \end{bmatrix}, \quad \hat{A}_d = \begin{bmatrix} A_d & 0 \\ 0 & 0 \end{bmatrix},$$

$$A_\rho = \begin{bmatrix} \rho_1 I & 0 \\ 0 & 0 \end{bmatrix}, \quad W_i = \begin{bmatrix} \bar{\alpha}_i(1 - \bar{\alpha}_i)I & 0 \\ 0 & 0 \end{bmatrix}, \quad T_i = \begin{bmatrix} \bar{\alpha}_i I & 0 \\ 0 & 0 \end{bmatrix}.$$

Proof. Let

$$\Theta_j(k) = t\{x(k - \tau_j(k)), x(k - \tau_j(k) + 1), \ldots, x(k)\} \quad (j = 1, 2, \ldots, q),$$

$$\chi(k) = \{\Theta_1(k) \cup \Theta_2(k) \cup \cdots \cup \Theta_q(k)\} = \bigcup_{j=1}^{q} \Theta_j(k).$$

Choose the following Lyapunov functional for system (3.9):

$$V(\chi(k)) = \sum_{i=1}^{3} V_i(k),$$

where

$$V_1(k) = \bar{x}^T(k)P\bar{x}(k), \quad V_2(k) = \sum_{j=1}^{q} \sum_{i=k-\tau_j(k)}^{k-1} \bar{x}^T(i)Q_j\bar{x}(i),$$

$$V_3(k) = \sum_{j=1}^{q} \sum_{m=-d_M+1}^{-d_m} \sum_{i=k+m}^{k-1} \bar{x}^T(i)Q_j\bar{x}(i),$$

with $P > 0$ and $Q_j > 0$ $(j = 1, 2, \ldots, q)$ being matrices to be determined. Then, along the trajectory of system (3.9) with $v(k) = 0$, we have

$$
\begin{aligned}
\mathbb{E}\{\Delta V \mid x(k)\} &= \mathbb{E}\{V(\chi(k+1)) \mid \chi(k)\} - V(\chi(k)) \\
&= \mathbb{E}\{V(\chi(k+1)) - V(\chi(k)) \mid \chi(k)\} \\
&= \sum_{i=1}^{3} \mathbb{E}\{\Delta V_i \mid \chi(k)\}
\end{aligned}
\tag{3.15}
$$

From (3.9), we can obtain that

$$
\begin{aligned}
\mathbb{E}\{\Delta V_1 \mid \chi(k)\} &= \mathbb{E}\{(\bar{x}^{\mathrm{T}}(k+1)P\bar{x}(k+1) - \bar{x}^{\mathrm{T}}(k)P\bar{x}(k)) \mid \chi(k)\} \\
&= \mathbb{E}\Bigg\{\Bigg(\bar{x}^{\mathrm{T}}(k)(\bar{A}^{\mathrm{T}}P\bar{A} + \tilde{A}^{\mathrm{T}}P\tilde{A} - P)\bar{x}(k) + 2\bar{x}^{\mathrm{T}}(k) \\
&\quad \times \bar{A}^{\mathrm{T}}P\left(\sum_{i=1}^{q}\bar{A}_{di}\bar{x}(k-\tau_i(k))\right) + 2\bar{x}^{\mathrm{T}}(k)\bar{A}^{\mathrm{T}}P\bar{F} \\
&\quad \times f(x(k))\Bigg) + \sum_{i=1}^{q}\bar{x}^{\mathrm{T}}(k-\tau_i(k))\tilde{A}_{di}^{\mathrm{T}}P\tilde{A}_{di}\bar{x}(k-\tau_i(k)) \\
&\quad + \left(\sum_{i=1}^{q}\bar{A}_{di}\bar{x}(k-\tau_i(k))\right)^{\mathrm{T}}P\left(\sum_{i=1}^{q}\bar{A}_{di}\bar{x}(k-\tau_i(k))\right) \\
&\quad + 2\left(\sum_{i=1}^{q}\bar{A}_{di}\bar{x}(k-\tau_i(k))\right)^{\mathrm{T}}P\bar{F}f(x(k)) + \bar{D}^{\mathrm{T}}P\bar{D} \\
&\quad + f^{\mathrm{T}}(x(k))\bar{F}^{\mathrm{T}}P\bar{F}f(x(k))) \mid \chi(k)\Bigg\},
\end{aligned}
$$

$$
\begin{aligned}
\mathbb{E}\{\tilde{A}_{di}^{\mathrm{T}}P\tilde{A}_{di}\} &= \mathbb{E}\left\{\begin{bmatrix}\tilde{\alpha}_i(k)A_d & 0 \\ 0 & 0\end{bmatrix}^{\mathrm{T}}P\begin{bmatrix}\tilde{\alpha}_i(k)A_d & 0 \\ 0 & 0\end{bmatrix}\right\} \\
&= \bar{\alpha}_i(1-\bar{\alpha}_i)\begin{bmatrix}A_d & 0 \\ 0 & 0\end{bmatrix}^{\mathrm{T}}P\begin{bmatrix}A_d & 0 \\ 0 & 0\end{bmatrix} \\
&= \bar{\alpha}_i(1-\bar{\alpha}_i)\hat{A}_d^{\mathrm{T}}P\hat{A}_d,
\end{aligned}
$$

$$
\begin{aligned}
\mathbb{E}\{\bar{D}^{\mathrm{T}}P\bar{D}\} &\leq \lambda_1\rho_2\left(\sum_{i=1}^{q}T_i\bar{x}(k-\tau_i(k))\right)^{\mathrm{T}}\left(\sum_{i=1}^{q}T_i\bar{x}(k-\tau_i(k))\right) \\
&\quad + \lambda_1\rho_2\sum_{i=1}^{q}\bar{x}^{\mathrm{T}}(k-\tau_i(k))W_i\bar{x}(k-\tau_i(k)) \\
&\quad + \lambda_1\bar{x}^{\mathrm{T}}(k)A_\rho\bar{x}(k).
\end{aligned}
\tag{3.16}
$$

Taking (3.15) and (3.16) into consideration, we have

$$
\mathbb{E}\left\{\Delta V_1 \mid \chi(k)\right\} \leq \bar{x}^{\mathrm{T}}(k)\left(\bar{A}^{\mathrm{T}}P\bar{A} + \sum_{j=1}^{r}\sigma_j^2 \bar{C}_j^{\mathrm{T}}P\bar{C}_j + \lambda_1 A_\rho - P\right)\bar{x}(k)
$$

$$
+2\bar{x}^{\mathrm{T}}(k)\bar{A}^{\mathrm{T}}P\left(\sum_{i=1}^{q}\bar{A}_{di}\bar{x}(k - \tau_i(k))\right) + 2\bar{x}^{\mathrm{T}}(k)\bar{A}^{\mathrm{T}}
$$

$$
\times P\bar{F}f(x(k)) + \sum_{i=1}^{q}\sum_{j=1}^{q}\bar{x}^{\mathrm{T}}(k - \tau_i(k))\left(\bar{A}_{di}^{\mathrm{T}}P\bar{A}_{dj}\right)
$$

$$
+\lambda_1\rho_2 T_i^{\mathrm{T}}T_j\right)\bar{x}(k - \tau_j(k)) + \sum_{i=1}^{q}\bar{x}^{\mathrm{T}}(k - \tau_i(k))
$$

$$
\times\left(\bar{\alpha}_i(1 - \bar{\alpha}_i)\hat{A}_d^{\mathrm{T}}P\hat{A}_d + \lambda_1\rho_2 W_i\right)\bar{x}(k - \tau_i(k))
$$

$$
+2\left(\sum_{i=1}^{q}\bar{A}_{di}\bar{x}(k - \tau_i(k))\right)^{\mathrm{T}}P\bar{F}f(x(k))
$$

$$
+f^{\mathrm{T}}(x(k))\bar{F}^{\mathrm{T}}P\bar{F}f(x(k)). \tag{3.17}
$$

Next, it can be derived that

$$
\mathbb{E}\left\{\Delta V_2 \mid \chi(k)\right\} \leq \mathbb{E}\left\{\sum_{j=1}^{q}\left(\bar{x}^{\mathrm{T}}(k)Q_j\bar{x}(k) - \bar{x}^{\mathrm{T}}(k - \tau_j(k))Q_j\right.\right.
$$

$$
\left.\left.\times\bar{x}(k - \tau_j(k)) + \sum_{i=k-d_M+1}^{k-d_m}\bar{x}^{\mathrm{T}}(i)Q_j\bar{x}(i)\right)\right|\chi(k)\right\},
$$

$$
\mathbb{E}\left\{\Delta V_3 \mid \chi(k)\right\} = \mathbb{E}\left\{\sum_{j=1}^{q}\left((d_M - d_m)\bar{x}^{\mathrm{T}}(k)Q_j\bar{x}(k)\right.\right.
$$

$$
\left.\left.-\sum_{i=k-d_M+1}^{k-d_m}\bar{x}^{\mathrm{T}}(i)Q_j\bar{x}(i)\right)\right|\chi(k)\right\}. \tag{3.18}
$$

Letting

$$
\xi(k) = \left[\bar{x}^{\mathrm{T}}(k)\bar{x}^{\mathrm{T}}(k - \tau_1(k))\cdots\bar{x}^{\mathrm{T}}(k - \tau_q(k))f^{\mathrm{T}}(x(k))\right]^{\mathrm{T}},
$$

the combination of (3.17) and (3.18) results in

$$
\mathbb{E}\left\{\Delta V \mid x(k)\right\} \leq \xi^{\mathrm{T}}(k)\Omega_1\xi(k), \tag{3.19}
$$

where

$$\Omega_1 = \begin{bmatrix} \Omega_{11} + \lambda_2 G^{\mathrm{T}} \tilde{R}_1 G & * & * \\ \hat{Z}^{\mathrm{T}} P \bar{A} & \Omega_{22} & * \\ \bar{F}^{\mathrm{T}} P \bar{A} & \bar{F}^{\mathrm{T}} P \hat{Z} & \bar{F}^{\mathrm{T}} P \bar{F} \end{bmatrix}.$$

Notice that (3.1) implies

$$\begin{bmatrix} \bar{x}(k) \\ f(x(k)) \end{bmatrix}^{\mathrm{T}} \begin{bmatrix} G^{\mathrm{T}} \tilde{R}_1 G & G^{\mathrm{T}} \tilde{R}_2 \\ \tilde{R}_2^{\mathrm{T}} G & I \end{bmatrix} \begin{bmatrix} \bar{x}(k) \\ f(x(k)) \end{bmatrix} \leq 0. \tag{3.20}$$

From (3.19) and (3.20), it follows that

$$\mathbb{E}\left\{\Delta V \mid x(k)\right\} \leq \xi^{\mathrm{T}}(k) \Omega \xi(k).$$

According to Theorem 3.2.3, we have $\Omega < 0$. Hence, for all $\xi(k) \neq 0$, $\mathbb{E}\{\Delta V \mid x(k)\} \leq \xi^{\mathrm{T}}(k)\Omega\xi(k) < 0$. Furthermore, from Theorem 1 in [25], the robustly exponential stability of system (3.9) can be confirmed in the mean-square sense. The proof is complete. $\square$

Next, we will analyze the H_∞ performance of the filtering error system (3.9).

Theorem 3.2.4 *Let the filter parameters A_{f}, B_{f} and C_{f} be given and γ be a prespecified positive constant. Then the filtering error system (3.9) is robustly exponentially stable in the mean square for $v(k) = 0$ and satisfies $\|\bar{z}(k)\|_2 \leq \gamma \|v(k)\|_2$ under the zero initial condition for any nonzero $v(k) \in l_2 [0, +\infty)$, if there exist matrices $P > 0$, $Q_j > 0$ $(j = 1, 2, \ldots, q)$ and positive constant scalars λ_1, λ_2 satisfying*

$$\Phi < 0, \tag{3.21}$$

$$P \leq \lambda_1 I, \tag{3.22}$$

where

$$\Phi = \begin{bmatrix} \Phi_{11} & * & * & * \\ \hat{Z}^{\mathrm{T}} P \bar{A} & \Omega_{22} & * & * \\ \Omega_{31} & \bar{F}^{\mathrm{T}} P \hat{Z} & \Omega_{33} & * \\ \bar{D}_1^{\mathrm{T}} P \bar{A} & \bar{D}_1^{\mathrm{T}} P \hat{Z} & \bar{D}_1^{\mathrm{T}} P \bar{F} & \bar{D}_1^{\mathrm{T}} P \bar{D}_1 - \gamma^2 I \end{bmatrix},$$

$$\Phi_{11} = \Omega_{11} + \bar{L}^{\mathrm{T}} \bar{L},$$

with Ω_{11}, Ω_{22}, Ω_{31}, Ω_{33}, $\bar{C}_j$, $\hat{A}_d$, A_ρ, W_i, T_i, $\hat{Z}$, $\hat{Z}_a$, $\hat{R}_1$, $\hat{R}_2$, and $\tilde{A}_i$ being defined as in Theorem 3.2.3.

Proof. It is clear that $\Phi < 0$ implies $\Omega < 0$. According to Theorem 3.2.3, the filtering error system (3.9) is robustly exponentially stable in the mean square.

Let us now deal with the H_∞ performance of the system (3.9). Construct the same Lyapunov–Krasovskii functional as in Theorem 3.2.3. A similar calculation as in the proof of Theorem 3.2.3 leads to

$$\mathbb{E}\{\Delta V \mid \chi(k)\} \le \xi_0^{\mathrm{T}}(k)\Omega_2\xi_0(k), \tag{3.23}$$

where

$$\xi_0(k) = \left[\bar{x}^{\mathrm{T}}(k)\bar{x}^{\mathrm{T}}(k-\tau_1(k))\cdots\bar{x}^{\mathrm{T}}(k-\tau_q(k))f^{\mathrm{T}}(x(k))v^{\mathrm{T}}(k)\right]^{\mathrm{T}},$$

$$\Omega_2 = \begin{bmatrix} \Omega_{11} + \lambda_2 G^{\mathrm{T}}\tilde{R}_1 G & * & * & * \\ \hat{Z}^{\mathrm{T}}P\bar{A} & \Omega_{22} & * & * \\ \bar{F}^{\mathrm{T}}P\bar{A} & \bar{F}^{\mathrm{T}}P\hat{Z} & \bar{F}^{\mathrm{T}}P\bar{F} & * \\ \bar{D}_1^{\mathrm{T}}P\bar{A} & \bar{D}_1^{\mathrm{T}}P\hat{Z} & \bar{D}_1^{\mathrm{T}}P\bar{F} & \bar{D}_1^{\mathrm{T}}P\bar{D}_1 \end{bmatrix}.$$

In order to deal with the H_∞ performance of the filtering system (3.9), we introduce the following index:

$$J(n) = \mathbb{E}\sum_{k=0}^{\infty}[\bar{z}^{\mathrm{T}}(k)\bar{z}(k) - \gamma^2 v^{\mathrm{T}}(k)v(k)], \tag{3.24}$$

where n is a nonnegative integer. Obviously, our aim is to show $J(n) < 0$ under the zero initial condition. From (3.20), (3.23), and (3.24), one has

$$J(n) = \mathbb{E}\sum_{k=0}^{n}[\bar{z}^{\mathrm{T}}(k)\bar{z}(k) - \gamma^2 v^{\mathrm{T}}(k)v(k) + \Delta V(\chi(k))] - \mathbb{E}V(\chi(n+1))$$

$$\le \mathbb{E}\sum_{k=0}^{n}\left\{\bar{x}^{\mathrm{T}}(k)\bar{L}^{\mathrm{T}}\bar{L}\bar{x}(k) - \gamma^2 v^{\mathrm{T}}(k)v(k) + \xi_0^{\mathrm{T}}(k)\Omega_2\xi_0(k)\right.$$

$$\left. -\lambda_2\begin{bmatrix}\bar{x}(k)\\f(x(k))\end{bmatrix}^{\mathrm{T}}\begin{bmatrix}G^{\mathrm{T}}\tilde{R}_1 G & G^{\mathrm{T}}\tilde{R}_2\\\tilde{R}_2^{\mathrm{T}}G & I\end{bmatrix}\begin{bmatrix}c\bar{x}(k)\\f(x(k))\end{bmatrix}\right\}$$

$$= \xi_0^{\mathrm{T}}(k)\Phi\xi_0(k).$$

According to Theorem 3.2.4, we have $J(n) \le 0$. Letting $n \to \infty$, we obtain

$$\|\bar{z}(k)\|_2 \le \gamma\|v(k)\|_2,$$

which completes the proof of Theorem 3.2.4. □

3.3 Robust H_∞ Filter Design

In this section, we aim at solving the H_∞ filter design problem for the system (3.7); that is, we are interested in determining the filter matrices in (3.8) such that the filtering error system in (3.9) is exponentially stable with a guaranteed $\mathcal{H}_\infty$ performance. The following theorem provides sufficient conditions for the existence of such H_∞ filters for system (3.9).

Theorem 3.3.1 *Let $\gamma > 0$ be a given positive constant and the admissible conditions hold. Then, for the nonlinear system (3.7) with multiple communication delays and packet dropouts, there exists an admissible H_∞ filter of the form (3.8) such that the filtering error system (3.9) is robustly exponentially stable in the mean square for $v(k) = 0$ and also satisfies $\|\bar{z}(k)\|_2 \le \gamma \|v(k)\|_2$ under the zero initial condition for any nonzero $v(k) \in l_2\,[0, +\infty)$, if there exist positive-definite matrices P and $Q_j > 0$ $(j = 1, 2, \dots, q)$, positive constant scalars ε, λ_1, and λ_2, and matrices X and C_f satisfying*

$$\Lambda < 0, \tag{3.25}$$

$$P \le \lambda_1 I \tag{3.26}$$

where

$$\Lambda = \begin{bmatrix} \Lambda_1 & * & * & * \\ 0 & -\gamma^2 I & * & * \\ \Lambda_2 & \Lambda_3 & \Lambda_4 & * \\ 0 & 0 & \Lambda_5 & -\varepsilon I \end{bmatrix}, \quad \Lambda_1 = \begin{bmatrix} \Lambda_{11} & * & * \\ 0 & \Lambda_{22} & * \\ -\lambda_2 \tilde{R}_2^T G & 0 & -\lambda_2 I \end{bmatrix},$$

$$\Lambda_2 = \begin{bmatrix} \bar{X} & 0 & 0 \\ \hat{L}_0 + C_f \hat{R}_3 & 0 & 0 \\ P\hat{A}_0 + X\hat{R}_1 & P\hat{Z} & P\hat{F} \end{bmatrix}, \quad \Lambda_3 = [\,0 \quad 0 \quad (P\hat{D}_0 + X\hat{D}_1)^T\,]^T,$$

$$\Lambda_4 = \mathrm{diag}\{-\bar{P}, -I, -P\}, \quad \Lambda_5 = [\,0 \quad 0 \quad H_0^T P\,], \quad E_0 = \begin{bmatrix} E & 0 \end{bmatrix},$$

$$\Lambda_{11} = \lambda_1 A_\rho - P + \sum_{j=1}^{q}(d_M - d_m + 1)Q_j - \lambda_2 G^T \tilde{R}_1 G + \varepsilon E_0^T E_0,$$

$$\Lambda_{22} = \mathrm{diag}\{-Q_1 + \tilde{A}_1, -Q_2 + \tilde{A}_2, \dots - Q_q + \tilde{A}_q\} + \lambda_1 \rho_2 \hat{Z}_a^T \hat{Z}_a,$$

$$\bar{X} = [\,\sigma_1 \hat{R}_{21}^T X^T \quad \cdots \quad \sigma_r \hat{R}_{2r}^T X^T\,]^T, \quad \hat{L}_0 = [\,L \quad 0\,],$$

$$\hat{R}_{2j} = \begin{bmatrix} 0 & 0 \\ C_j & 0 \end{bmatrix}, \quad \hat{D}_1 = \begin{bmatrix} 0 \\ D_2 \end{bmatrix}, \quad \hat{A}_0 = \begin{bmatrix} A & 0 \\ 0 & 0 \end{bmatrix},$$

$$\hat{R}_3 = [\,0 \quad -I\,], \quad \hat{E} = [\,0 \quad I\,]^T, \quad \hat{D}_0 = [\,D_1^T \quad 0\,]^T,$$

$$\hat{R}_1 = \begin{bmatrix} 0 & I \\ \Xi C & 0 \end{bmatrix}, \quad H_0 = [\,H^T \quad 0\,]^T, \quad \bar{P} = \mathrm{diag}\{\underbrace{P, \dots, P}_{r}\}.$$

Furthermore, if $(P, Q_j, X, C_f, \varepsilon, \lambda_1, \lambda_2)$ is a feasible solution of (3.25) and (3.26), then the system matrices of the admissible H_∞ filter in the form of (3.8) can be obtained by means of the matrices X and C_f, where

$$[A_f \quad B_f] = [\hat{E}^T P \hat{E}]^{-1} \hat{E}^T X. \tag{3.27}$$

Proof. From Theorem 3.2.4, we know that there exists an admissible filter in the form of (3.8) such that the filtering error system (3.9) is robustly exponentially stable with a guaranteed H_∞ performance γ if there exist matrices $P > 0$, $Q_j > 0$ $(j = 1, 2, \dots, q)$, and positive constant scalars λ_1, λ_2 satisfying (3.21) and (3.22). By the Schur complement lemma, (3.21) is equivalent to

$$
\begin{bmatrix}
\lambda_1 A_\rho + \check{\Omega}_{11} & * & * & * & * & * & * \\
0 & \Lambda_{22} & * & * & * & * & * \\
-\lambda_2 \tilde{R}_2^{\mathrm{T}} G & 0 & -\lambda_2 I & * & * & * & * \\
0 & 0 & 0 & -\gamma^2 I & * & * & * \\
\bar{P}\hat{C} & 0 & 0 & 0 & -\bar{P} & * & * \\
\bar{L} & 0 & 0 & 0 & 0 & -I & * \\
P\bar{A} & P\hat{Z} & P\bar{F} & P\bar{D}_1 & 0 & 0 & -P
\end{bmatrix} < 0, \qquad (3.28)
$$

where

$$
\check{\Omega}_{11} = -P + \sum_{j=1}^{q}(d_M - d_m + 1)Q_j - \lambda_2 G^{\mathrm{T}}\tilde{R}_1 G,
$$

$$
\hat{C} = [\,\sigma_1 \bar{C}_1^{\mathrm{T}} \quad \sigma_2 \bar{C}_2^{\mathrm{T}} \quad \cdots \quad \sigma_r \bar{C}_r^{\mathrm{T}}\,]^{\mathrm{T}}.
$$

In order to avoid partitioning the positive-definite matrices P and Q_j, we rewrite the parameters in Theorem 3.2.4 in the following form:

$$
\bar{A} = \hat{A}_0 + H_0 F(k) E_0 + \hat{E} K \hat{R}_1, \quad \bar{C}_j = \hat{E} K \hat{R}_{2j},
$$

$$
\bar{L} = \hat{L}_0 + C_{\mathrm{f}} \hat{R}_3, \quad \bar{D}_1 = \hat{D}_0 + \hat{E} K \hat{D}_1,
$$

$$
K = [\,A_{\mathrm{f}} \quad B_{\mathrm{f}}\,], \quad X = P\hat{E}K,
$$

and therefore we can get (3.27). Then, from Lemma 3.2.2, we can obtain (3.25). The proof is now complete. $\qquad\square$

Remark 3.3 *In Theorem 3.3.1, the robust H_∞ filtering problem is solved for a class of discrete-time nonlinear networked systems with multiple stochastic communication delays and multiple packet dropouts by using a linear matrix inequality (LMI) approach. Obviously, our main results can be easily specialized to many special cases; for example, the cases when there are no nonlinearities, or no stochastic disturbances, or no parameter uncertainties, and so on. These specialized results are not listed here to keep the exposition concise. It is also worth pointing out that the main results in this chapter can be easily extended to the delayed jumping systems with sensor nonlinearities [116] and other more complicated systems. Note that we mainly focus on the effects brought by multiple stochastic communication delays and packet dropouts, which are two of the most important network-induced characteristics.*

Remark 3.4 *Lemma 3.2.2 is used to tackle the norm-bounded parameter uncertainties in the proof of Theorem 3.3.1. Comparing with existing literature, the system we consider is more comprehensive since the random delays, partial measurement missing, sector nonlinearities, parameter uncertainties, and stochastic disturbances are simultaneously taken into account. For deterministic time-delay system, a lot of research attention has been paid to the selection of*

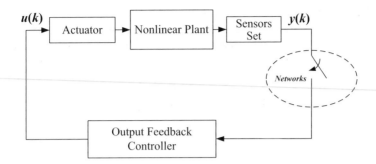

Figure 3.1 The framework of output feedback control systems over networks environments

Lyapunov functionals to reduce the conservatism; for example, see He et al. [153]. Similarly, for the discrete-time stochastic system considered in this chapter, we could further reduce the conservatism of the main results by making an effort towards the construction of more general Lyapunov functionals (e.g., the one used by Liu et al. [154]), which leaves a relatively minor research issue for further investigation.

3.4 Robust H_∞ Fuzzy Control

3.4.1 Problem Formulation

In this section we consider the output feedback control problem for discrete-time fuzzy systems in NCSs, where the framework is shown in Figure 3.1. The sensors are connected to the controller via a network which is shared by other NCSs and subject to communication delays and missing measurements (packet dropouts). The fuzzy systems with multiple stochastic communication delays and uncertain parameters can be described as follows:

△ **Plant Rule i:** IF $\theta_1(k)$ is M_{i1} and $\theta_2(k)$ is M_{i2} and $\cdots$ and $\theta_p(k)$ is M_{ip} THEN

$$x(k+1) = A_i(k)x(k) + A_{di} \sum_{m=1}^{h} \alpha_m(k)x(k - \tau_m(k)) + B_{1i}u(k) + D_{1i}v(k),$$

$$\tilde{y}(k) = C_i x(k) + D_{2i}v(k),$$

$$z(k) = C_{zi}(k)x(k) + B_{2i}u(k) + D_{3i}v(k),$$

$$x(k) = \phi(k), \ \forall k \in Z^-, \ i = 1, \ldots, r. \tag{3.29}$$

where M_{ij} is the fuzzy set, r is the number of IF–THEN rules, and $\theta(k) = [\theta_1(k), \theta_2(k), \ldots, \theta_p(k)]$ is the premise variable vector. It is assumed that the premise variables do not depend on the input variables $u(k)$, which is needed to avoid a complicated defuzzification process of fuzzy controllers. $x(k) \in \mathbb{R}^n$ represents the state vector; $u(k) \in \mathbb{R}^m$ is the control input; $\tilde{y}(k) \in \mathbb{R}^s$ is the process output; $z(k) \in \mathbb{R}^q$ is the controlled output; $v(k) \in \mathbb{R}^p$ is a vector of exogenous inputs which belongs to $l_2[0, \infty)$ (such as reference signals, disturbance signals, sensor noise, etc.); $\tau_m(k) \ (m = 1, 2, \ldots, h)$ are the communication delays that occur according to the stochastic variables $\alpha_m(k)$; $\alpha_m(k)$ satisfies the condition (3.5) and $\phi(k) \ (\forall k \in Z^-)$ is the initial state.

The matrices $A_i(k) = A_i + \Delta A_i(k)$, $C_{zi}(k) = C_{zi} + \Delta C_{zi}(k)$, and A_i, A_{di}, B_{1i}, B_{2i}, C_i, C_{zi}, D_{1i}, D_{2i}, and D_{3i} are known constant matrices with compatible dimensions. The matrices $\Delta A_i(k)$ and $\Delta C_{zi}(k)$ represent time-varying norm-bounded parameter uncertainties that satisfy

$$\begin{bmatrix} \Delta A_i(k) \\ \Delta C_{zi}(k) \end{bmatrix} = \begin{bmatrix} H_{ai} \\ H_{ci} \end{bmatrix} F(k)E, \tag{3.30}$$

where H_{ai}, H_{ci} and E are constant matrices of appropriate dimensions, and $F(k)$ is an unknown matrix function satisfying

$$F^{\mathrm{T}}(k)F(k) \leq I, \forall k. \tag{3.31}$$

The parameter uncertainties $\Delta A_i(k)$ and $\Delta C_{zi}(k)$ are said to be admissible if both (3.30) and (3.31) hold.

In this section, the missing measurement (packet dropout) phenomenon constitutes another focus of our present research. The multiple missing measurements are described by

$$y(k) = \Xi C_i x(k) + D_{2i} v(k)$$
$$= \sum_{l=1}^{s} \beta_l C_{il} x(k) + D_{2i} v(k), \tag{3.32}$$

where $y(k) \in \mathbb{R}^s$ is the *actual* measurement signal of (3.4.1), $\Xi := \mathrm{diag}\{\beta_1, \ldots, \beta_s\}$ with β_l $(l = 1, \ldots, s)$ being s unrelated random variables which are also unrelated with $\alpha_m(k)$. It is assumed that β_l has the probabilistic density function $q_l(s)$ $(l = 1, \ldots, s)$ on the interval $[0, 1]$ with mathematical expectation μ_l and variance σ_l^2. C_{il} is defined by

$$C_{il} := \mathrm{diag}\{\underbrace{0, \ldots, 0}_{l-1}, 1, \underbrace{0, \ldots, 0}_{s-l}\}C_i.$$

In the following, we denote $\bar{\Xi} = \mathbb{E}\{\Xi\}$.

For a given pair of $(x(k), u(k))$, the final output of the fuzzy system is inferred as follows:

$$x(k+1) = \sum_{i=1}^{r} h_i(\theta(k)) \Bigg[A_i(k)x(k) + B_{1i}u(k)$$
$$+ A_{di} \sum_{m=1}^{h} \alpha_m(k)x(k - \tau_m(k)) + D_{1i}v(k) \Bigg],$$

$$y(k) = \sum_{i=1}^{r} h_i(\theta(k))[\Xi C_i x(k) + D_{2i}v(k)],$$

$$z(k) = \sum_{i=1}^{r} h_i(\theta(k))[C_{zi}(k)x(k) + B_{2i}u(k) + D_{3i}v(k)], \tag{3.33}$$

where the fuzzy basis functions are given by

$$h_i(\theta(k)) = \frac{\vartheta_i(\theta(k))}{\sum\limits_{i=1}^{r} \vartheta_i(\theta(k))}, \quad \vartheta_i(\theta(k)) = \prod_{j=1}^{p} M_{ij}(\theta_j(k)),$$

with $M_{ij}(\theta_j(k))$ representing the grade of membership of $\theta_j(k)$ in M_{ij}. Here, $\vartheta_i(\theta(k))$ has the following basic property:

$$\vartheta_i(\theta(k)) \geq 0, i = 1, 2, \ldots, r, \quad \sum_{i=1}^{r} \vartheta_i(\theta(k)) > 0, \forall k,$$

and therefore

$$h_i(\theta(k)) \geq 0, i = 1, 2, \ldots, r, \quad \sum_{i=1}^{r} h_i(\theta(k)) = 1, \forall k.$$

In what follows, we define $h_i := h_i(\theta(k))$ for brevity.

In this section, by the parallel distributed compensation, we consider the following *fuzzy dynamic output feedback controller* for the fuzzy system (3.33):

△ **Controller Rule i:** IF $\theta_1(k)$ is M_{i1} and $\theta_2(k)$ is M_{i2} and $\cdots$ and $\theta_p(k)$ is M_{ip} THEN

$$\begin{cases} x_c(k+1) = A_{ki}x_c(k) + B_{ki}y(k) \\ u(k) = C_{ki}x_c(k), \quad i = 1, 2, \ldots, r \end{cases} \tag{3.34}$$

where $x_c(k) \in R^n$ is the controller state; A_{ki}, B_{ki}, and C_{ki} are controller parameters to be determined. Then, the overall fuzzy output feedback controller is given by

$$\begin{cases} x_c(k+1) = \sum\limits_{i=1}^{r} h_i \left[A_{ki}x_c(k) + B_{ki}y(k) \right] \\ u(k) = \sum\limits_{i=1}^{r} h_i C_{ki}x_c(k), \quad i = 1, 2, \ldots, r \end{cases} \tag{3.35}$$

From (3.33) and (3.35), the closed-loop system can be obtained as

$$\bar{x}(k+1) = \sum_{i=1}^{r}\sum_{j=1}^{r} h_i h_j \left[\left(A_{ij}(k) + B_{ij} \right) \bar{x}(k) + D_{ij}v(k) \right.$$
$$\left. + \sum_{m=1}^{h} (\bar{A}_{dmi} + \tilde{A}_{dmi})\bar{x}(k - \tau_m(k)) \right], \tag{3.36}$$
$$z(k) = \sum_{i=1}^{r}\sum_{j=1}^{r} h_i h_j [\bar{C}_{ij}(k)\bar{x}(k) + D_{3i}v(k)],$$

where

$$\bar{x}(k) = \begin{bmatrix} x(k) \\ x_c(k) \end{bmatrix}, \quad A_{ij}(k) = \begin{bmatrix} A_i(k) & B_{1i}C_{kj} \\ B_{ki}\bar{\Xi}C_j & A_{ki} \end{bmatrix},$$

$$B_{ij} = \begin{bmatrix} 0 & 0 \\ B_{ki}(\Xi - \bar{\Xi})C_j & 0 \end{bmatrix}, \quad D_{ij} = \begin{bmatrix} D_{1i} \\ B_{ki}D_{2j} \end{bmatrix},$$

$$\bar{A}_{dmi} = \begin{bmatrix} \bar{\alpha}_m A_{di} & 0 \\ 0 & 0 \end{bmatrix}, \quad \tilde{A}_{dmi} = \begin{bmatrix} \tilde{\alpha}_m(k)A_{di} & 0 \\ 0 & 0 \end{bmatrix},$$

$$\bar{C}_{ij}(k) = [C_{zi}(k) \quad B_{2i}C_{kj}],$$

with $\tilde{\alpha}_m(k) = \alpha_m(k) - \bar{\alpha}_m$. It is clear that $\mathbb{E}\{\tilde{\alpha}_m(k)\} = 0$ and $\mathbb{E}\{\tilde{\alpha}_m^2(k)\} = \bar{\alpha}_m(1 - \bar{\alpha}_m)$.

Our aim in this section is to develop techniques to deal with the robust H_∞ dynamic output feedback control problem for the discrete-time fuzzy system (3.36) such that, for all admissible multiple stochastic communication delays, multiple missing measurements, and uncertain parameters, the following two requirements are satisfied simultaneously:

(R1) The fuzzy system (3.36) is exponentially stable in the mean square.
(R2) Under zero initial condition, the controlled output $z(k)$ satisfies

$$\sum_{k=0}^{\infty} \mathbb{E}\{\|z(k)\|^2\} \leqslant \gamma^2 \sum_{k=0}^{\infty} \mathbb{E}\{\|v(k)\|^2\} \tag{3.37}$$

for all nonzero $v(k)$, where $\gamma > 0$ is a prescribed scalar.

3.4.2   Performance Analysis

Lemma 3.4.1   [155] For any real matrices X_{ij} for $i, j = 1, 2, \ldots, r$ and $n > 0$ with appropriate dimensions, we have

$$\sum_{i=1}^{r}\sum_{j=1}^{r}\sum_{k=1}^{r}\sum_{l=1}^{r} h_i h_j h_k h_l X_{ij}^{\mathrm{T}} \Lambda X_{kl} \leq \sum_{i=1}^{r}\sum_{j=1}^{r} h_i h_j X_{ij}^{\mathrm{T}} \Lambda X_{ij}. \tag{3.38}$$

For convenience of presentation, we first discuss the nominal system of (3.36) (i.e., without parameter uncertainties $\Delta A_i(k)$ and $\Delta C_{zi}(k)$) and will eventually extend our main results to the general case. We have the following analysis result that serves as a theoretical basis for the subsequent design problem.

By using similar analysis techniques to those in Section 3.4.1, some parallel results are derived and listed as follows.

Theorem 3.4.2   *Consider the nominal fuzzy system of (3.36) with given controller parameters and a prescribed H_∞ performance $\gamma > 0$. Then, the nominal fuzzy system of (3.36) is* *exponentially stable with disturbance attenuation level γ if there exist matrices $P > 0$ and $Q_k > 0$ $(k = 1, 2, \ldots, h)$ satisfying*

$$\begin{bmatrix} \Pi_i & * \\ 0.5\Sigma_{ii} & \Lambda \end{bmatrix} < 0, \tag{3.39}$$

$$\begin{bmatrix} 4\Pi_i & * \\ \Sigma_{ij} & \Lambda \end{bmatrix} < 0, \tag{3.40}$$

where

$$\Pi_i = \mathrm{diag}\{\Upsilon_k, F_i, -\gamma^2 I\}, \quad \Lambda = \mathrm{diag}\{-\check{P}, -P, -I\},$$

$$\Sigma_{ij} = \begin{bmatrix} \Phi_{11ij} & \Phi_{12ij} \\ \Phi_{21ij} & D_{3i} + D_{3j} \end{bmatrix}, \quad \Phi_{11ij} = \begin{bmatrix} \check{C}_{ij} + \check{C}_{ji} & 0 \\ P(A_{ij} + A_{ji}) & P(\hat{Z}_{mi} + \hat{Z}_{mj}) \end{bmatrix},$$

$$\Phi_{12ij} = \begin{bmatrix} 0 \\ P(D_{ij} + D_{ji}) \end{bmatrix}, \quad \Phi_{21ij} = [\bar{C}_{ij} + \bar{C}_{ji} \quad 0],$$

$$\Upsilon_k = \sum_{k=1}^{h}(d_T - d_t + 1)Q_k - P, \quad \bar{C}_{ij} = [C_{zi} \quad B_{2i}C_{kj}],$$

$$\hat{C}_{lij} = \begin{bmatrix} 0 & 0 \\ B_{ki}C_{jl} & 0 \end{bmatrix}, \quad A_{ij} = \begin{bmatrix} A_i & B_{1i}C_{kj} \\ B_{ki}\bar{\Xi}C_j & A_{ki} \end{bmatrix},$$

$$\check{C}_{ij} = [\sigma_1\hat{C}_{1ij}^{\mathrm{T}}P, \ldots, \sigma_s\hat{C}_{sij}^{\mathrm{T}}P]^{\mathrm{T}}, \quad \hat{A}_{di} = \begin{bmatrix} A_{di} & 0 \\ 0 & 0 \end{bmatrix},$$

$$\check{P} = \mathrm{diag}\{\underbrace{P, \ldots, P}_{s}\}, \quad \check{P} = \mathrm{diag}\{\underbrace{P, \ldots, P}_{h}\},$$

$$\check{A}_{di} = \mathrm{diag}\{\underbrace{\hat{A}_{di}, \ldots, \hat{A}_{di}}_{h}\}, \quad \hat{Z}_{mi} = [\bar{A}_{d1i}, \ldots, \bar{A}_{dhi}],$$

$$\hat{\alpha} = \mathrm{diag}\{\bar{\alpha}_1(1 - \bar{\alpha}_1), \ldots, \bar{\alpha}_h(1 - \bar{\alpha}_h)\},$$

$$F_i = \hat{\alpha}\check{A}_{di}^{\mathrm{T}}\check{P}\check{A}_{di} - \hat{Q}, \quad \hat{Q} = \mathrm{diag}\{Q_1, \ldots, Q_h\}. \tag{3.41}$$

3.4.3 Controller Design

In this section, we aim at solving the robust H_∞ fuzzy output feedback controller design problem for the system (3.36). That is, we are interested in determining the controller parameters in (3.35) such that the closed-loop fuzzy system in (3.36) is exponentially stable with a guaranteed H_∞ performance. The following theorem provides sufficient conditions for the existence of such an H_∞ fuzzy controller for the nominal fuzzy system of (3.36).

Theorem 3.4.3 *Consider the nominal fuzzy system of (3.36). For a prescribed constant $\gamma > 0$, if there exist positive-definite matrices $P > 0$, $L > 0$, and $Q_k > 0$ $(k = 1, 2, \ldots, h)$,* *matrices K_i and $\bar{C}_{ki}$ such that*

$$\Gamma_1 \triangleq \begin{bmatrix} \Pi_i & * \\ 0.5\bar{\Sigma}_{ii} & \bar{\Lambda} \end{bmatrix} < 0 \quad i = 1, 2, \ldots, r, \tag{3.42}$$

$$\Gamma_2 \triangleq \begin{bmatrix} 4\Pi_i & * \\ \bar{\Sigma}_{ij} & \bar{\Lambda} \end{bmatrix} < 0, \quad 1 \le i < j \le r, \tag{3.43}$$

$$PL = I \tag{3.44}$$

hold, then the nominal system of (3.36) is exponentially stable with disturbance attenuation γ, where Π_i is defined in Theorem 3.4.2 and

$$\bar{\Lambda} = \text{diag}\{-\bar{L}, -L, -I\}, \quad \bar{\Sigma}_{ij} = \begin{bmatrix} \bar{\Phi}_{11ij} + \bar{\Phi}_{11ji} & \bar{\Phi}_{12ij} + \bar{\Phi}_{12ji} \\ \bar{\Phi}_{21ij} + \bar{\Phi}_{21ji} & D_{3i} + D_{3j} \end{bmatrix},$$

$$\bar{\Phi}_{11ij} = \begin{bmatrix} \check{C}_{ij} & 0 \\ \bar{A}_i + \bar{E}K_i\bar{R}_j + \bar{B}_{1i}\bar{C}_{kj} & \hat{Z}_{mi} \end{bmatrix}, \quad \bar{\Phi}_{12ij} = \begin{bmatrix} 0 \\ \bar{D}_{1i} + \bar{E}K_i\bar{D}_{2j} \end{bmatrix},$$

$$\bar{\Phi}_{21ij} = [\bar{C}_{zi} + B_{2i}\bar{C}_{kj} \quad 0], \quad \bar{A}_i = \begin{bmatrix} A_i & 0 \\ 0 & 0 \end{bmatrix}, \quad \bar{E} = \begin{bmatrix} 0 \\ I \end{bmatrix}, \quad \bar{B}_{1i} = \begin{bmatrix} B_{1i} \\ 0 \end{bmatrix},$$

$$R_{il} = \begin{bmatrix} 0 & 0 \\ C_{il} & 0 \end{bmatrix}, \quad \bar{D}_{1i} = \begin{bmatrix} D_{1i} \\ 0 \end{bmatrix}, \quad \bar{D}_{2i} = \begin{bmatrix} 0 \\ D_{2i} \end{bmatrix}, \quad K_i = [A_{ki} \quad B_{ki}],$$

$$\bar{C}_{ki} = [0 \quad C_{ki}], \quad \bar{C}_{zi} = [C_{zi} \quad 0], \quad \bar{L} = \text{diag}\{\underbrace{L, \ldots, L}_{s}\},$$

$$\bar{R}_i = \begin{bmatrix} 0 & I \\ \bar{\Xi}C_i & 0 \end{bmatrix}, \quad \check{C}_{ij} = \begin{bmatrix} \sigma_1\bar{E}K_i\hat{R}_{j1} \\ \cdots \\ \sigma_s\bar{E}K_i\hat{R}_{js} \end{bmatrix}. \tag{3.45}$$

Furthermore, if $(P, L, Q_k, K_i, \bar{C}_{ki})$ is a feasible solution of (3.42)–(3.44), then the controller parameters in the form of (3.35) can be easily obtained by K_i and $\bar{C}_{ki}$.

Having established the main results for nominal systems, we are now in a position to show that the robust H_∞ controller parameters can be determined based on the results of Theorem 3.4.3.

Theorem 3.4.4   *Consider the uncertain fuzzy system (3.36). For a prescribed constant $\gamma > 0$, if there exist positive-definite matrices $P > 0$, $L > 0$, and $Q_k > 0$ ($k = 1, 2, \ldots, h$), matrices K_i and $\bar{C}_{ki}$, and a positive constant scalar $\varepsilon > 0$ such that*

$$\begin{bmatrix} \Pi_i & * & * \\ 0.5\bar{\Sigma}_{ii} & \bar{\Lambda} & * \\ \hat{\Sigma} & 0.5\tilde{\Sigma}_{ii} & \Xi \end{bmatrix} < 0 \quad i = 1, 2, \ldots, r, \tag{3.46}$$

$$\begin{bmatrix} 4\Pi_i & * & * \\ \bar{\Sigma}_{ij} & \bar{\Lambda} & * \\ \hat{\Sigma} & \tilde{\Sigma}_{ij} & \Xi \end{bmatrix} < 0 \quad 1 \le i < j \le r, \tag{3.47}$$

$$PL = I \tag{3.48}$$

hold, where Π_i is defined in Theorem 3.4.2 and

$$\hat{\Sigma} = [\,\hat{\Phi} \quad 0\,], \quad \hat{\Phi} = \begin{bmatrix} 0 \\ \varepsilon\hat{E} \end{bmatrix}, \quad \Xi = \text{diag}\{-\varepsilon I, -\varepsilon I\},$$

$$\tilde{\Sigma}_{ij} = \begin{bmatrix} 0 & \bar{H}_{ai}^{\mathrm{T}} + \bar{H}_{aj}^{\mathrm{T}} & \bar{H}_{ci}^{\mathrm{T}} + \bar{H}_{cj}^{\mathrm{T}} \\ 0 & 0 & 0 \end{bmatrix},$$

then the system (3.36) is exponentially stable with disturbance attenuation γ. Furthermore, if $(P, L, Q_k, K_i, \bar{C}_{ki})$ is a feasible solution of (3.46)–(3.48), then the controller parameters in the form of (3.35) can be obtained directly from K_i and $\bar{C}_{ki}$.

Note that there is a matrix equality in Theorem 3.4.4 which gives rise to significant difficulty in numerical computation. Nevertheless, such difficulty can be overcome by using the CCL algorithm proposed in Refs [97, 156]. In view of this observation, we put forward the following nonlinear minimization problem involving LMI conditions instead of the original nonconvex feasibility problem formulated in Theorem 3.4.4.

The nonlinear minimization problem: $\min \text{tr}(PL)$ subject to (3.46) and (3.47) and

$$\begin{bmatrix} P & I \\ I & L \end{bmatrix} \geq 0. \tag{3.49}$$

If the solution of $\min \text{tr}(PL)$ subject to (3.46), (3.47), and (3.49) exists and $\min \text{tr}(PL) = n$, then the conditions in Theorem 3.4.4 are solvable.

Finally, the following algorithm is suggested to solve the above problem.

Algorithm H_∞ Fuzzy Control (HinfFC)

Step 1. Find a feasible set $(P_{(0)}, L_{(0)}, Q_{k(0)}, K_{i(0)}, \bar{C}_{ki(0)})$ satisfying (3.46), (3.47), and (3.49). Set $q = 0$.

Step 2. According to (3.46), (3.47), and (3.49), solve the LMI problem: $\min \text{tr}(PL_{(q)} + P_{(q)}L)$.

Step 3. Substitute the matrix variables $(P, L, Q_k, K_i, \bar{C}_{ki})$ obtained into (3.39) and (3.40). If conditions (3.39) and (3.40) are satisfied with $|\text{tr}(PL) - n| < \delta$ for some sufficiently small scalar $\delta > 0$, then output the feasible solutions. Exit.

Step 4. If $q > N$, where N is the maximum number of iterations allowed, then output the feasible solutions $(P, L, Q_k, K_i, \bar{C}_{ki})$ and exit. Else, set $q = q + 1$ and go to Step 2.

3.5 Illustrative Examples

In this section, some simulation examples are presented to demonstrate the theory presented in this chapter.

3.5.1   Example 1

In this example, we consider robust H_∞ filtering for a class of nonlinear networked systems with multiple stochastic communication delays and packet dropouts.

The system data of (3.2) are as follows:

$$A = \begin{bmatrix} 0.2 & 0 & 0.1 \\ 0.1 & -0.3 & 0.1 \\ 0.1 & 0 & -0.2 \end{bmatrix}, \quad H = \begin{bmatrix} 0.1 \\ 0.2 \\ 0.1 \end{bmatrix},$$

$$F(k) = \sin(0.6k), \quad E = \begin{bmatrix} 0.1 & 0.1 & 0.1 \end{bmatrix},$$

$$A_d = \begin{bmatrix} 0.2 & 0 & 0.1 \\ 0.1 & -0.3 & 0.1 \\ 0.1 & 0 & -0.2 \end{bmatrix}, \quad F = \begin{bmatrix} 0.2 & 0 & 0.1 \\ 0.1 & 0.3 & 0.1 \\ 0.1 & 0 & 0.2 \end{bmatrix},$$

$$D_1 = \begin{bmatrix} -0.2 & 0 & 0.1 \\ -0.1 & 0.1 & 0.1 \\ 0 & 0.2 & 0.1 \end{bmatrix}, \quad C = \begin{bmatrix} 1 & 0.8 & 0.7 \\ -0.6 & 0.9 & 0.6 \\ 0.2 & 0.1 & 0.1 \end{bmatrix},$$

$$D_2 = \begin{bmatrix} 0.9 & -0.6 & 0.1 \\ 0.5 & 0.8 & 0.1 \\ 0.2 & 0.3 & 0.1 \end{bmatrix}, \quad L = \begin{bmatrix} -0.1 & 0 & 0.1 \end{bmatrix}.$$

Let $\gamma = 0.9$, $f(x(k)) = 0.4\sin(x(k))$, and $g(x(k), \tilde{x}(k), k) = 0.5x(k) + 0.5\tilde{x}(k)$. Then assume that the time-varying communication delays satisfy $2 \le \tau_i(k) \le 6$ $(i = 1, 2)$ and

$$\bar{\alpha}_1 = \mathbb{E}\{\alpha_1(k)\} = 0.8, \quad \bar{\alpha}_2 = \mathbb{E}\{\alpha_2(k)\} = 0.6.$$

Suppose that the probabilistic density functions of β_1, β_2, and β_3 in [0, 1] are described by

$$q_1(s_1) = \begin{cases} 0s_1 = 0 \\ 0.1s_1 = 0.5 \\ 0.9s_1 = 1 \end{cases}, \quad q_2(s_2) = \begin{cases} 0.1s_2 = 0 \\ 0.1s_2 = 0.5 \\ 0.8s_2 = 1 \end{cases}, \quad q_3(s_3) = \begin{cases} 0s_3 = 0 \\ 0.2s_3 = 0.5 \\ 0.8s_3 = 1 \end{cases}$$

from which the expectations and variances can be easily calculated as $\mu_1 = 0.95$, $\mu_2 = 0.85$, $\mu_3 = 0.9$, $\sigma_1 = 0.15$, $\sigma_2 = 0.32$, and $\sigma_3 = 0.2$. The initial condition is set to be $x_0 = \begin{bmatrix} 1 & 0 & -1 \end{bmatrix}^T$, $\hat{x}_0 = \begin{bmatrix} 0 & 0 & 0 \end{bmatrix}^T$ and the external disturbance v_k is described by

$$v_k = \begin{cases} 0.1, & 20 \le k \le 50 \\ -0.1, & 70 \le k \le 100 \\ 0, & \text{else} \end{cases}$$

We would like to design a filter in the form of (3.8) so that the filtering error system in (3.9) is exponentially stable with a guaranteed H_∞ norm bound γ. By applying Theorem 3.3.1 with

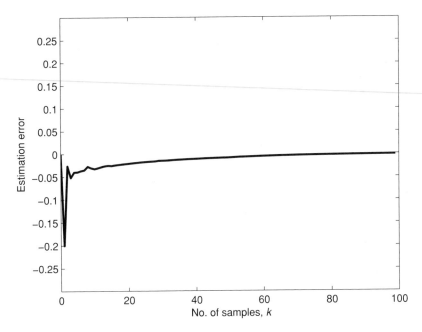

Figure 3.2 Estimation error $z(k) - \hat{z}(k)$

help from the Matlab LMI toolbox, we can obtain the desired H_∞ filter parameters as follows (other matrices are omitted to save space):

$$A_f = \begin{bmatrix} 0.3170 & 0.2021 & 0.1123 \\ 0.3169 & 0.3169 & 0.3169 \\ 0.3170 & 0.1106 & 0.3170 \end{bmatrix}, \quad B_f = \begin{bmatrix} -0.0079 & -0.0407 & 0.0944 \\ -0.0080 & -0.0408 & 0.0948 \\ -0.0079 & -0.0406 & 0.0942 \end{bmatrix},$$

$$C_f = [\, 0.3280 \quad 0.1220 \quad 0.4231 \,].$$

The simulation results are shown in Figures 3.2–3.7. Figure 3.2 plots the estimation error $\bar{z}(k)$. The actual state response $x_i(k)$ and the estimate $\hat{x}_i(k)$ $(i = 1, 2, 3)$ are depicted in Figures 3.3–3.5. Figure 3.6 shows the time-varying delays $\tau_i(k)$ $(i = 1, 2)$. The Bernoulli sequences $\alpha_i(k)$ $(i = 1, 2)$ are drawn in Figure 3.7. All the simulations have confirmed our theoretical analysis for the problems of robust H_∞ filtering for discrete nonlinear networked systems with multiple time-varying random communication delays and multiple packet dropouts.

3.5.2 Example 2

In this example, we consider robust H_∞ fuzzy output-feedback control with multiple probabilistic delays and multiple missing measurements.

Consider a T–S fuzzy model (3.4.1) with multiple communication delays and multiple missing measurements. The rules are as follows:

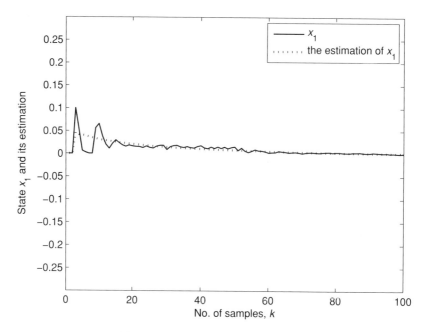

Figure 3.3 $x_1(k)$ and its estimate $\hat{x}_1(k)$

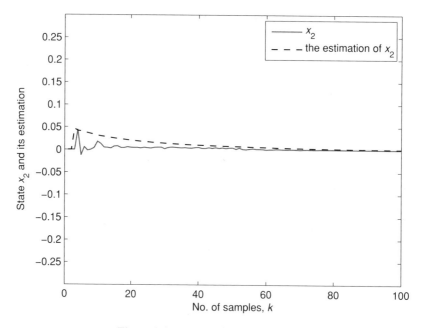

Figure 3.4 $x_2(k)$ and its estimate $\hat{x}_2(k)$

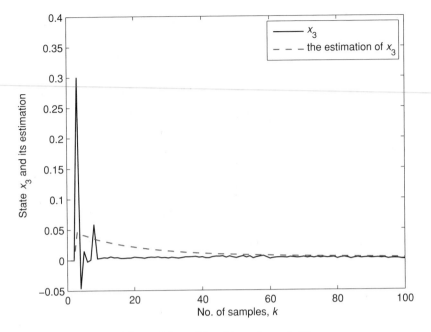

Figure 3.5 $x_3(k)$ and its estimate $\hat{x}_3(k)$

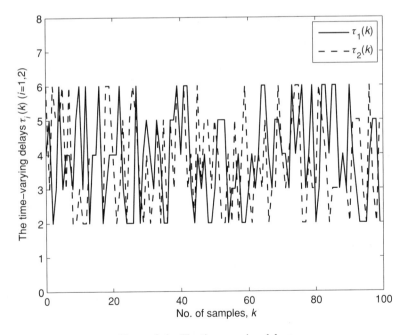

Figure 3.6 The time-varying delays

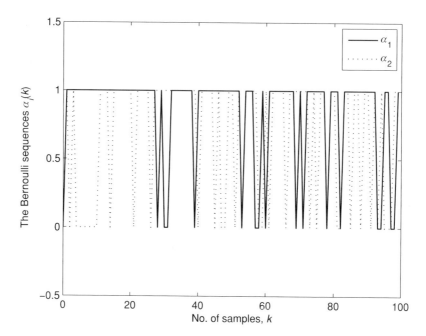

Figure 3.7 The Bernoulli sequences $\alpha_i(k)$

Plant Rule 1: IF $x_1(k)$ is $h_1(x_1(k))$ THEN

$$\begin{cases} x(k+1) = A_1(k)x(k) + A_{d1} \sum_{m=1}^{h} \alpha_m(k)x(k - \tau_m(k)) + B_{11}u(k) + D_{11}v(k), \\ y(k) = \Xi C_1 x(k) + D_{21}v(k), \\ z(k) = C_{z1}(k)x(k) + B_{21}u(k) + D_{31}v(k). \end{cases} \tag{3.50}$$

Plant Rule 2: IF $x_1(k)$ is $h_2(x_1(k))$ THEN

$$\begin{cases} x(k+1) = A_2(k)x(k) + A_{d2} \sum_{m=1}^{h} \alpha_m(k)x(k - \tau_m(k)) + B_{12}u(k) + D_{12}v(k), \\ y(k) = \Xi C_2 x(k) + D_{22}v(k), \\ z(k) = C_{z2}(k)x(k) + B_{22}u(k) + D_{32}v(k). \end{cases} \tag{3.51}$$

The model parameters are as follows:

$$A_1 = \begin{bmatrix} 1 & 0.2 & 0 \\ 0.1 & 0.1 & 0.1 \\ 0.1 & 0.2 & 0.2 \end{bmatrix}, \quad D_{11} = \begin{bmatrix} 0.1 \\ 0 \\ 0 \end{bmatrix}, \quad A_{d1} = \begin{bmatrix} 0.03 & 0 & -0.01 \\ 0.02 & 0.03 & 0 \\ 0.04 & 0.05 & -0.1 \end{bmatrix},$$

$$B_{11} = \begin{bmatrix} 1 & 1 \\ 0.4 & 1 \\ 0 & 1 \end{bmatrix}, \quad D_{31} = \begin{bmatrix} -0.1 \\ 0 \\ 0.1 \end{bmatrix}, \quad C_1 = \begin{bmatrix} 1 & 0.8 & 0.7 \\ -0.6 & 0.9 & 0.6 \end{bmatrix},$$

$$D_{21} = \begin{bmatrix} 0.15 \\ 0 \end{bmatrix}, \quad C_2 - \begin{bmatrix} 0.1 & 0.8 & 0.7 \\ -0.6 & 0.9 & 0.6 \end{bmatrix}, \quad D_{22} = \begin{bmatrix} 0.1 \\ 0 \end{bmatrix},$$

$$C_{z1} = \begin{bmatrix} 0.2 & 0 & 0 \\ 0 & 0 & 0 \\ 0 & 0 & 0.1 \end{bmatrix}, \quad B_{21} = \begin{bmatrix} 1 & 1 \\ 0 & 1 \\ 0 & 1 \end{bmatrix}, \quad H_{a1} = \begin{bmatrix} 0.1 \\ 0.1 \\ 0.1 \end{bmatrix}, \quad H_{c1} = \begin{bmatrix} 0.1 \\ 0 \\ 0.1 \end{bmatrix},$$

$$H_{a2} = \begin{bmatrix} 0.1 \\ 0 \\ 0.1 \end{bmatrix}, \quad E = \begin{bmatrix} 0.1 \\ 0.1 \\ 0.1 \end{bmatrix}^{\mathrm{T}}, \quad H_{c2} = \begin{bmatrix} 0.1 \\ 0 \\ 0.5 \end{bmatrix}, \quad D_{32} = \begin{bmatrix} 0.1 \\ 0 \\ 0.1 \end{bmatrix},$$

$$A_2 = \begin{bmatrix} 1 & -0.38 & 0 \\ -0.2 & 0 & 0.21 \\ 0.1 & 0 & -0.55 \end{bmatrix}, \quad B_{12} = \begin{bmatrix} 1 & 0 \\ 1 & 1 \\ 0 & 1 \end{bmatrix}, \quad B_{22} = \begin{bmatrix} 1 & 0 \\ 0 & 1 \\ 1 & 1 \end{bmatrix},$$

$$A_{d2} = \begin{bmatrix} 0 & 0.01 & -0.01 \\ 0.02 & 0.03 & 0 \\ 0.04 & 0.05 & -0.1 \end{bmatrix}, \quad D_{12} = \begin{bmatrix} 0.1 \\ 0 \\ 0.1 \end{bmatrix}, \quad C_{z2} = \begin{bmatrix} 0.1 & 0 & 0 \\ 0.2 & 0 & 0.2 \\ 0 & 0.1 & 0.2 \end{bmatrix}.$$

Assume that the time-varying communication delays satisfy $2 \le \tau_m(k) \le 6$ ($m = 1, 2$) and

$$\bar{\alpha}_1 = \mathbb{E}\{\alpha_1(k)\} = 0.8, \quad \bar{\alpha}_2 = \mathbb{E}\{\alpha_2(k)\} = 0.6.$$

Let the probabilistic density functions of β_1 and β_2 in $[0, 1]$ be described by

$$q_1(s_1) = \begin{cases} 0 & s_1 = 0 \\ 0.1 & s_1 = 0.5 \\ 0.9 & s_1 = 1 \end{cases} \quad \text{and} \quad q_2(s_2) = \begin{cases} 0.1 & s_2 = 0 \\ 0.1 & s_2 = 0.5 \\ 0.8 & s_2 = 1 \end{cases}, \tag{3.52}$$

from which the expectations and variances can be easily calculated as $\mu_1 = 0.95$, $\mu_2 = 0.85$, $\sigma_1 = 0.15$, and $\sigma_2 = 0.32$.

The membership function is assumed to be

$$h_1 = \begin{cases} 1, & x_0(1) = 0, \\ |\sin(x_0(1))|/x_0(1), & \text{else,} \end{cases} \tag{3.53}$$

$$h_2 = 1 - h_1.$$

Our aim is to design a dynamic output feedback paralleled controller in the form of (3.35) such that the system (3.36) is exponentially stable with a guaranteed H_∞ norm bound γ.

Letting $\gamma = 0.9$ and applying Theorem 3.4.4 with help from Algorithm HinfFC (see Section 3.4.3), we can obtain the desired H_∞ controller parameters as follows (other matrices are omitted to save space):

$$A_{k1} = \begin{bmatrix} -0.3671 & 0.0015 & 0.1389 \\ -0.2568 & 0.0028 & 0.1000 \\ -0.1402 & -0.0063 & 0.0417 \end{bmatrix}, \quad B_{k1} = \begin{bmatrix} 0.0245 & 0.1445 \\ 0.0187 & 0.1091 \\ 0.0031 & 0.0246 \end{bmatrix},$$

$$A_{k2} = \begin{bmatrix} -0.5428 & -0.0071 & 0.2040 \\ -0.4209 & -0.0074 & 0.1557 \\ -0.0515 & 0.0069 & 0.0292 \end{bmatrix}, \quad B_{k2} = \begin{bmatrix} 0.1098matrix & -0.1159 \\ 0.0426 & -0.1057 \\ 0.1741 & 0.0509 \end{bmatrix},$$

$$C_{k1} = \begin{bmatrix} -0.8083 & -0.0014 & 0.3029 \\ 0.2878 & 0.0006 & -0.1093 \end{bmatrix}, \quad C_{k2} = \begin{bmatrix} -0.4780 & 0.0039 & 0.1819 \\ 0.6162 & 0.0004 & -0.2310 \end{bmatrix}.$$

For simulation purposes, we set the initial condition as

$$x(0) = [1 \quad 0 \quad -1]^{\mathrm{T}}, \quad x_{\mathrm{c}}(0) = [0 \quad 0 \quad 0]^{\mathrm{T}} \tag{3.54}$$

and the external disturbance as $v(k) \equiv 0$. Figure 3.8 gives the state evolutions for the uncontrolled fuzzy systems, which are apparently unstable. Figure 3.9 gives the state simulation results of the closed-loop fuzzy system, from which we can see that the closed-loop system is exponentially stable.

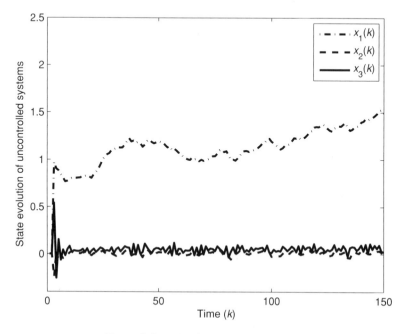

Figure 3.8 $x(k)$ of uncontrolled systems

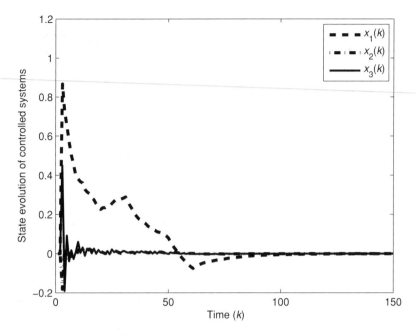

Figure 3.9 $x(k)$ of controlled systems

Next, to illustrate the disturbance attenuation performance, the initial condition is chosen as $x(0) = 0$, $x_c(0) = 0$, and the external disturbance $v(k)$ is assumed to be

$$v(k) = \begin{cases} 0.3, & 20 \leq k \leq 30, \\ -0.2, & 50 \leq k \leq 60, \\ 0, & \text{else.} \end{cases} \tag{3.55}$$

Figure 3.10 shows the controller output, Figure 3.11 plots the controller state, and Figure 3.12 depicts the disturbance input $v(k)$ and controlled output $z(k)$. Figure 3.13 shows the time-varying delays $\tau_m(k)$ ($m = 1, 2$). All the simulation results confirm our theoretical analysis for the robust H_∞ fuzzy control problem for discrete-time fuzzy systems with multiple time-varying random communication delays and multiple missing measurements.

3.5.3 Example 3

In this example, we consider an uncertain nonlinear mass–spring–damper mechanical system [157] controlled through a network, whose dynamic equation is

$$\ddot{x}(t) = c(t)\dot{x}(t) - 0.02x(t) - 0.67x^3(t) + u(t). \tag{3.56}$$

Assume that $x(t) \in [-1.5, 1.5]$, $\dot{x}(t) \in [-1.5, 1.5]$, and $c(t)\dot{x}(t) = -0.1\dot{x}^3(t)$, where $c(t)$ is the uncertain term and $c(t) \in [-0.225, 0]$.

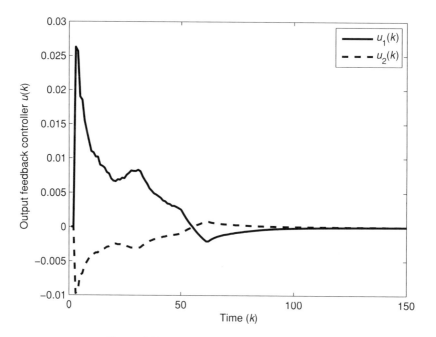

Figure 3.10   Output feedback controllers $u(k)$

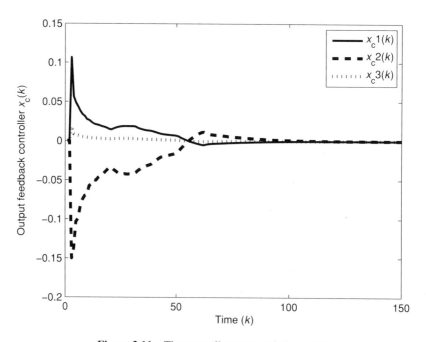

Figure 3.11   The controller state evolution $x_c(k)$

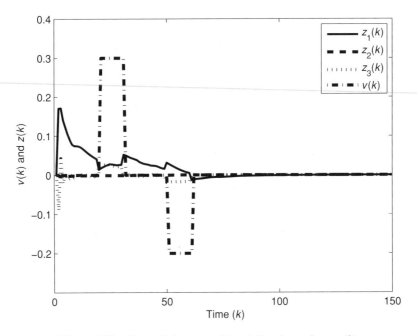

Figure 3.12   Controlled output $z(k)$ and disturbance input $v(k)$

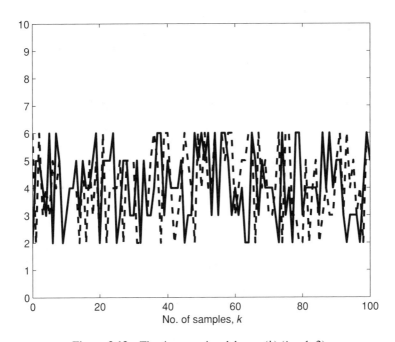

Figure 3.13   The time-varying delays $\tau_i(k)$ $(i = 1, 2)$

Consider the following controlled and measurement outputs:

$$z(t) = [\, x^T(t) \quad u^T(t) \quad v^T(t)\,]^T, \quad y(t) = [\, x^T(t) \quad v^T(t)\,]^T. \tag{3.57}$$

Using the same procedure as in [157], the nonlinear term $-0.67x^3(t)$ can be represented as

$$-0.67x^3 = M_1(x) \cdot 0 \cdot x - M_2(x) \cdot 1.5075x,$$

where $M_1(x), M_2(x) \in [0, 1]$, and $M_1(x) + M_2(x) = 1$. By solving the equations, $M_1(x)$ and $M_2(x)$ are obtained as follows:

$$M_1(x) = 1 - \frac{x^2(t)}{2.25}, \quad M_2(x) = \frac{x^2(t)}{2.25}.$$

M_1 and M_2 can be interpreted as membership functions of fuzzy sets. By using these fuzzy sets and set sampling time $T = 0.02$, the uncertain nonlinear system (3.56) and (3.57) can be represented by the following T–S fuzzy model:

Plant Rule 1: IF $x(k)$ is $M_1(x)$ THEN

$$\begin{cases} x(k + 1) = (A_1 + \Delta A_1(k))\, x(k) + B_{11}u(k) + D_{11}v(k) \\[2mm] \qquad + A_{d1} \sum_{m=1}^{h} \alpha_m(k)x(k - \tau_m(k)), \\[2mm] y(k) = \Xi C_1 x(k) + D_{21}v(k), \\[2mm] z(k) = (C_{z1} + \Delta C_{z1}(k)t)x(k) + B_{21}u(k) + D_{31}v(k). \end{cases} \tag{3.58}$$

Plant Rule 2: IF $x(k)$ is $M_2(x)$ THEN

$$\begin{cases} x(k + 1) = (A_2 + \Delta A_2(k))\, x(k) + B_{12}u(k) + D_{12}v(k) \\[2mm] \qquad + A_{d2} \sum_{m=1}^{h} \alpha_m(k)x(k - \tau_m(k)), \\[2mm] y(k) = \Xi C_2 x(k) + D_{22}v(k), \\[2mm] z(k) = (C_{z2} + \Delta C_{z2}(k))x(k) + B_{22}u(k) + D_{32}v(k). \end{cases} \tag{3.59}$$

The model parameters are

$$A_1 = \begin{bmatrix} -0.1125 & -0.02 \\ 1 & 0 \end{bmatrix}, \quad A_2 = \begin{bmatrix} -0.1125 & -1.527 \\ 1 & 0 \end{bmatrix},$$

$$C_{z1} = C_{z2} = \begin{bmatrix} 0 & 1 \\ 0 & 0 \end{bmatrix}, \quad C_1 = C_2 = [\,0 \quad 1\,],$$

$$A_{d1} = A_{d2} = 0, \quad B_{21} = B_{22} = [\,0 \quad 1\,]^T,$$

$$D_{21} = D_{22} = 0.5, \quad B_{11} = B_{12} = [\,1 \quad 0\,]^T,$$

$$D_{11} = D_{12} = 0, \quad D_{31} = D_{32} = [\,-0.1 \quad 0\,]^T,$$

and $\Delta A_1(k)$, $\Delta A_2(k)$ and $\Delta C_{z1}(k)$, $\Delta C_{z2}(k)$ can be represented in the form of (3.30) with

$$H_{a1} = H_{a2} = \begin{bmatrix} -0.1125 \\ 0 \end{bmatrix}, \quad E = \begin{bmatrix} 0.1 \\ 0.1 \end{bmatrix}^{\mathrm{T}}, \quad H_{c1} = H_{c2} = \begin{bmatrix} 0.1 \\ 0 \end{bmatrix}.$$

Let the probabilistic density functions of β_1 in $[0, 1]$ be described by

$$q(s_1) = \begin{cases} 0 & s_1 = 0 \\ 0.1 & s_1 = 0.5 \\ 0.9 & s_1 = 1 \end{cases}, \tag{3.60}$$

from which the expectations and variances can be easily calculated as $\mu = 0.95$ and $\sigma = 0.15$.
Letting $\gamma = 0.8$ and applying Theorem 3.4.4 with help from Algorithm HinfFC (see Section 3.4.3), we obtain the solution as follows:

$$A_{k1} = \begin{bmatrix} 0.6442 & 0.0125 \\ -0.2583 & 0.0457 \end{bmatrix}, \quad B_{k1} = \begin{bmatrix} 0.1205 \\ -0.0177 \end{bmatrix},$$

$$A_{k2} = \begin{bmatrix} -0.1098 & 0.1147 \\ -0.1765 & 0.0508 \end{bmatrix}, \quad B_{k2} = \begin{bmatrix} -0.1159 \\ 0.0509 \end{bmatrix},$$

$$C_{k1} = [-0.6145 \quad 0.2876], \quad C_{k2} = [0.1039 \quad -0.1765].$$

First, we assume

$$x(0) = [1 \quad 0]^{\mathrm{T}}, \quad x_c(0) = [0 \quad 0.5]^{\mathrm{T}}, \tag{3.61}$$

Figure 3.14 gives the state evolutions of the closed-loop fuzzy system when the external disturbance $v(k) = 0$, from which we can see that the two states converge to zero.

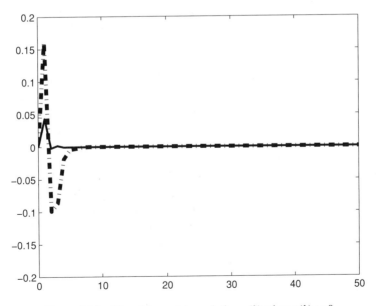

Figure 3.14 Closed-loop state evolution $x(k)$ when $v(k) \equiv 0$

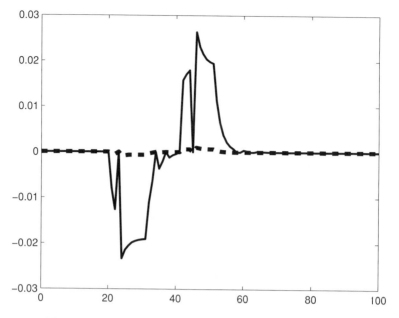

Figure 3.15   Closed-loop state evolution $x(k)$ with $v(k)$ in (3.55)

Next, to illustrate the disturbance attenuation performance, we assume zero initial condition and the external disturbance $v(k)$ is as in (3.55). Figure 3.15 shows the changing curves of the state variables with which, according to Theorem 3.4.4, the addressed uncertain discrete-time fuzzy systems with multiple time-varying random communication delays and multiple missing measurements is exponentially stable in the mean square and the effect of the disturbance input on the controlled output is constrained to the given level.

3.6   Summary

In this chapter, we have studied the robust H_∞ filtering and fuzzy output feedback control problem for nonlinear networked systems with multiple time-varying random communication delays and multiple packet dropouts. First, the H_∞ filtering problem has been considered for the systems involves parameter uncertainties, state-dependent stochastic disturbances (multiplicative noises or Itô-type noises), multiple stochastic time-varying delays, sector-bounded nonlinearities, and multiple packet dropouts. Sufficient conditions for the robustly exponential stability of the filtering error dynamics have been obtained and, at the same time, the prescribed H_∞ disturbance rejection attenuation level has been guaranteed. Then, some parallel results have also been derived for a class of uncertain discrete-time fuzzy systems with both multiple probabilistic delays and multiple missing measurements by using similar analysis techniques. Finally, the results of this chapter have been demonstrated by some simulation examples.

4

Filtering and Control for Systems with Repeated Scalar Nonlinearities

In this chapter, the H_∞ filtering and control problems are investigated for systems with repeated scalar nonlinearities and missing measurements. The nonlinear system is described by a discrete-time state equation involving a repeated scalar nonlinearity which typically appears in recurrent neural networks. The H_∞ filtering problem in the presence of missing measurements is first considered. The communication links, existing between the plant and filter, are assumed to be imperfect and a stochastic variable satisfying the Bernoulli random binary distribution is utilized to model the phenomenon of the missing measurements. The stable full- and reduced-order filters are designed such that the filtering process is stochastically stable and the filtering error satisfies the H_∞ performance constraint for all admissible missing observations and nonzero exogenous disturbances under the zero initial condition. Sufficient conditions are obtained for the existence of admissible filters. Since these conditions involve matrix equalities, the cone complementarity linearization procedure is employed to cast the nonconvex feasibility problem into a sequential minimization problem subject to linear matrix inequalities, which can be readily solved by using standard numerical software. Moreover, the multiple missing measurements are included to model the randomly intermittent behaviors of the individual sensors, where the missing probability for each sensor/actuator is governed by a random variable satisfying a certain probabilistic distribution on the interval [0, 1]. By using similar analysis techniques, the observer-based H_∞ control problem is also studied for systems with repeated scalar nonlinearities and multiple packet losses, and a set of parallel results is derived. Finally, some simulation examples are given to illustrate the main results of this chapter.

Filtering, Control and Fault Detection with Randomly Occurring Incomplete Information, First Edition.
Hongli Dong, Zidong Wang, and Huijun Gao.

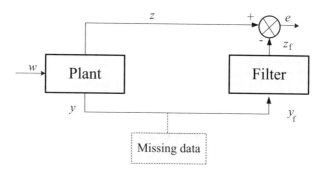

Figure 4.1   The filtering problem with unreliable communication links

4.1 Problem Formulation for Filter Design

The filtering problem with unreliable communication links is shown in Figure 4.1, where the physical plant is of the characteristic of repeated scalar nonlinearities, and the data packet dropout phenomenon occurs from the plant to the filter.

4.1.1 The Physical Plant

As in Gao *et al.* [158], we consider the discrete-time nonlinear system described as follows:

$$\mathcal{S}: \begin{cases} x_{k+1} = Af(x_k) + Bw_k, \\ y_k = Cf(x_k) + Dw_k, \\ z_k = Hf(x_k), \end{cases} \tag{4.1}$$

where $x_k \in \mathbb{R}^n$ represents the state vector, $y_k \in \mathbb{R}^m$ is the measured output, $z_k \in \mathbb{R}^p$ is the signal to be estimated, $w_k \in \mathbb{R}^l$ is the disturbance input which belongs to $l_2[0, \infty)$, (A, B, C, D, H) are system matrices with compatible dimensions, and f is a nonlinear function satisfying the following assumption as in Chu and Glover [159].

Assumption 4.1   *The nonlinear function $f : \mathbb{R} \to \mathbb{R}$ in system (4.1) satisfies*

$$\forall a, b \in \mathbb{R} \quad |f(a) + f(b)| \leq |a + b|. \tag{4.2}$$

In the following, for the vector $x = [x_1 x_2 \cdots x_n]^{\mathrm{T}}$, we denote

$$f(x) \triangleq =[f(x_1) f(x_2) \cdots f(x_n)]^{\mathrm{T}}.$$

The model (4.1) is called a system with a repeated scalar nonlinearity [159]. The block diagram of (4.1) is drawn in Figure 4.2, where λ denotes the time delay and $\circ$ denotes composition. This type of nonlinearity is analogous to an upper linear fractional transformation

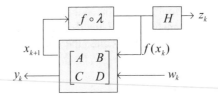

Figure 4.2 The block-diagram of systems with repeated scalar nonlinearities

with respect to a repeated scalar block δI [160], but the upper block is now nonlinear instead of linear. As a result, such an uncertain model could find applications in many practical situations, such as recurrent artificial neural networks.

Remark 4.1 *[159, 161] Note that f is odd (by putting $b = -a$) and 1-Lipschitz (by putting $b = -b$). Therefore, some typical classes of nonlinearities can be described by f, such as the semilinear function (i.e., the standard saturation* sat$(s) := s$ *if* $|s| \leq 1$ *and* sat$(s) := $ sgn(s) *if $|s| > 1$), the hyperbolic tangent function that has been extensively used for the activation function in neural networks, the sine function, and so on.*

Remark 4.2 *The plant model structure (4.1) can be used to describe a broad class of real-time dynamical systems, such as digital control systems having saturation-type nonlinearities on the state or on the controller [162, 163], recurrent artificial neural networks (e.g., see Cao and Wang [164] and references cited therein), neural networks defined on hypercubes [165], n-stand cold-rolling mills [166], fixed-point state-space digital filters using saturation overflow arithmetic [159, 167], manufacturing systems for decision-making [168], marketing and production control problems [166].*

4.1.2 The Communication Link

In a perfect world, the measurement out of (4.1) is y_k. However, in reality, such as network-based communication systems, the data may be lost during their transmission from the sensor to the filter. Let the *actual* measurement signal of (4.1) (i.e., the *actual* input available to the filter) be denoted by y_{fk}. Clearly, y_{fk} may not be equal to the *ideal* output y_k of the plant due to the lossy signal transmission. In this chapter, we model the missing data phenomenon via a stochastic Bernoulli approach:

$$y_{fk} = \beta_k y_k, \tag{4.3}$$

where y_{fk} is the actual available input to the filter and β_k is a stochastic variable taking values on either 1 or 0. Obviously, $\beta_k = 0$ holds when the communication link fails (i.e., data are lost), and $\beta_k = 1$ means successful transmission. A natural assumption on the sequence $\{\beta_k\}$ can be made as follows:

$$\text{Prob}\{\beta_k = 1\} = \mathbb{E}\{\beta_k\} = \bar{\beta}, \quad \text{Prob}\{\beta_k = 0\} = 1 - \bar{\beta}.$$

Note that the missing probability could be estimated through statistical tests. Such a stochastic Bernoulli approach has been extensively used for dealing with data missing problems; for example, see Wang *et al.* [169] and references cited therein.

4.1.3 The Filter

In this chapter, the filter is of the following structure:

$$\mathcal{E}: \begin{cases} \hat{x}_{k+1} = A_F f(\hat{x}_k) + B_F y_{fk}, \\ \hat{z}_k = C_F f(\hat{x}_k), \end{cases} \tag{4.4}$$

where $\hat{x}_k \in \mathbb{R}^k$ is the state estimate, $\hat{z}_k \in \mathbb{R}^p$ is the output signal of the filter which is used for an estimation of z_k, $y_{fk} \in \mathbb{R}^m$ is the *actual* measurement signal of (4.1) (i.e., the *actual* input to the filter), and A_F, B_F, and C_F are appropriately dimensioned filter matrices to be determined. It is clear from (4.3) and (4.4) that

$$\mathcal{F}: \begin{cases} \hat{x}_{k+1} = A_F f(\hat{x}_k) + B_F \beta_k y_k, \\ \hat{z}_k = C_F f(\hat{x}_k). \end{cases} \tag{4.5}$$

Our aim in this chapter is to design both the full-order (when $k = n$) and reduced-order (when $1 \leq k < n$) filters within the same framework.

4.1.4 The Filtering Error Dynamics

Augmenting the model of $\mathcal{S}$ to include the states of the filter $\mathcal{F}$, the filtering error system is given by

$$\mathcal{G}: \begin{cases} \bar{x}_{k+1} = \bar{A}_1 f(\bar{x}_k) + \tilde{\beta}_k \bar{A}_2 f(\bar{x}_k) + \bar{B}_1 w_k + \tilde{\beta}_k \bar{B}_2 w_k, \\ \bar{z}_k = \bar{C} f(\bar{x}_k), \end{cases} \tag{4.6}$$

where

$$\bar{x}_k = [x_k^T \quad \hat{x}_k^T]^T, \quad \bar{z}_k = z_k - \hat{z}_k,$$

$$\bar{A}_1 = \begin{bmatrix} A & 0 \\ \bar{\beta} B_F C & A_F \end{bmatrix}, \quad \bar{B}_1 = \begin{bmatrix} B \\ \bar{\beta} B_F D \end{bmatrix}, \quad \bar{C} = [H \quad -C_F],$$

$$\bar{A}_2 = \begin{bmatrix} 0 & 0 \\ B_F C & 0 \end{bmatrix}, \quad \bar{B}_2 = \begin{bmatrix} 0 \\ B_F D \end{bmatrix} \tag{4.7}$$

and $\tilde{\beta}_k = \beta_k - \bar{\beta}$. It is clear that $\mathbb{E}\{\tilde{\beta}_k\} = 0$ and $\mathbb{E}\{\tilde{\beta}_k^2\} = \bar{\beta}(1 - \bar{\beta})$.

Before proceeding further, we first introduce the following definitions and lemmas.

Definition 4.1.1   *The solution $\bar{x}_k = 0$ of the filter error system in (4.6) with $w_k \equiv 0$ is said to be stochastically stable if, for any $\varepsilon > 0$, there exists a $\delta > 0$ such that*

$$\mathbb{E}\{\|\bar{x}_k\|\} < \varepsilon, \tag{4.8}$$

whenever $k \in \mathbb{I}^+$ and $\|\bar{x}_0\| < \delta$.

Definition 4.1.2   *A square matrix $P \triangleq =[p_{ij}] \in \mathbb{R}^{n \times n}$ is called diagonally dominant if for all $i = 1, \ldots, n$*

$$p_{ii} \geqslant \sum_{j \neq i} |p_{ij}|. \tag{4.9}$$

Lemma 4.1.3   [159] If $P > 0$ is diagonally dominant, then for all nonlinear functions f satisfying (4.2) the following inequality holds for all $\bar{x}_k \in \mathbb{R}^{n+k}$:

$$f^{\mathrm{T}}(\bar{x}_k) P f(\bar{x}_k) \leq \bar{x}_k^{\mathrm{T}} P \bar{x}_k. \tag{4.10}$$

Remark 4.3   *It will be seen later that the purpose of requiring the matrix P to satisfy (4.10) is to admit the quadratic Lyapunov function $V(\bar{x}_k) = \bar{x}_k^{\mathrm{T}} P \bar{x}_k$.*

Lemma 4.1.4   [170] If there exist a Lyapunov function $V(\bar{x}_k)$ and a function $\phi(x) \in OL$ satisfying the conditions

$$V(0) = 0, \tag{4.11}$$

$$\phi(\|\bar{x}_k\|) \leq V(\bar{x}_k), \tag{4.12}$$

$$\mathbb{E}\{V(\bar{x}_{k+1})\} - \mathbb{E}\{V(\bar{x}_k)\} < 0, \quad k \in \mathbb{I}^+, \tag{4.13}$$

then the solution $\bar{x}_k = 0$ of the filtering error system in (4.6) with $w_k \equiv 0$ is stochastically stable.

Consider the filtering problem in the presence of a missing data phenomenon and suppose the parameter $\bar{\beta}$ describing intermittent transmission is known. We are now in a position to state the problem of nonlinear stochastic H_∞ filtering with data loss as follows.

Problem H_∞ filtering with data loss (HFDL): Given a scalar $\gamma > 0$, design a filter in the form of (4.4) such that:

(i) (stochastic stability) the filtering error system in (4.6) is stochastically stable in the sense of Definition 4.1.1;
(ii) (H_∞ performance) under zero initial condition, the error output $\bar{z}_k$ satisfies

$$\|\bar{z}\|_{\mathbb{E}} \leq \gamma \|w\|_2, \tag{4.14}$$

where

$$\|\bar{z}\|_{\mathbb{E}} \triangleq \mathbb{E}\left\{\sqrt{\sum_{k=0}^{\infty} \bar{z}_k^T \bar{z}_k}\right\}.$$

If the above two conditions are satisfied, the filtering error system is said to be stochastically stable with a guaranteed H_∞ performance γ, and the problem HFDL is solved.

4.2   Filtering Performance Analysis

In this section, the problem HFDL formulated in Section 4.1.4 will be tackled via a quadratic approach described in the following theorem.

Theorem 4.2.1   *Consider system $\mathcal{S}$ in (4.1) and suppose the filter matrices (A_F, B_F, C_F) of $\mathcal{E}$ in (4.4) are given. The filtering error system $\mathcal{G}$ in (4.6) is stochastically stable with a given H_∞ performance γ if there exists a positive diagonally dominant matrix P satisfying*

$$\begin{bmatrix} \bar{A}_1 & \bar{B}_1 \\ g\bar{A}_2 & g\bar{B}_2 \\ \bar{C} & 0 \end{bmatrix}^T \begin{bmatrix} P & 0 & 0 \\ 0 & P & 0 \\ 0 & 0 & I \end{bmatrix} \begin{bmatrix} \bar{A}_1 & \bar{B}_1 \\ g\bar{A}_2 & g\bar{B}_2 \\ \bar{C} & 0 \end{bmatrix} - \begin{bmatrix} P & 0 \\ 0 & \gamma^2 I \end{bmatrix} < 0, \qquad (4.15)$$

where $g = \sqrt{\bar{\beta}(1 - \bar{\beta})}$.

Proof.   Define the following Lyapunov function candidate:

$$V(\bar{x}_k) = \bar{x}_k^T P \bar{x}_k. \qquad (4.16)$$

When $w_k \equiv 0$, (4.6) becomes

$$\bar{x}_{k+1} = \bar{A}_1 f(\bar{x}_k) + \tilde{\beta}_k \bar{A}_2 f(\bar{x}_k),$$
$$\bar{z}_k = \bar{C} f(\bar{x}_k),$$

and then the difference of the Lyapunov function is calculated as

$$\begin{aligned} \Delta V(\bar{x}_k) &= \mathbb{E}\{V(\bar{x}_{k+1}) \mid \bar{x}_k\} - V(\bar{x}_k) \\ &= \mathbb{E}\{V(\bar{x}_{k+1})\} - \mathbb{E}\{V(\bar{x}_k)\} \\ &= \mathbb{E}\{f^T(\bar{x}_k)(\bar{A}_1^T + \tilde{\beta}_k \bar{A}_2^T)P(\bar{A}_1 + \tilde{\beta}_k \bar{A}_2)f(\bar{x}_k) \mid \bar{x}_k\} - \bar{x}_k^T P \bar{x}_k \\ &= f^T(\bar{x}_k)(\bar{A}_1^T P \bar{A}_1 + g^2 \bar{A}_2^T P \bar{A}_2)f(\bar{x}_k) - \bar{x}_k^T P \bar{x}_k. \end{aligned}$$

According to Lemma 4.1.3, we have

$$\Delta V(\bar{x}_k) \leq f^T(\bar{x}_k)(\bar{A}_1^T P \bar{A}_1 + g^2 \bar{A}_2^T P \bar{A}_2 - P)f(\bar{x}_k).$$

Note that (4.15) implies

$$\bar{A}_1^T P \bar{A}_1 + g^2 \bar{A}_2^T P \bar{A}_2 + \bar{C}^T \bar{C} - P < 0,$$

and subsequently

$$\Phi = \bar{A}_1^T P \bar{A}_1 + g^2 \bar{A}_2^T P \bar{A}_2 - P < 0.$$

Thus, we have

$$\mathbb{E}\{V(\bar{x}_{k+1})\} - \mathbb{E}\{V(\bar{x}_k)\} < f^T(\bar{x}_k)\Phi f(\bar{x}_k) < 0,$$

which satisfies (4.13). Taking $\phi(\bar{x}_k) = \lambda_{\min}(P)\bar{x}_k^2$ such that $\phi(\bar{x}_k) \in OL$, we obtain

$$\phi(\|\bar{x}_k\|) = \lambda_{\min}(P)\|\bar{x}_k\|^2 = \lambda_{\min}(P)\bar{x}_k^T \bar{x}_k \leq \bar{x}_k^T P \bar{x}_k = V(\bar{x}_k),$$

which satisfies (4.12). Considering $V(0) = 0$, it follows readily from Lemma 4.1.4 that the filtering error system in (4.6) with $w_k \equiv 0$ is stochastically stable.

Next, the H_∞ performance criteria for the filter error system in (4.6) will be established. Assuming zero initial conditions, an index is introduced as follows:

$$\bar{J} = \mathbb{E}\{V(\bar{x}_{k+1}) \mid \bar{x}_k\} + \bar{z}_k^T \bar{z}_k - \gamma^2 w_k^T w_k - f^T(\bar{x}_k) P f(\bar{x}_k).$$

Defining

$$\eta_k = [\, f^T(\bar{x}_k) \quad w_k^T \,]^T,$$

we have

$$\mathbb{E}\{V(\bar{x}_{k+1}) \mid \bar{x}_k\}$$
$$= \mathbb{E}\left\{ \eta_k^T \left(\begin{bmatrix} \bar{A}_1^T \\ \bar{B}_1^T \end{bmatrix} P [\, \bar{A}_1 \quad \bar{B}_1 \,] + g^2 \begin{bmatrix} \bar{A}_2^T \\ \bar{B}_2^T \end{bmatrix} P [\, \bar{A}_2 \quad \bar{B}_2 \,] \right) \eta_k \,\middle|\, \bar{x}_k \right\},$$

and

$$\bar{z}_k^T \bar{z}_k = \eta_k^T \begin{bmatrix} \bar{C}^T \\ 0 \end{bmatrix} [\, \bar{C} \quad 0 \,] \eta_k.$$

It then follows that

$$\bar{J} = \eta_k^T \left(\begin{bmatrix} \bar{A}_1^T \\ \bar{B}_1^T \end{bmatrix} P [\, \bar{A}_1 \quad \bar{B}_1 \,] + g^2 \begin{bmatrix} \bar{A}_2^T \\ \bar{B}_2^T \end{bmatrix} P [\, \bar{A}_2 \quad \bar{B}_2 \,] \right.$$
$$\left. + \begin{bmatrix} \bar{C}^T \\ 0 \end{bmatrix} [\, \bar{C} \quad 0 \,] - \begin{bmatrix} P & 0 \\ 0 & \gamma^2 I \end{bmatrix} \right) \eta_k,$$

which, from (4.15), indicates that $\bar{J} \leq 0$ or

$$\mathbb{E}\{V(\bar{x}_{k+1}) \mid \bar{x}_k\} + \bar{z}_k^{\mathrm{T}} \bar{z}_k - \gamma^2 w_k^{\mathrm{T}} w_k - f^{\mathrm{T}}(\bar{x}_k) P f(\bar{x}_k) \leq 0.$$

According to Lemma 4.1.3, we have

$$\mathbb{E}\{V(\bar{x}_{k+1}) \mid \bar{x}_k\} + \bar{z}_k^{\mathrm{T}} \bar{z}_k - \gamma^2 w_k^{\mathrm{T}} w_k - \bar{x}_k^{\mathrm{T}} P \bar{x}_k \leq 0.$$

Taking mathematical expectation on both sides, we obtain

$$\mathbb{E}\{V(\bar{x}_{k+1})\} - \mathbb{E}\{V(\bar{x}_k)\} + \mathbb{E}\{\bar{z}_k^{\mathrm{T}} \bar{z}_k\} - \gamma^2 w_k^{\mathrm{T}} w_k \leq 0.$$

For $k = 0, 1, 2, \ldots, \infty$, summing up both sides under zero initial condition and considering $\mathbb{E}\{V_k\} \geq 0$, we arrive at

$$\mathbb{E}\left\{\sum_{k=0}^{\infty} \bar{z}_k^{\mathrm{T}} \bar{z}_k\right\} - \sum_{k=0}^{\infty} \gamma^2 w_k^{\mathrm{T}} w_k \leq 0,$$

which is equivalent to (4.14). The proof is now complete. □

4.3 Filter Design

In this section, we aim at designing a filter in the form of (4.4) based on Theorem 4.2.1. That is, we are interested in determining the filter matrices in (4.4) such that the filtering error system in (4.6) is stochastically stable with a guaranteed H_∞ performance. The following theorem provides sufficient conditions for the existence of such H_∞ filters for system $\mathcal{S}$.

Theorem 4.3.1   *Consider the system $\mathcal{S}$ in (4.1). Then, an admissible H_∞ filter of the form $\mathcal{E}$ in (4.4) exists if there exist matrices $0 < P \triangleq [p_{ij}] \in \mathbb{R}^{(n+k)\times(n+k)}$, $L > 0$, K, C_F, $R = R^{\mathrm{T}} \triangleq [r_{ij}] \in \mathbb{R}^{(n+k)\times(n+k)}$ satisfying*

$$\begin{bmatrix} -L & 0 & 0 & \bar{A}_0 + EKR_1 & \bar{B}_0 + EKS_1 \\ * & -L & 0 & gEKR_2 & gEKS_2 \\ * & * & -I & \bar{C}_0 + C_F T & 0 \\ * & * & * & -P & 0 \\ * & * & * & * & -\gamma^2 I \end{bmatrix} < 0, \tag{4.17}$$

$$p_{ii} - \sum_{j \neq i}(p_{ij} + 2r_{ij}) \geq 0, \tag{4.18}$$

$$r_{ij} \geq 0 \quad \forall i \neq j, \tag{4.19}$$

$$p_{ij} + r_{ij} \geq 0 \quad \forall i \neq j, \tag{4.20}$$

$$PL = I, \tag{4.21}$$

where

$$\bar{A}_0 = \begin{bmatrix} A & 0 \\ 0 & 0 \end{bmatrix}, \quad \bar{B}_0 = \begin{bmatrix} B \\ 0 \end{bmatrix}, \quad \bar{C}_0 = [H \quad 0],$$

$$E = \begin{bmatrix} 0 \\ I \end{bmatrix}, \quad R_1 = \begin{bmatrix} 0 & I \\ \bar{\beta}C & 0 \end{bmatrix}, \quad R_2 = \begin{bmatrix} 0 & 0 \\ C & 0 \end{bmatrix},$$

$$S_1 = \begin{bmatrix} 0 \\ \bar{\beta}D \end{bmatrix}, \quad S_2 = \begin{bmatrix} 0 \\ D \end{bmatrix}, \quad T = [0 \quad -I]. \tag{4.22}$$

Furthermore, if (P, L, K, C_F, R) *is a feasible solution of (4.17)–(4.21), then the system matrices of an admissible* H_∞ *filter in the form of (4.4) can be obtained by means of the matrices* K *and* C_F, *where*

$$K = [A_F \quad B_F]. \tag{4.23}$$

Proof. From Theorem 4.2.1, we know that there exists an admissible filter $\mathcal{E}$ in the form of (4.4) such that the filtering error system $\mathcal{G}$ in (4.6) is stochastically stable with a guaranteed H_∞ performance γ if there exists a positive diagonally dominant matrix P satisfying (4.15). By the Schur complement lemma, (4.15) is equivalent to

$$\begin{bmatrix} -P^{-1} & 0 & 0 & \bar{A}_1 & \bar{B}_1 \\ * & -P^{-1} & 0 & g\bar{A}_2 & g\bar{B}_2 \\ * & * & -I & \bar{C} & 0 \\ * & * & * & -P & 0 \\ * & * & * & * & -\gamma^2 I \end{bmatrix} < 0. \tag{4.24}$$

Rewrite (4.7) in the following form:

$$\bar{A}_1 = \bar{A}_0 + E[A_F \quad B_F]R_1, \quad \bar{A}_2 = E[A_F \quad B_F]R_2,$$
$$\bar{B}_1 = \bar{B}_0 + E[A_F \quad B_F]S_1, \quad \bar{B}_2 = E[A_F \quad B_F]S_2,$$
$$\bar{C} = \bar{C}_0 + C_F T \tag{4.25}$$

where $\bar{A}_0, \bar{B}_0, \bar{C}_0, E, R_1, R_2, S_1, S_2,$ and T are defined in (4.22). Noticing (4.25), (4.24) can be rewritten as

$$\begin{bmatrix} -P^{-1} & 0 & 0 & \bar{A}_0 + E[A_F \quad B_F]R_1 & \bar{B}_0 + E[A_F \quad B_F]S_1 \\ * & -P^{-1} & 0 & gE[A_F \quad B_F]R_2 & gE[A_F \quad B_F t]S_2 \\ * & * & -I & \bar{C}_0 + C_F T & 0 \\ * & * & * & -P & 0 \\ * & * & * & * & -\gamma^2 I \end{bmatrix} < 0,$$

which, by noticing (4.21) and (4.23), is equivalent to (4.17). Furthermore, from (4.18)–(4.20), we have

$$p_{ii} \geq \sum_{j \neq i}(p_{ij} + 2r_{ij}) = \sum_{j \neq i}(|p_{ij} + r_{ij}| + |-r_{ij}|) \geq \sum_{j \neq i}|p_{ij}|,$$

which guarantees the positive-definite matrix P to be diagonally dominant, and the proof is then complete.   □

It is worth noting that, by far, we are unable to apply the LMI approach in the design of filters because of the matrix equality in Theorem 4.3.1. Fortunately, this problem can be addressed with help from the CCL algorithm proposed in El Ghaoui *et al.* [171]. The basic idea in the CCL algorithm is that if the LMI

$$\Omega(P, L) = \begin{bmatrix} P & I \\ I & L \end{bmatrix} \geq 0$$

is feasible in the $n \times n$ matrix variables $L > 0$ and $P > 0$, then tr(PL) $\geq n$; and tr(PL) $= n$ if and only if $PL = I$. Based on this, it is likely to be able to solve the equalities in (4.21) by using of CCL algorithm.

According to Gao *et al.* [158], if the solution of min tr(PL) subject to (4.17)–(4.20) and $\Omega(P, L) \geq 0$ is $n + k$, then the conditions in Theorem 4.3.1 are solvable. In view of this observation, we put forward the following nonlinear minimization problem involving LMI conditions instead of the original nonconvex feasibility problem formulated in Theorem 4.3.1.

Algorithm H_∞ Filtering (HinfF)

Step 1. Find a feasible set $(P_{(0)}, L_{(0)}, K_{(0)}, C_{F(0)}, R_{(0)})$ satisfying (4.17)–(4.20) and $\Omega(P, L) \geq 0$, if there are none, exit. Set $q = 0$.

Step 2. According to (4.17)–(4.20) and $\Omega(P, L)$, solve the LMI problem: min tr($PL_{(q)} + P_{(q)}L$).

Step 3. If the stopping criterion is satisfied, then output the feasible solutions (P, L, K, C_F, R) and exit. Else, set $q = q + 1$ and go to Step 2.

Remark 4.4   *The proposed Algorithm HinfF can be used to solve the feasibility problem in Theorem 4.3.1 for a given constant γ. It should be pointed out that the technique used here for solving the filtering problem is very different from many existing results in the literature concerning filtering problems. The techniques used in many publications fall into the variable linearization category, where the original nonlinear matrix inequality is transformed into an LMI by performing congruence transformations and by defining new matrix variables. Since the linearization techniques involve the partition of the positive matrix P, they will lead to the dilation difficulty if they were to be used to deal with the problem addressed here because the positive matrix P is restricted to be diagonally dominant.*

Remark 4.5   *Our main results are based on the LMI conditions. The LMI Control Toolbox implements state-of-the-art interior-point LMI solvers. While these solvers are significantly*

*faster than classical convex optimization algorithms, it should be kept in mind that the com-
plexity of LMI computations remains higher than that of solving, say, a Riccati equation. For
instance, problems with a thousand design variables typically take over an hour on today's
workstations. However, research on LMI optimization is a very active area in the applied math,
optimization, and operations research community, and substantial speed-ups can be expected
in the future.*

Remark 4.6 *The CCL procedure is sometimes time consuming in execution. Nevertheless,
the system considered in this chapter is not time varying and, therefore, the filter design can
be implemented in an offline manner. Therefore, the running-time issue is not a concern.
We would, of course, like to consider other design approaches in the future that will not
involve equality constraints and, therefore, the CCL is no longer needed. Other possible
future research directions include real-time applications of the proposed filtering theory in
telecommunications, and further extensions of the present results to more complex systems
with unreliable communication links, such as sampled-data systems, bilinear systems, and a
class of nonlinear systems.*

4.4 Observer-Based H_∞ Control with Multiple Packet Losses

4.4.1 Problem Formulation

The Physical Plant

Consider the discrete-time system with repeated scalar nonlinearities described as following:

$$\begin{cases} x_{k+1} = Af(x_k) + B_2u_k + B_1w_k, \\ z_k = C_1f(x_k) + D_1w_k, \\ y_{ck} = C_2x_k + D_2w_k, \end{cases} \tag{4.26}$$

where $x_k \in \mathbb{R}^n$ represents the state vector, $u_k \in \mathbb{R}^m$ is the control input, $z_k \in \mathbb{R}^r$ is the controlled
output, $y_{ck} \in \mathbb{R}^p$ is the process output, $w_k \in \mathbb{R}^q$ is the disturbance input which belongs to
$l_2[0, \infty)$, and A, B_1, B_2, C_1, C_2, D_1, and D_2 are known real matrices with appropriate
dimensions. f is a nonlinear function satisfying the assumption described in (4.2).

The Controller

The dynamic observer-based control scheme for the system (4.26) is described by

$$\begin{cases} \hat{x}_{k+1} = Af(\hat{x}_k) + B_2u_k + L(y_k - \hat{y}_k), \\ \hat{y}_k = C_2\hat{x}_k, \\ \hat{u}_k = K\hat{x}_k, \end{cases} \tag{4.27}$$

where $\hat{x}_k \in \mathbb{R}^n$ is the state estimate of the system (4.26), $y_k \in \mathbb{R}^p$ is the measured output,
$\hat{u}_k \in \mathbb{R}^m$ is the control input without transmission missing, and $L \in \mathbb{R}^{n \times p}$ and $K \in \mathbb{R}^{m \times n}$ are
the observer and controller gains, respectively.

The Communication Links

It should be pointed out that, owing to the existence of the communication links, the phenomenon of data packet dropout will inevitably induce missing observations. That is, the process output is probably not equivalent to the measured output (i.e., $y_{ck} \neq y_k$). In this section, the measurement with multiple communication packet loss is described by

$$y_k = \Xi y_{ck} = \sum_{i=1}^{p} \alpha_i (C_{2i} x_k + D_{2i} w_k), \tag{4.28}$$

where $\Xi := \mathrm{diag}\{\alpha_1, \ldots, \alpha_p\}$ with α_i $(i = 1, \ldots, p)$ being p unrelated random variables which are also unrelated to w_k. It is assumed that α_i has the probabilistic density function $q_i(s)$ $(i = 1, \ldots, p)$ on the interval $[0, 1]$ with mathematical expectation μ_i and variance σ_i^2. C_{2i} and D_{2i} are defined by

$$C_{2i} := \mathrm{diag}\{\underbrace{0, \ldots, 0}_{i-1}, 1, \underbrace{0, \ldots, 0}_{p-i}\} C_2, \quad D_{2i} := \mathrm{diag}\{\underbrace{0, \ldots, 0}_{i-1}, 1, \underbrace{0, \ldots, 0}_{p-i}\} D_2.$$

α_i could satisfy any discrete probabilistic distribution on the interval $[0, 1]$, which includes the widely used Bernoulli distribution as a special case. In the following, we denote $\bar{\Xi} = \mathbb{E}\{\Xi\}$.

Similarly, the control input with multiple communication packet loss is described by

$$u_k = \Omega \hat{u}_k = \sum_{j=1}^{m} \beta_j K_j \hat{x}_k, \tag{4.29}$$

where $\Omega = \mathrm{diag}\{\beta_1, \ldots, \beta_m\}$ with β_j $(j = 1, \ldots, m)$ being m unrelated random variables and

$$K_j = \mathrm{diag}\{\underbrace{0, \ldots, 0}_{j-1}, 1, \underbrace{0, \ldots, 0}_{m-j}\} K.$$

It is assumed that β_j has the probabilistic density function $m_j(s)$ on the interval $[0, 1]$ with mathematical expectation ϑ_j and variance ξ_j^2. We define $\bar{\Omega} = \mathbb{E}\{\Omega\}$.

Remark 4.7 *It can be noted from (4.28) and (4.29) that the diagonal matrices Ξ and Ω, which consist of random variables, are introduced to reflect the random multiple packet losses in, respectively, the sensor-to-controller and controller-to-actuator channels. The random packet-loss mode in the sensor output has recently been studied in many chapters on NCSs, most of which were concerned with the linear system with single packet-loss. To the best of our knowledge, there has been little research so far on the control problem for nonlinear systems in the presence of multiple packet losses in both sensor-to-controller and controller-to-actuator channels.*

Remark 4.8 *In real systems, the measurement data may be transferred through multiple sensors and actuators. For different sensors or actuators, the data missing probability may be different. In this sense, it would be more reasonable to assume that the data missing* *law for each individual sensor/actuator satisfies an individual probabilistic distribution. In (4.28), the diagonal matrix Ξ represents the whole missing status, where the random variable α_i corresponds to the ith sensor. We note that the data loss (also called packet dropout or measurement missing) phenomenon has been extensively studied and several models have been introduced. The Bernoulli distributed model is arguably the most popular one in which 0 is used to stand for an entire missing of signals and 1 denotes the intactness. However, for various reasons, such as sensor aging and sensor temporal failure, the data missing at one moment might be partial, and therefore the missing probability cannot be simply described by 0 or 1. In (4.28), α_i can take a value on the interval $[0, 1]$ and the probability for α_i to take different values may differ from each other. It is easy to see that the Bernoulli distribution is included as a special case. Similar discussion can be applied to the diagonal matrix Ω.*

The Closed-Loop System

Letting the estimation error be

$$e_k := x_k - \hat{x}_k, \tag{4.30}$$

the closed-loop system can be obtained as follows by substituting (4.27)–(4.29) into (4.26) and (4.30):

$$
\begin{cases}
x_{k+1} = Af(x_k) + B_2\bar{\Omega}Kx_k + B_2(\Omega - \bar{\Omega})Kx_k - B_2\bar{\Omega}Ke_k \\
\qquad\quad - B_2(\Omega - \bar{\Omega})Ke_k + B_1w_k, \\
e_{k+1} = A[f(x_k) - f(\hat{x}_k)] + (LC_2 - L\bar{\Xi}C_2)x_k - L(\Xi - \bar{\Xi})C_2x_k \\
\qquad\quad - LC_2e_k + \left[(B_1 - L\bar{\Xi}D_2) - L(\Xi - \bar{\Xi})D_2\right]w_k,
\end{cases}
\tag{4.31}
$$

or, in a compact form,

$$\varsigma_{k+1} = \check{A}\eta_k + \bar{A}\varsigma_k + \psi_k\hat{A}\varsigma_k + \bar{B}w_k, \tag{4.32}$$

where

$$\varsigma_k = [\,x_k^{\mathrm{T}} \;\; e_k^{\mathrm{T}}\,]^{\mathrm{T}}, \quad \eta_k = [\,f^{\mathrm{T}}(x_k) \;\; f^{\mathrm{T}}(x_k) - f^{\mathrm{T}}(\hat{x}_k)\,]^{\mathrm{T}}, \quad \check{A} = \begin{bmatrix} A & 0 \\ 0 & A \end{bmatrix},$$

$$\bar{A} = \begin{bmatrix} B_2\bar{\Omega}K & -B_2\bar{\Omega}K \\ LC_2 - L\bar{\Xi}C_2 & -LC_2 \end{bmatrix}, \quad \psi_k = \begin{bmatrix} B_2(\Omega - \bar{\Omega}) & 0 \\ 0 & L(\Xi - \bar{\Xi}) \end{bmatrix},$$

$$\hat{A} = \begin{bmatrix} K & -K \\ -C_2 & 0 \end{bmatrix}, \quad \bar{B} = \begin{bmatrix} B_1 \\ (B_1 - L\bar{\Xi}D_2) - L(\Xi - \bar{\Xi})D_2 \end{bmatrix}.$$

It should be pointed out that, in the closed-loop system (4.32), the stochastic matrices Ξ and Ω appear, which make the difference from (1) the traditional deterministic system without random packet losses and (2) the system with single random packet loss. Before proceeding further, we introduce the following definition, assumption, and lemmas, which will be needed for the derivation of our main results.

Definition 4.4.1   *[170] The solution $\varsigma_k = 0$ of the closed-loop system in (4.32) with $w_k \equiv 0$ is said to be stochastically stable if, for any $\varepsilon > 0$, there exists a $\delta > 0$ such that $\mathbb{E}\{\|\varsigma_k\|\} < \varepsilon$ whenever $k \in \mathbb{I}^+$ and $\|\varsigma_0\| < \delta$.*

Assumption 4.2   *[172] The matrix B_2 is of full column rank; that is, $\mathrm{rank}(B_2) = m$.*

Remark 4.9   *For the matrix B_2 of full column rank, there always exist two orthogonal matrices $U \in \mathbb{R}^{n \times n}$ and $V \in \mathbb{R}^{m \times m}$ such that*

$$\tilde{B}_2 = U B_2 V = \begin{bmatrix} U_1 \\ U_2 \end{bmatrix} B_2 V = \begin{bmatrix} \Sigma \\ 0 \end{bmatrix}, \tag{4.33}$$

where $U_1 \in \mathbb{R}^{m \times n}$ and $U_2 \in \mathbb{R}^{(n-m) \times n}$, and $\Sigma = \mathrm{diag}\{\tau_1, \tau_2, \ldots, \tau_m\}$, where τ_i ($i = 1, 2, \ldots, m$) are nonzero singular values of B_2.

Lemma 4.4.2   [172] For the matrix $B_2 \in \mathbb{R}^{n \times m}$ with full column rank, if matrix P_1 is of the structure

$$P_1 = U^{\mathrm{T}} \begin{bmatrix} P_{11} & 0 \\ 0 & P_{22} \end{bmatrix} U = U_1^{\mathrm{T}} P_{11} U_1 + U_2^{\mathrm{T}} P_{22} U_2, \tag{4.34}$$

where $P_{11} \in \mathbb{R}^{m \times m} > 0$ and $P_{22} \in \mathbb{R}^{(n-m) \times (n-m)} > 0$, and U_1 and U_2 are defined in (4.33), then there exists a nonsingular matrix $P \in \mathbb{R}^{m \times m}$ such that $B_2 P = P_1 B_2$.

Remark 4.10   *The purpose of Lemma 4.4.2 is to find a solution P to $B_2 P = P_1 B_2$, which will later facilitate our development of the LMI approach to the controller design. The assumption of B_2 being full column rank is just for presentation convenience, which does not lose any generality, as we can always conduct the congruence transformation on B_2. If the condition (4.34) holds, then P exists, but it may not be unique unless B_2 is square and nonsingular.*

We aim to design the controller (4.27) for the system (4.26) such that, in the presence of multiple random packet losses, the closed-loop system (4.32) is stochastically stable and the H_∞ performance constraint is satisfied. To be more specific, we describe the problem as follows.

Problem H_∞ Control with Multiple Data Losses (HCMDL)

For given the communication link parameters $\bar{\Xi}$ and $\bar{\Omega}$ and the scalar $\gamma > 0$, design the controller (4.27) for the system (4.26) such that the closed-loop system satisfies the following two performance requirements:

(i) (stochastic stability) the closed-loop system in (4.32) is stochastically stable in the sense of Definition 4.4.1;

(ii) (H_∞ performance) under zero initial condition, the controlled output z_k satisfies $\|z\|_{\mathrm{E}} \leq \gamma \|w\|_2$, where

$$\|\bar{z}\|_{\mathrm{E}} \triangleq \mathbb{E}\left\{ \sqrt{\sum_{k=0}^{\infty} z_k^{\mathrm{T}} z_k} \right\}, \tag{4.35}$$

and $\|\cdot\|_2$ stands for the usual l_2 norm.

If the above two conditions are satisfied, the closed-loop system is said to be stochastically stable with a guaranteed H_∞ performance γ, and the problem HCMDL is solved.

4.4.2   Main Results

By using similar analysis techniques, some parallel results are derived and listed as follows.

Theorem 4.4.3   *Suppose that both the controller gain matrix K and the observer gain matrix L are given. The closed-loop system in (4.32) is stochastically stable with a guaranteed H_∞ performance γ if there exist positive-definite matrices P_1, P_2 and two scalars $\rho_1 > 0$, $\rho_2 > 0$ satisfying*

$$\begin{bmatrix} \Lambda + \Lambda_1 & \Lambda_2 \\ \Lambda_2^{\mathrm{T}} & \Lambda_3 \end{bmatrix} < 0, \tag{4.36}$$

$$P_1 \leq \rho_1 I, \qquad P_2 \leq \rho_2 I \tag{4.37}$$

where

$$\Lambda = 2 \begin{bmatrix} B_2 \bar{\Omega} K & -B_2 \bar{\Omega} K \\ LC_2 - L\bar{\Xi}C_2 & -LC_2 \end{bmatrix}^{\mathrm{T}} \begin{bmatrix} P_1 & 0 \\ 0 & P_2 \end{bmatrix} \begin{bmatrix} B_2 \bar{\Omega} K & -B_2 \bar{\Omega} K \\ LC_2 - L\bar{\Xi}C_2 & -LC_2 \end{bmatrix}$$

$$+ \begin{bmatrix} 2\rho_1 \lambda_1 I - P_1 & 0 \\ 0 & 2\rho_2 \lambda_1 I - P_2 \end{bmatrix} + \sum_{j=1}^{m} \xi_j^2 \bar{B}_j^{\mathrm{T}} P_1 \bar{B}_j + \sum_{i=1}^{p} \sigma_i^2 \bar{C}_i^{\mathrm{T}} P_2 \bar{C}_i,$$

$$\Lambda_1 = \begin{bmatrix} \rho_1 \lambda_1 I + 2\lambda_2 I & 0 \\ 0 & \rho_2 \lambda_1 I \end{bmatrix}, \qquad \Lambda_2 = \begin{bmatrix} \bar{\Lambda}_{21} \\ -\bar{\Lambda}_{211} - (LC_2)^{\mathrm{T}} P_2 (B_1 - L\bar{\Xi}D_2) \end{bmatrix},$$

$$\bar{\Lambda}_{21} = \bar{\Lambda}_{211} + (LC_2 - L\bar{\Xi}C_2)^{\mathrm{T}} P_2 (B_1 - L\bar{\Xi}D_2) + \sum_{i=1}^{p} \sigma_i^2 (LC_{2i})^{\mathrm{T}} P_2 (LD_{2i}),$$

$$\bar{\Lambda}_{211} = (B_2 \bar{\Omega} K)^{\mathrm{T}} P_1 B_1, \qquad \lambda_1 = \lambda_{\max}(A^{\mathrm{T}} A), \qquad \lambda_2 = \lambda_{\max}(C_1^{\mathrm{T}} C_1),$$

$$\Lambda_3 = 2B_1^{\mathrm{T}} P_1 B_1 + 2(B_1 - L\bar{\Xi}D_2)^{\mathrm{T}} P_2 (B_1 - L\bar{\Xi}D_2) + \sum_{i=1}^{p} \sigma_i^2 (LD_{2i})^{\mathrm{T}} P_2 (LD_{2i})$$

$$+ 2D_1^{\mathrm{T}} D_1 - \gamma^2 I, \qquad \bar{B}_j = [\, B_2 K_j \quad -B_2 K_j \,], \qquad \bar{C}_i = [\, -LC_{2i} \quad 0 \,]. \tag{4.38}$$

In the following, we will deal with the controller design problem and derive the explicit expression of the controller parameters; that is, determine the controller parameters in (4.27) such that the closed-loop system in (4.32) is stochastically stable and the controlled output z_k satisfies (4.35).

Theorem 4.4.4 *Consider the system (4.26). There exists a dynamic observer-based controller in the form of (4.27) such that the closed-loop system in (4.32) is stochastically stable with a guaranteed H_∞ performance γ if there exist positive-definite matrices $P_{11} \in \mathbb{R}^{m \times m}$, $P_{22} \in \mathbb{R}^{(n-m) \times (n-m)}$, $P_2 \in \mathbb{R}^{n \times n}$, real matrices $M_j \in \mathbb{R}^{m \times n}$ $(j = 1, \ldots, m)$, $N \in \mathbb{R}^{n \times p}$ and two scalars $\rho_1 > 0$, $\rho_2 > 0$ satisfying*

$$\begin{bmatrix} \Pi_1 & \Pi_2^{\mathrm{T}} \\ \Pi_2 & \Pi_3 \end{bmatrix} < 0, \tag{4.39}$$

$$P_1 \leq \rho_1 I, \tag{4.40}$$

$$P_2 \leq \rho_2 I, \tag{4.41}$$

where

$$\Pi_1 = \mathrm{diag}\{-P_1 + 3\rho_1\lambda_1 I + 2\lambda_2 I, -P_2 + 3\rho_2\lambda_1 I, -\gamma^2 I + 2D_1^{\mathrm{T}} D_1\},$$

$$\Pi_3 = \mathrm{diag}\{-P_1, -P_2, -P_1, -P_2, -P_1, -P_2, -\hat{P}_1, -\hat{P}_2\},$$

$$\hat{B} = [\xi_1 M_1^{\mathrm{T}} B_2^{\mathrm{T}}, \ldots, \xi_m M_m^{\mathrm{T}} B_2^{\mathrm{T}}]^{\mathrm{T}}, \quad \hat{C} = [-\sigma_1 C_{21}^{\mathrm{T}} N^{\mathrm{T}}, \ldots, -\sigma_p C_{2p}^{\mathrm{T}} N^{\mathrm{T}}]^{\mathrm{T}},$$

$$\hat{P}_1 = \mathrm{diag}\{\underbrace{P_1, \ldots, P_1}_{m}\}, \quad \hat{P}_2 = \mathrm{diag}\{\underbrace{P_2, \ldots, P_2}_{p}\},$$

$$\hat{D} = [-\sigma_1 D_{21}^{\mathrm{T}} N^{\mathrm{T}}, \ldots, -\sigma_p D_{2p}^{\mathrm{T}} N^{\mathrm{T}}]^{\mathrm{T}}, \quad P_1 := U_1^{\mathrm{T}} P_{11} U_1 + U_2^{\mathrm{T}} P_{22} U_2,$$

$$\Pi_2 = \begin{bmatrix} \sum_{j=1}^{m} \vartheta_j B_2 M_j & -\sum_{j=1}^{m} \vartheta_j B_2 M_j & P_1 B_1 \\ NC_2 - N\bar{\Xi}C_2 & -NC_2 & P_2 B_1 - N\bar{\Xi}D_2 \\ \sum_{j=1}^{m} \vartheta_j B_2 M_j & -\sum_{j=1}^{m} \vartheta_j B_2 M_j & 0 \\ NC_2 - N\bar{\Xi}C_2 & -NC_2 & 0 \\ 0 & 0 & P_1 B_1 \\ 0 & 0 & P_2 B_1 - N\bar{\Xi}D_2 \\ \hat{B} & -\hat{B} & 0 \\ \hat{C} & 0 & \hat{D} \end{bmatrix}. \tag{4.42}$$

Furthermore, the controller parameters are given by

$$K = \sum_{j=1}^{m} V\Sigma^{-1} P_{11}^{-1} \Sigma V^{\mathrm{T}} M_j, \quad L = P_2^{-1} N. \tag{4.43}$$

Remark 4.11 *As we can see from Theorem 4.4.3, in the presence of multiple random packet losses, the H_∞ control problem is solved for systems with repeated scalar nonlinearities, and* *an observer-based feedback controller is designed to stochastically stabilize the networked system and also achieve the prescribed H_∞ disturbance rejection attenuation level. Other possible future research directions include real-time applications of the proposed filtering theory in telecommunications and further extensions of the present results to more complex systems with unreliable communication links, such as sampled-data systems, bilinear systems, and a class of nonlinear systems.*

4.5 Illustrative Examples

In this section, some simulation examples are presented to demonstrate the theory presented in this chapter.

4.5.1 Example 1

In this example, we consider H_∞ filter design with repeated scalar nonlinearities (in the presence of missing measurements).

Consider the following system:

$$
x_{k+1} = \begin{bmatrix} 0 & -0.5 \\ 1 & 1 \end{bmatrix} f(x_k) + \begin{bmatrix} -6 & 0 \\ 1 & 0 \end{bmatrix} w_k,
$$

$$
y_k = [-100 \quad 10] f(x_k) + [0 \quad 1] w_k, \tag{4.44}
$$

$$
z_k = [1 \quad 0] f(x_k).
$$

Here, the nonlinear function $f(x_k) = \sin(x_k)$ satisfies (4.2). We like to design full-order ($k = 2$) and reduced-order ($k = 1$) filters in the form of (4.4) so that the filtering error system in (4.6) is stochastically stable with a guaranteed H_∞ norm-bound γ.

Let $\bar{\beta} = 0.8$. By applying Theorem 4.3.1 with help from Algorithm HinfF (Section 4.3), we can easily obtain admissible H_∞ filters and their associated matrix parameters as follows.

Case 1: Full-order **H_∞** *filter design with* $\gamma^* = 0.5$

$$
P = \begin{bmatrix} 1.0286 & 0.0288 & 0.4629 & 0.4629 \\ 0.0288 & 0.0979 & 0.0218 & 0.0218 \\ 0.4629 & 0.0218 & 6.5325 & -5.9677 \\ 0.4629 & 0.0218 & -5.9677 & 6.5325 \end{bmatrix}, \quad A_F = \begin{bmatrix} -0.2881 & -0.2961 \\ 0.0386 & 0.0125 \end{bmatrix},
$$

$$
B_F = \begin{bmatrix} -0.0004 \\ -0.0020 \end{bmatrix}, \quad C_F = [-0.4607 \quad -0.4607].
$$

In this case, we assume $w_k \equiv 0$ and let the initial condition be $x_0 = [0.3 \quad -0.8]^T$, $\hat{x}_0 = [0 \quad 0]^T$. Figure 4.3 shows that the estimation error converges to zero, which confirms that the filtering error system in (4.6) is stochastically stable in the sense of Definition 4.1.1. To further illustrate the performance of the designed filter, we now assume the zero initial conditions and set the external disturbance w_k by

$$
w_k = \begin{cases} 1, & 20 \le k \le 50, \\ -1, & 70 \le k \le 100, \\ 0, & \text{else.} \end{cases} \tag{4.45}
$$

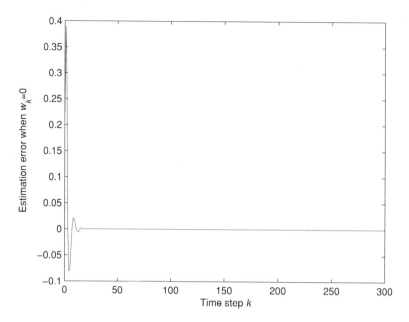

Figure 4.3   Estimation error when $w_k = 0$

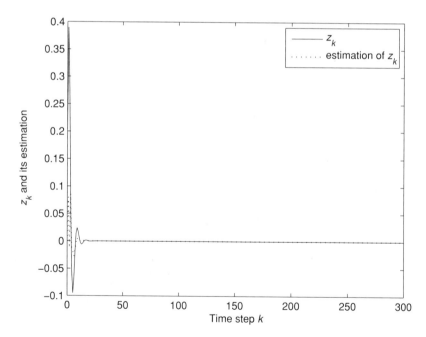

Figure 4.4 z_k and its estimation when $w_k = 0$

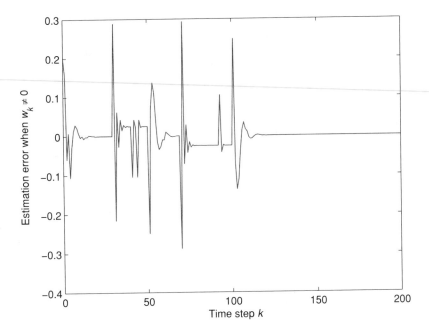

Figure 4.5   Estimation error

Figure 4.5 shows the response of signal $\bar{z}_k$, and Figure 4.6 gives the simulation results of z_k and $\hat{z}_k$. By calculation, we obtain that $\|\bar{z}\|_2^2 = 2.2483$ and $\|w\|_2^2 = 52$, and subsequently $\gamma = 0.2079$ (below the minimum $\gamma^* = 0.5$). Therefore, the HFDL presented in (4.31) is solved, which shows the effectiveness of the H_∞ filter design.

Case 2: First-order H_∞ filter design with $\gamma^* = 0.5$
Similar to Case 1, we assume the zero initial conditions and set the external disturbance w_k as that of (4.45):

$$P = \begin{bmatrix} 1.7403 & 0.4665 & -0.3250 \\ 0.4665 & 0.6514 & -0.0172 \\ -0.3250 & -0.0172 & 0.3929 \end{bmatrix}, \quad A_F = -0.0941,$$

$$B_F = -0.0044, \quad C_F = 0.3047,$$

and the simulation can be carried out in the same way. Figure 4.7 shows the response of the signal $\bar{z}_k$, and Figure 4.8 gives the simulation results of z_k and $\hat{z}_k$. Similarly, we have $\gamma = 0.2059$, which confirms the effectiveness of the H_∞ filter design.

4.5.2   Example 2

Following Mahmoud [166], we consider a factory that produces two kinds of products $(j = 1, 2)$ sharing common resources and raw materials, like color TV and black-and-white TV, PC and laptop computer, and so on. The information transmission is conducted through

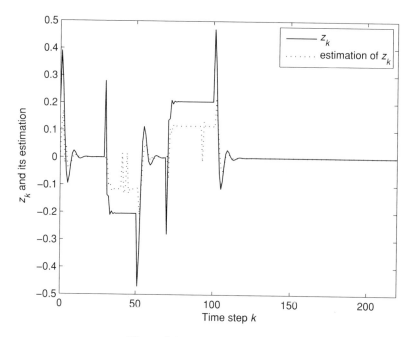

Figure 4.6 z_k and its estimation

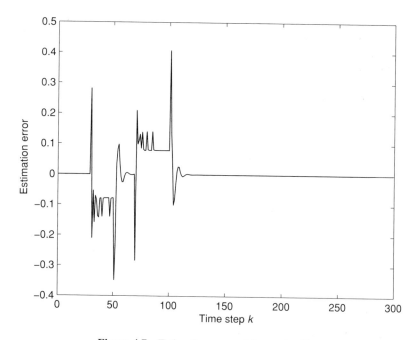

Figure 4.7   Estimation error of first-order filter

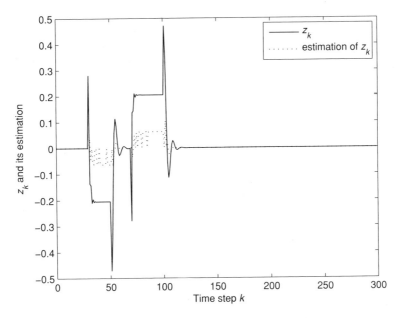

Figure 4.8 z_k and its estimation of first-order filter

networks which are subject to possible packet losses. During the kth period (quarter or season), we define

s_{jk}: amount of sales of product j;

a_{jk}: advertisement cost spent for product j;

i_{jk}: amount of inventory of product j;

p_{jk}: production of product j.

Let

$$x_k = \begin{bmatrix} s_{1k} \\ s_{2k} \\ i_{1k} \\ i_{2k} \end{bmatrix}, \quad u_k = \begin{bmatrix} p_{1,k+1} \\ p_{2,k+1} \\ a_{1k} \\ a_{2k} \end{bmatrix}.$$

The effect of advertisements on sales in the marketing process and the interlink between inventory and production in the production process can then be expressed dynamically by the following form:

$$x_{k+1} = Af(x_k) + Bw_k + (E + \Delta E)u_k,$$

where $f(x_k)$ is a saturation nonlinearity function, $\Delta E u_k$ denotes the uncertain changes in production and advertisements costs, and y_k denotes the measured amount of inventory of

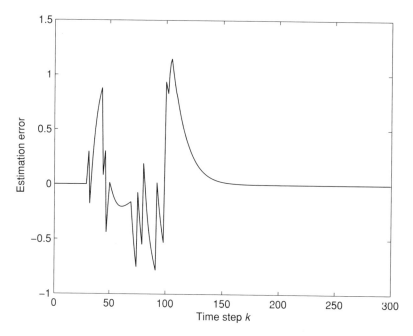

Figure 4.9   Estimation error of Example 2

product. The purpose is to design an H_∞ filter to estimate the amount of sales of product. It is easily seen that the above model fits the model (4.1) nicely when $E + \Delta E = 0$.

Let us now consider a specific example for the above combined marketing and production filtering problem, where

$$A = \begin{bmatrix} 0.7 & 0 & 0 & 0 \\ 0 & 0.5 & 0 & 0 \\ -0.7 & 0 & 0.9 & 0 \\ 0 & -0.5 & 0 & 0.9 \end{bmatrix}, \quad B = \begin{bmatrix} 0 & 0.2 & 0 & 0 \\ 0.1 & 0 & 0 & 0 \\ 0 & -0.2 & 0 & 0 \\ -0.1 & 0 & 0 & 0 \end{bmatrix}, \quad E + \Delta E = 0,$$

$$C = \begin{bmatrix} 1 & 0 & 0 & 0 \end{bmatrix}, \quad D = \begin{bmatrix} 0.5 & 0 & 0 & 0 \end{bmatrix}, \quad H = \begin{bmatrix} 0 & 0 & 1 & 0 \end{bmatrix}, \quad \gamma^* = 3.$$

In the simulation, the probability for network-induced data packet dropouts is set as 10%. Assume the zero initial conditions and let the external disturbance w_k be the same as (4.45). Figure 4.9 depicts the estimation error of the sales amount, and Figure 4.10 depicts the sales amount of product and its estimation. It can be calculated that $\gamma = 0.6674$, which is less than the minimum $\gamma^* = 3$. We can see that the designed filter produces a satisfactory estimate of z_k, showing the effectiveness of the H_∞ filter design.

4.5.3   Example 3

In this example, we consider observer-based H_∞ control for systems with repeated scalar nonlinearities and multiple packet losses.

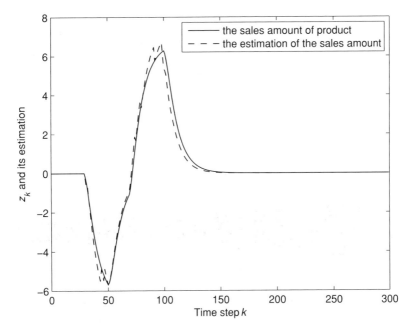

Figure 4.10 The sales amount of product and its estimation

The system data of (4.31) are given as follows:

$$A = \begin{bmatrix} -1 & 0 & 0 \\ 0 & 0.8 & 0.5 \\ 0.5 & 0 & 0 \end{bmatrix}, \quad B_1 = \begin{bmatrix} 0.1 & 1 & 0.4 \\ 0.2 & -0.2 & 1 \\ 0.2 & -0.1 & 0.1 \end{bmatrix}, \quad B_2 = \begin{bmatrix} 1 & 0 & 0.6 \\ 0 & 0.3 & 0.5 \\ 1 & 1.2 & -0.5 \end{bmatrix},$$

$$C_1 = \begin{bmatrix} 0.5 & 0 & 0 \\ 0.4 & 0.5 & 0.5 \\ 1 & 0.2 & 0.2 \end{bmatrix}, \quad C_2 = \begin{bmatrix} 1 & 1 & 5 \\ 0.2 & -0.4 & 0.4 \\ 0 & 0 & 0.3 \end{bmatrix}, \quad D_1 = \begin{bmatrix} 0.1 & -1 & 0.3 \\ 0.2 & -0.3 & 0.3 \\ 1 & 0.1 & -0.1 \end{bmatrix},$$

$$D_2 = \begin{bmatrix} -0.2 & 0.1 & 0.2 \\ 0 & 0.2 & 1 \\ 1 & 0.4 & 0 \end{bmatrix}.$$

Assuming that the probabilistic density functions of α_1, α_2, and α_3 in $[0, 1]$ are described by

$$q_1(s_1) = \begin{cases} 0 & s_1 = 0 \\ 0.1 & s_1 = 0.5 \\ 0.9 & s_1 = 1 \end{cases}, \quad q_2(s_2) = \begin{cases} 0.1 & s_2 = 0 \\ 0.1 & s_2 = 0.5 \\ 0.8 & s_2 = 1 \end{cases},$$

$$q_3(s_3) = \begin{cases} 0 & s_3 = 0 \\ 0.2 & s_3 = 0.5 \\ 0.8 & s_3 = 1 \end{cases},$$

$$(4.46)$$

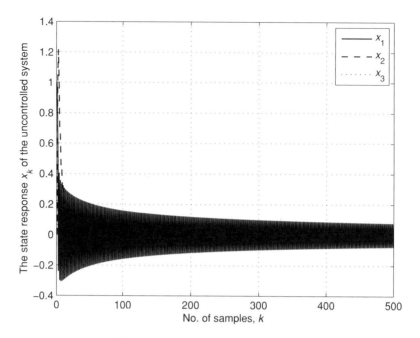

Figure 4.11 x_k of uncontrolled system

from which the expectations and variances can be easily calculated as $\mu_1 = 0.95$, $\mu_2 = 0.85$, $\mu_3 = 0.9$, $\sigma_1 = 0.15$, $\sigma_2 = 0.32$, and $\sigma_3 = 0.2$. In the same way, we assume the probabilistic density functions of β_1, β_2, and β_3 in [0, 1] to be

$$m_1(s_1) = \begin{cases} 0 & s_1 = 0 \\ 0.4 & s_1 = 0.5 \\ 0.6 & s_1 = 1 \end{cases}, \quad m_2(s_2) = \begin{cases} 0.05 & s_2 = 0 \\ 0.15 & s_2 = 0.5 \\ 0.8 & s_2 = 1 \end{cases},$$

$$m_3(s_3) = \begin{cases} 0 & s_3 = 0 \\ 0.2 & s_3 = 0.5 \\ 0.8 & s_3 = 1 \end{cases}, \tag{4.47}$$

from which we can calculate that $\vartheta_1 = 0.8$, $\vartheta_2 = 0.875$, $\vartheta_3 = 0.9$, $\xi_1 = 0.245$, $\xi_2 = 0.268$, and $\xi_3 = 0.2$. By applying Theorem 4.4.3, we can obtain an admissible solution as follows:

$$K = \begin{bmatrix} -0.0364 & 0.0708 & -0.3856 \\ -0.0242 & 0.0550 & -0.3159 \\ 0.0255 & -0.0579 & 0.3356 \end{bmatrix}, \quad L = \begin{bmatrix} 0.0509 & 0.3040 & 0.8950 \\ 0.1242 & 0.4020 & 0.3154 \\ 0.0040 & 0.0247 & 0.0488 \end{bmatrix}.$$

For the purpose of simulation, we let the initial conditions be $x_0 = [1 \quad 0 \quad 0]^T$, $\hat{x}_0 = [0 \quad 0 \quad 0]^T$, and the disturbance input be $w_k = [k^{-2} \quad k^{-2} \quad k^{-2}]^T$. Figure 4.11 displays the state evolutions of the uncontrolled system, which are apparently unstable. Figure 4.12 shows the state simulation results of the closed-loop system, from which we can see that the desired objective is achieved.

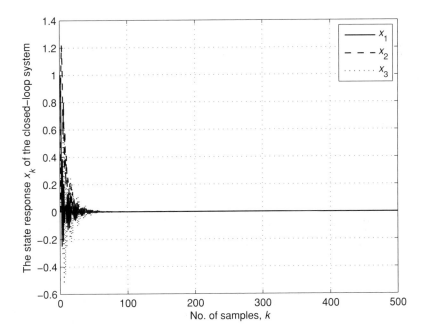

Figure 4.12 x_k of controlled system

4.5.4 Example 4

In this example, we aim to illustrate the effectiveness of our results for different measurement missing cases. Here,

$$A = \begin{bmatrix} -1 & 0 & -0.9 \\ 2 & 0.8 & 0.5 \\ 0.5 & 0 & 1.2 \end{bmatrix},$$

and the other system data of (4.31) is the same as in Example 3. First, we assume the probabilistic density functions of α_1, α_2, α_3 and β_1, β_2, β_3 are the same as (4.46) and (4.47), respectively, and obtain an admissible solution as follows:

$$K = \begin{bmatrix} -0.0276 & 0.0551 & -0.3031 \\ -0.0087 & 0.0295 & -0.1799 \\ 0.0095 & -0.0318 & 0.1967 \end{bmatrix}, \quad L = \begin{bmatrix} 0.0514 & 0.3017 & 0.8770 \\ 0.1217 & 0.3971 & 0.3288 \\ 0.0042 & 0.0236 & 0.0414 \end{bmatrix},$$

for which the simulation result of the state responses is given in Figure 4.13 that confirms the realization of our design goal.

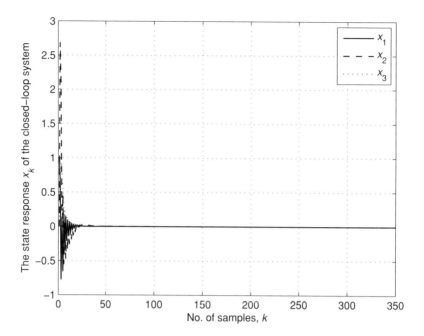

Figure 4.13 x_k when the packet-loss probability is lower

Next, let us consider the case when the multiple packet-loss probability becomes higher. Take the probabilistic density functions of α_1, α_2, and α_3 in [0, 1] as

$$q_1(s_1) = \begin{cases} 0 & s_1 = 0 \\ 0.3 & s_1 = 0.5 \\ 0.7 & s_1 = 1 \end{cases}, \quad q_2(s_2) = \begin{cases} 0 & s_2 = 0 \\ 0.3 & s_2 = 0.5 \\ 0.7 & s_2 = 1 \end{cases},$$

$$q_3(s_3) = \begin{cases} 0 & s_3 = 0 \\ 0.5 & s_3 = 0.5 \\ 0.5 & s_3 = 1 \end{cases},$$

and the probabilistic density functions of β_1, β_2, and β_3 in [0, 1] as

$$m_1(s_1) = \begin{cases} 0.8 & s_1 = 0 \\ 0.1 & s_1 = 0.5 \\ 0.1 & s_1 = 1 \end{cases}, \quad m_2(s_2) = \begin{cases} 0.2 & s_2 = 0 \\ 0.1 & s_2 = 0.5 \\ 0.7 & s_2 = 1 \end{cases},$$

$$m_3(s_3) = \begin{cases} 0.2 & s_3 = 0 \\ 0.1 & s_3 = 0.5 \\ 0.7 & s_3 = 1 \end{cases}.$$

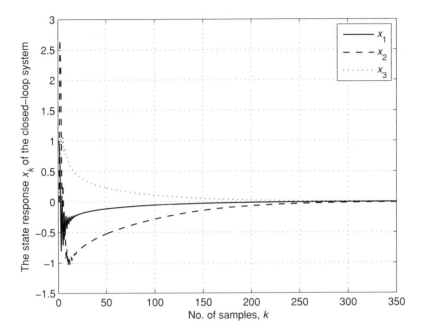

Figure 4.14 x_k when the packet-loss probability is higher

By calculating the expectations and variances of the random variables, we have arrived at the following solution:

$$
K = \begin{bmatrix} -0.0135 & 0.0676 & -0.2887 \\ 0.0424 & 0.0102 & 0.1361 \\ -0.0465 & -0.0465 & -0.1507 \end{bmatrix}, \quad L = \begin{bmatrix} 0.0570 & 0.3216 & 1.0009 \\ 0.1433 & 0.4253 & 0.3136 \\ 0.0071 & 0.0317 & 0.0495 \end{bmatrix}.
$$

Again, the simulation result of the state responses are depicted in Figure 4.14. As we can see from Figures 4.13 and 4.14, when the packet losses are more severe, the dynamical behavior of the NCSs takes longer to converge and, furthermore, the robustness of the closed-loop system is rather degraded.

4.6   Summary

In this chapter, we have investigated the H_∞ filtering and control problems for systems with repeated scalar nonlinearities and missing measurements. The nonlinear system has been described by a discrete-time state equation involving a repeated scalar nonlinearity which typically appears in recurrent neural networks. The missing measurements have been modeled by a stochastic variable satisfying the Bernoulli random binary distribution. The quadratic Lyapunov function has been used to design both full- and reduced-order H_∞ filters such that, for the admissible random missing measurement and repeated scalar nonlinearities, the filtering error system is stochastically stable and preserves a guaranteed H_∞ performance. Furthermore, a novel H_∞ control problem has been considered for systems with repeated scalar nonlinearities under multiple missing measurements. The random communication packet losses have been allowed to occur, simultaneously, in the communication channels from the sensor to the controller and from the controller to the actuator, and the missing probability for each sensor is governed by an individual random variable satisfying a certain probabilistic distribution on the interval [0, 1]. In the presence of random packet losses, an observer-based feedback controller has been designed to stochastically stabilize the networked system. Both the stability analysis and controller synthesis problems have been investigated in detail. Finally, the results of this chapter have been demonstrated by some simulation examples.

5

Filtering and Fault Detection for Markov Systems with Varying Nonlinearities

This chapter addresses the filtering and fault detection problems for discrete-time MJSs with incomplete knowledge of TPs, RVNs and sensor saturations. The RVNs and the sensor saturations are introduced to reflect the limited capacity of the communication networks resulting from the noisy environment, probabilistic communication failures, measurements of limited amplitudes, and so on. Two kinds of TP matrices for the Markovian process are considered: one with polytopic uncertainties and one with partially unknown entries. First, the robust H_∞ filtering problem is considered for a class of discrete time-varying MJSs with randomly occurring nonlinearities and sensor saturation. The main purpose is to design a robust filter, over a given finite horizon, such that the H_∞ disturbance attenuation level is guaranteed for the time-varying MJSs in the presence of both the randomly occurring nonlinearities and the sensor saturation. Sufficient conditions are established for the existence of the desired filter satisfying the H_∞ performance constraint in terms of a set of RLMIs. Considering the fact that the fault detection problem can often be converted into an auxiliary H_∞ filtering problem, the other research focus of this chapter is to investigate the fault detection problem for discrete-time MJSs with incomplete knowledge of TPs, RVNs, and sensor saturations. Two energy norm indices are used for the fault detection problem: one in order to account for the restraint of disturbance and the other the sensitivity of faults. The purpose of the problem addressed is to design an optimized fault detection filter such that (1) the fault detection dynamics is stochastically stable, (2) the effect from the exogenous disturbance on the residual is attenuated with respect to a minimized H_∞-norm, and (3) the sensitivity of the residual to the fault is enhanced by means of a maximized H_∞-norm. The characterization of the gains of the desired fault detection filters is derived in terms of the solution to a convex optimization problem that can be easily solved by using the semi-definite program method. Finally, simulation examples are employed to show the effectiveness of the main results of this chapter.

Filtering, Control and Fault Detection with Randomly Occurring Incomplete Information, First Edition.
Hongli Dong, Zidong Wang, and Huijun Gao.
© 2013 John Wiley & Sons, Ltd. Published 2013 by John Wiley & Sons, Ltd.

5.1   Problem Formulation for Robust H_∞ Filter Design

Let $r(k)$ $(k \in [0, N])$ be a Markov chain taking values in a finite state space $S = \{1, 2, \ldots, s\}$ with TP matrix $\hat{\Psi} = [\lambda_{ij}]$ given by

$$\text{Prob}\{r(k + 1) = j | r(k) = i\} = \lambda_{ij}, \ \forall i, j \in S,$$

where $\lambda_{ij} \geq 0$ $(i, j \in S)$ is the TP from i to j and $\sum_{j=1}^{s} \lambda_{ij} = 1, \ \forall i \in S$.

In this chapter, we consider the following two cases where the TP matrix $\hat{\Psi} = [\lambda_{ij}]$ is imperfectly known.

Case 1:

The TP matrix $\hat{\Psi}$ belongs to a given polytope, namely $\hat{\Psi} \in \mathfrak{R}$, where $\mathfrak{R}$ is a given convex-bounded polyhedral domain described by v vertices as follows:

$$\mathfrak{R} := \left\{ \hat{\Psi} \ \middle| \ \hat{\Psi} = \sum_{r=1}^{v} \psi_r \hat{\Psi}^{(r)}, \sum_{r=1}^{v} \psi_r = 1, \psi_r \geq 0, \ r = 1, 2, \ldots, v \right\} \tag{5.1}$$

and $\hat{\Psi}^{(r)} = [\lambda_{ij}^{(r)}]$ $(i, j = 1, \ldots, s, r = 1, \ldots, v)$ are given TP matrices. It is easy to see that the convex combination of these TP matrices is also a possible TP matrix.

Case 2:

Some elements in matrix $\hat{\Psi}$ are unknown; for example, the TP matrix $\hat{\Psi}$ may be

$$\hat{\Psi} = \begin{bmatrix} \lambda_{11} & ? & ? \\ ? & \lambda_{22} & ? \\ \lambda_{31} & \lambda_{32} & \lambda_{33} \end{bmatrix},$$

where "?" represents the unknown entries. For notation clarity, for any $i \in S$, we denote that

$$\hat{\Psi}_k^i := \{j : \lambda_{ij} \text{ is known}\}, \ \hat{\Psi}_{uk}^i := \{j : \lambda_{ij} \text{ is unknown}\}.$$

We consider the following class of uncertain discrete stochastic nonlinear time-varying MJSs in the presence of sensor saturation defined on $k \in [0, N]$:

$$\begin{cases} x(k + 1) = (A(k, r(k)) + \Delta A(k, r(k)))x(k) + \alpha(k)f(k, x(k)) \\ \qquad\qquad + D_1(k, r(k))w(k), \\ y(k) = \sigma(y_s(k, r(k))) + \beta(k)g(k, x(k)) + D_2(k, r(k))w(k), \\ z(k) = (L(k, r(k)) + \Delta L(k, r(k)))x(k), \\ y_s(k, r(k)) = C(k, r(k))x(k, \\ x(0) = \varphi_0, \end{cases} \tag{5.2}$$

where $x(k) \in \mathbb{R}^n$ represents the state vector, $y(k) \in \mathbb{R}^r$ is the measurement output, $z(k) \in \mathbb{R}^m$ is a linear combination of the state variables to be estimated, $w(k) \in \mathbb{R}^p$ is the disturbance input which belongs to $l_2[0, \infty)$, and φ_0 is a given real initial value. For fixed-system mode, $A(k, r(k))$, $D_1(k, r(k))$, $C(k, r(k))$, $D_2(k, r(k))$, and $L(k, r(k))$ are known, real, time-varying matrices with appropriate dimensions. $\Delta A(k, r(k))$ and $\Delta L(k, r(k))$ are unknown matrices representing the time-varying parameter uncertainties of the form

$$\begin{bmatrix} \Delta A(k, r(k)) \\ \Delta L(k, r(k)) \end{bmatrix} = \begin{bmatrix} H_1(r(k)) \\ H_2(r(k)) \end{bmatrix} F(k, r(k)) N(r(k)), \forall r(k) \in S,$$

where, for fixed-system mode, $H_1(r(k))$, $H_2(r(k))$, and $N(r(k))$ are known, real, constant matrices of appropriate dimensions which characterize how the uncertain parameter in $F(k, r(k))$ enters the nominal matrices $A(k, r(k))$ and $L(k, r(k))$, and $F(k, r(k))$ is an unknown time-varying matrix satisfying

$$F^{\mathrm{T}}(k, r(k)) F(k, r(k)) \leq I, \ \forall r(k) \in S. \tag{5.3}$$

The parameter uncertainties $\Delta A(k, r(k))$ and $\Delta L(k, r(k))$ are said to be admissible if (5.3) holds.

The nonlinear functions $f(k, x(k)) : [0, N] \times \mathbb{R}^n \to \mathbb{R}^n$ and $g(k, x(k)) : [0, N] \times \mathbb{R}^r \to \mathbb{R}^r$ satisfy the following conditions:

$$\begin{aligned} \| f(k, x(k)) \|^2 &\leq \varepsilon_1(k) \| E_1(k) x(k) \|^2, \\ \| g(k, x(k)) \|^2 &\leq \varepsilon_2(k) \| E_2(k) x(k) \|^2 \end{aligned} \tag{5.4}$$

for all $k \in [0, N]$, where $\varepsilon_1(k) > 0$ and $\varepsilon_2(k) > 0$ are known positive scalars, and $E_1(k)$ and $E_2(k)$ are known constant matrices.

The stochastic variables $\alpha(k)$ and $\beta(k)$ are two independent Bernoulli sequences which account for the phenomena of randomly occurring nonlinearities. A natural assumption on the sequences $\alpha(k)$ and $\beta(k)$ can be made as follows:

$$\begin{aligned} \mathrm{Prob}\{\alpha(k) = 1\} &= \mathbb{E}\{\alpha(k)\} = \bar{\alpha} \quad \mathrm{Prob}\{\alpha(k) = 0\} = 1 - \bar{\alpha} \\ \mathrm{Prob}\{\beta(k) = 1\} &= \mathbb{E}\{\beta(k)\} = \bar{\beta} \quad \mathrm{Prob}\{\beta(k) = 0\} = 1 - \bar{\beta} \end{aligned} \tag{5.5}$$

where $\bar{\alpha} \in [0 \quad 1]$ and $\bar{\beta} \in [0 \quad 1]$ are known constants. We also assume that the $r(k)$, $\alpha(k)$, and $\beta(k)$ are mutually independent.

Remark 5.1 *As described in (5.2), the nonlinear functions $f(k, x(k))$ and $g(k, x(k))$ could occur independently and randomly according to individual probability distributions specified a priori through statistical tests.*

The saturation function $\sigma(\cdot) : \mathbb{R}^r \mapsto \mathbb{R}^r$ is defined as

$$\sigma(v) = \begin{bmatrix} \sigma_1^{\mathrm{T}}(v_1) & \sigma_2^{\mathrm{T}}(v_2) & \cdots & \sigma_r^{\mathrm{T}}(v_r) \end{bmatrix}^{\mathrm{T}} \tag{5.6}$$

with $\sigma_i(v_i) = \text{sign}(v_i) \min\{V_{i,\max}, |V_i|\}$, where $V_{i,\max}$ is the ith element of the vector $V_{\max}$, the saturation level.

Definition 5.1.1   *[148] A nonlinearity $\Psi : \mathbb{R}^m \mapsto \mathbb{R}^m$ is said to satisfy a sector condition if*

$$(\Psi(v) - \bar{H}_1 v)^{\mathsf{T}}(\Psi(v) - \bar{H}_2 v) \leq 0, \ \forall v \in \mathbb{R}^r \tag{5.7}$$

for some real matrices $\bar{H}_1, \bar{H}_2 \in \mathbb{R}^{r \times r}$, where $\bar{H} = \bar{H}_2 - \bar{H}_1$ is a positive-definite symmetric matrix. In this case, we say that Ψ belongs to the sector $[\bar{H}_1, \bar{H}_2]$.

If we assume that there exist two diagonal matrices K_1 and K_2 such that $0 \leq K_1 < I \leq K_2$, then the saturation function $\sigma(y_s(k, r(k)))$ in (5.2) can be decomposed into a linear and a nonlinear part as

$$\sigma(y_s(k, r(k))) = K_1 C(k, r(k))x(k) + \Psi(y_s(k, r(k))), \tag{5.8}$$

where $\Psi(y_s(k, r(k)))$ is a nonlinear vector-valued function satisfying a sector condition with $\bar{H}_1 = 0$, $\bar{H}_2 = K$, and can be described as follows:

$$\Psi^{\mathsf{T}}(y_s(k, r(k)))(\Psi(y_s(k, r(k))) - K y_s(k, r(k))) \leq 0, \tag{5.9}$$

where $K = K_2 - K_1$.

In this chapter, the linear time-varying filter under consideration is of the following structure:

$$\begin{cases} \hat{x}(k+1) = A_{\mathrm{f}}(k, r(k))\hat{x}(k) + B_{\mathrm{f}}(k, r(k))y(k) \\ \hat{z}(k) = L_{\mathrm{f}}(k, r(k))\hat{x}(k) \end{cases} \tag{5.10}$$

where $\hat{x}(k) \in \mathbb{R}^n$ represents the state estimate and $z(k) \in \mathbb{R}^m$ is the estimated output. For fixed-system mode, the time-varying matrices $A_{\mathrm{f}}(k, r(k))$, $B_{\mathrm{f}}(k, r(k))$, and $L_{\mathrm{f}}(k, r(k))$ are the filter parameters to be designed.

Note that the set S comprises various operation modes of the system in (5.2) and (5.10), and the Markov chain $\{r(k), k \in [0, N]\}$ takes values in the finite set $S = \{1, 2, \ldots, s\}$. For presentation convenience, for each possible $r(k) = i$ ($i \in S$), a matrix $N(k, r(k))$ will be denoted by $N_i(k)$; for example, $A(k, r(k))$ is denoted by $A_i(k)$, and $E(r(k))$ by E_i, and so on.

Let us now work on the system mode $r(k) = i$, $\forall i \in S$. Setting $\bar{x}(k) = [x^{\mathsf{T}}(k) \ \hat{x}^{\mathsf{T}}(k)]^{\mathsf{T}}$ and $\bar{z}(k) = z(k) - \hat{z}(k)$, we obtain an augmented system from (5.2) and (5.10) as follows:

$$\begin{cases} \bar{x}(k+1) = \bar{A}_i(k)\bar{x}(k) + \bar{B}_i(k)\Psi(y_{si}(k)) + (G_i(k) + \bar{G}_i(k))h(k, x(k)) + \bar{D}_i(k)w(k) \\ \bar{z}(k) = \bar{L}_i(k)\bar{x}(k) \end{cases} \tag{5.11}$$

where

$$\bar{A}_i(k) = \begin{bmatrix} A_i(k) + \Delta A_i(k) & 0 \\ B_{\mathrm{f}i}(k)K_1 C_i(k) & A_{\mathrm{f}i}(k) \end{bmatrix}, \quad \bar{B}_i(k) = \begin{bmatrix} 0 \\ B_{\mathrm{f}i}(k) \end{bmatrix},$$

$$h(k, x(k)) = \begin{bmatrix} f(k, x(k)) \\ g(k, x(k)) \end{bmatrix}, \quad \bar{L}_i(k) = [\, L_i(k) + \Delta L_i(k) \quad -L_{\mathrm{f}i}(k)\,],$$

$$\bar{D}_i(k) = \begin{bmatrix} D_{1i}(k) \\ B_{\mathrm{f}i}(k)D_{2i}(k) \end{bmatrix}, \quad G_i(k) = \mathrm{diag}\{\bar{\alpha}I, \bar{\beta}B_{\mathrm{f}i}(k)\},$$

$$\bar{G}_i(k) = \mathrm{diag}\{(\alpha(k) - \bar{\alpha})I, (\beta(k) - \bar{\beta})B_{\mathrm{f}i}(k)\}. \tag{5.12}$$

Our aim in this chapter is to design a finite-horizon filter in the form of (5.10) such that, for the given disturbance attenuation level $\gamma > 0$, positive-definite matrices Q_i $(i = 1, 2, \ldots, s)$ and the initial state $\bar{x}(0)$, the H_∞ performance index satisfies the following inequality:

$$\mathbb{E}\{\|\bar{z}(k)\|_{[0,N]}^2\} \leq \gamma^2(\mathbb{E}\{\|w(k)\|_{[0,N]}^2\} + e^{\mathrm{T}}(0)Q_i e(0)) \tag{5.13}$$

where $e(0) = x(0) - \hat{x}(0)$.

The finite-horizon filtering problem in the presence of sensor saturation and Markovian jump parameters addressed above is referred to as the robust finite-horizon H_∞ filtering problem for uncertain nonlinear discrete time-varying Markovian jump stochastic systems with sensor saturation constraint.

5.2 Performance Analysis of Robust H_∞ Filter

Given the unknown TP matrices described in *Case 1* and *Case 2*, we first propose the following H_∞ performance analysis results with the desired time-varying filter (5.10). For presentation convenience, we denote

$$\Omega_{11i}(k) = 2\bar{A}_i^{\mathrm{T}}(k)\bar{P}_i(k+1)\bar{A}_i(k) - P_i(k) + \bar{L}_i^{\mathrm{T}}(k)\bar{L}_i(k) + \rho_i(k)\bar{E}(k),$$

$$\Omega_{21i}(k) = \bar{B}_i^{\mathrm{T}}(k)\bar{P}_i(k+1)\bar{A}_i(k) + \frac{1}{2}\tau_i(k)\tilde{C}_i(k),$$

$$\Omega_{22i}(k) = 2\bar{B}_i^{\mathrm{T}}(k)\bar{P}_i(k+1)\bar{B}_i(k) - \tau_i(k)I,$$

$$\Omega_{31i}(k) = \bar{D}_i^{\mathrm{T}}(k)\bar{P}_i(k+1)\bar{A}_i(k), \quad \Omega_{32i}(k) = \bar{D}_i^{\mathrm{T}}(k)\bar{P}_i(k+1)\bar{B}_i(k),$$

$$\Omega_{33i}(k) = 2\bar{D}_i^{\mathrm{T}}(k)\bar{P}_i(k+1)\bar{D}_i(k) - \gamma^2 I, \quad \bar{P}_i(k) = \sum_{j=1}^{s} \lambda_{ij}P_j(k), \; i, j = 1, \ldots, s,$$

$$\hat{G}_i(k) = \mathrm{diag}\left\{\sqrt{\bar{\alpha}(1-\bar{\alpha})}I, \sqrt{\bar{\beta}(1-\bar{\beta})}B_{\mathrm{f}i}(k)\right\}, \quad \tilde{C}_i(k) = [\, KC_i(k) \; 0\,],$$

$$E_\varepsilon(k) = \varepsilon_1(k)E_1^{\mathrm{T}}(k)E_1(k) + \varepsilon_2(k)E_2^{\mathrm{T}}(k)E_2(k), \quad \bar{E}(k) = \mathrm{diag}\{E_\varepsilon(k), 0\}. \tag{5.14}$$

Theorem 5.2.1   *Consider system (5.2) subject to randomly occurring nonlinearities (5.4) and (5.5) and sensor saturation (5.6). Let the disturbance attenuation level $\gamma > 0$, sets of positive scalars $\{\rho_i(k) > 0, i \in S\}_{0 \le k \le N}$, $\{\tau_i(k) > 0, i \in S\}_{0 \le k \le N}$, positive-definite matrices $Q_i > 0, i \in S$, and the filter parameters $\{A_{fi}(k)\}_{0 \le k \le N}$, $\{B_{fi}(k)\}_{0 \le k \le N}$, and $\{L_{fi}(k)\}_{0 \le k \le N}$ $(i \in S)$ be given. The $\mathcal{H}_\infty$ performance index defined in (5.13) is achieved for all nonzero $w(k)$ if, with the initial condition $P_i(0) \le \gamma^2 [I \quad -I]^{\mathrm{T}} Q_i [I \quad -I]$, there exists a family of positive-definite matrices $\{P_i(k)\}_{0 \le k \le N+1}$ $(i \in S)$ satisfying the following recursive matrix inequalities:*

$$
\begin{bmatrix}
-\rho_i(k)I & * & * \\
\hat{G}_i(k) & -\bar{P}_i^{-1}(k+1) & * \\
G_i(k) & 0 & -\frac{1}{4}\bar{P}_i^{-1}(k+1)
\end{bmatrix} \le 0 \tag{5.15}
$$

$$
\begin{bmatrix}
\Omega_{11i}(k) & * & * \\
\Omega_{21i}(k) & \Omega_{22i}(k) & * \\
\Omega_{31i}(k) & \Omega_{32i}(k) & \Omega_{33i}(k)
\end{bmatrix} \le 0 \tag{5.16}
$$

for all $0 \le k \le N$, where $\Omega_{11i}(k)$, $\Omega_{21i}(k)$, $\Omega_{22i}(k)$, $\Omega_{31i}(k)$, $\Omega_{32i}(k)$, $\Omega_{33i}(k)$, $\bar{P}_i(k)$, $\hat{G}_i(k)$, $\tilde{C}_i(k)$, $E_\varepsilon(k)$, and $\bar{E}(k)$ are defined in (5.14).

Proof.   For $r(k) = i$, define the following Lyapunov function:

$$
V(\bar{x}(k), r(k)) = \bar{x}^{\mathrm{T}}(k) P_i(k) \bar{x}(k), \tag{5.17}
$$

where $P_i(k) = \mathrm{diag}\{P_{1i}(k), P_{2i}(k)\} > 0$ are the solutions to (5.15) and (5.16). Then, for $r(k) = i$ and $r(k+1) = j$, one has from (5.11) that

$$
\begin{aligned}
\mathbb{E}&\{\Delta V(\bar{x}(k), r(k))\} \\
&= \mathbb{E}\{V(\bar{x}(k+1), r(k+1)) \mid (\bar{x}(k), r(k)) - V(\bar{x}(k), r(k))\} \\
&= \mathbb{E}\{\bar{x}^{\mathrm{T}}(k+1)\bar{P}_i(k+1)\bar{x}(k+1) \mid \bar{x}(k)\} - \bar{x}^{\mathrm{T}}(k)P_i(k)\bar{x}(k) \\
&= \mathbb{E}\{[\bar{A}_i(k)\bar{x}(k) + \bar{B}_i(k)\Psi(y_{si}(k)) + (G_i(k) + \bar{G}_i(k))h(k, x(k)) + \bar{D}_i(k)w(k)]^{\mathrm{T}} \\
&\quad \times \bar{P}_i(k+1)[\bar{A}_i(k)\bar{x}(k) + \bar{B}_i(k)\Psi(y_{si}(k)) + (G_i(k) + \bar{G}_i(k))h(k, x(k)) \\
&\quad + \bar{D}_i(k)w(k)] \mid \bar{x}(k)\} - \bar{x}^{\mathrm{T}}(k)P_i(k)\bar{x}(k) \\
&= \bar{x}^{\mathrm{T}}(k)(\bar{A}_i^{\mathrm{T}}(k)\bar{P}_i(k+1)\bar{A}_i(k) - P_i(k))\bar{x}(k) + 2\bar{x}^{\mathrm{T}}(k)\bar{A}_i^{\mathrm{T}}(k)\bar{P}_i(k+1)\bar{B}_i(k) \\
&\quad \times \Psi(y_{si}(k)) + 2\bar{x}^{\mathrm{T}}(k)\bar{A}_i^{\mathrm{T}}(k)\bar{P}_i(k+1)G_i(k)h(k, x(k)) + 2\bar{x}^{\mathrm{T}}(k)\bar{A}_i^{\mathrm{T}}(k)\bar{P}_i(k+1) \\
&\quad \times \bar{D}_i(k)w(k) + \Psi^{\mathrm{T}}(y_{si}(k))\bar{B}_i^{\mathrm{T}}(k)\bar{P}_i(k+1)\bar{B}_i(k)\Psi(y_{si}(k)) + 2\Psi^{\mathrm{T}}(y_{si}(k))\bar{B}_i^{\mathrm{T}}(k) \\
&\quad \times \bar{P}_i(k+1)G_i(k)h(k, x(k)) + 2\Psi^{\mathrm{T}}(y_{si}(k))\bar{B}_i^{\mathrm{T}}(k)\bar{P}_i(k+1)\bar{D}_i(k)w(k) \\
&\quad + 2h^{\mathrm{T}}(k, x(k))G_i^{\mathrm{T}}(k)\bar{P}_i(k+1)\bar{D}_i(k)w(k) + h^{\mathrm{T}}(k, x(k))(G_i^{\mathrm{T}}(k)\bar{P}_i(k+1)G_i(k) \\
&\quad + \hat{G}_i^{\mathrm{T}}(k)\bar{P}_i(k+1)\hat{G}_i(k))h(k, x(k)) + w^{\mathrm{T}}(k)\bar{D}_i^{\mathrm{T}}(k)\bar{P}_i(k+1)\bar{D}_i(k)w(k). \tag{5.18}
\end{aligned}
$$

From the elementary inequality $2a^{\mathrm{T}}b \leq a^{\mathrm{T}}\bar{P}_i(k)a + b^{\mathrm{T}}\bar{P}_i^{-1}(k)b$, it follows that

$$2\bar{x}^{\mathrm{T}}(k)\bar{A}_i^{\mathrm{T}}(k)\bar{P}_i\,(k+1)G_i(k)h(k, x(k))$$
$$\leq \bar{x}^{\mathrm{T}}(k)\bar{A}_i^{\mathrm{T}}(k)\bar{P}_i(k+1)\bar{A}_i(k)\bar{x}(k)$$
$$+h^{\mathrm{T}}(k, x(k))G_i^{\mathrm{T}}(k)\bar{P}_i(k+1)G_i(k)h(k, x(k)),$$
$$2\Psi^{\mathrm{T}}(y_{si}(k))\bar{B}_i^{\mathrm{T}}\,(k)\bar{P}_i(k+1)G_i(k)h(k, x(k))$$
$$\leq \Psi^{\mathrm{T}}(y_{si}(k))\bar{B}_i^{\mathrm{T}}(k)\bar{P}_i(k+1)\bar{B}_i(k)\Psi(y_{si}(k))$$
$$+h^{\mathrm{T}}(k, x(k))G_i^{\mathrm{T}}(k)\bar{P}_i(k+1)G_i(k)h(k, x(k)),$$
$$2h^{\mathrm{T}}(k, x(k))G_i^{\mathrm{T}}\,(k)\bar{P}_i(k+1)\bar{D}_i(k)w(k)$$
$$\leq w^{\mathrm{T}}(k)\bar{D}_i^{\mathrm{T}}(k)\bar{P}_i(k+1)\bar{D}_i(k)w(k)$$
$$+h^{\mathrm{T}}(k, x(k))G_i^{\mathrm{T}}(k)\bar{P}_i(k+1)G_i(k)h(k, x(k)).$$

It can be seen that (5.15) is equivalent to

$$4G_i^{\mathrm{T}}(k)\bar{P}_i(k+1)G_i(k) + \hat{G}_i^{\mathrm{T}}(k)\bar{P}_i(k+1)\hat{G}_i(k) \leq \rho_i(k)I$$

and therefore it follows from (5.4) that

$$h^{\mathrm{T}}\,(k, x(k))(4G_i^{\mathrm{T}}(k)\bar{P}_i(k+1)G_i(k) + \hat{G}_i^{\mathrm{T}}(k)\bar{P}_i(k+1)\hat{G}_i(k))h(k, x(k))$$
$$\leq \bar{x}^{\mathrm{T}}(k)\rho_i(k)\bar{E}(k)\bar{x}(k).$$

Adding the zero term $\mathbb{E}\{\bar{z}^{\mathrm{T}}(k)\bar{z}(k) - \gamma^2\omega^{\mathrm{T}}(k)\omega(k) - \bar{z}^{\mathrm{T}}(k)\bar{z}(k) + \gamma^2\omega^{\mathrm{T}}(k)\omega(k)\}$ to $\mathbb{E}\{\Delta V(\bar{x}(k), r(k))\}$ results in

$$\mathbb{E}\{\Delta V(\bar{x}(k), r(k))\}$$

$$\leq \mathbb{E}\left\{\begin{bmatrix} \bar{x}(k) \\ \Psi(y_{si}(k)) \\ \omega(k) \end{bmatrix}^{\mathrm{T}} \Lambda_k \begin{bmatrix} \bar{x}(k) \\ \Psi(y_{si}(k)) \\ \omega(k) \end{bmatrix} - \bar{z}^{\mathrm{T}}(k)\bar{z}(k) + \gamma^2\omega^{\mathrm{T}}(k)\omega(k)\right\} \quad (5.19)$$

$$= \mathbb{E}\{\eta^{\mathrm{T}}(k)\Lambda_k\eta(k) - \bar{z}^{\mathrm{T}}(k)\bar{z}(k) + \gamma^2\omega^{\mathrm{T}}(k)\omega(k)\},$$

where

$$\Lambda_k = \begin{bmatrix} \Omega_{11i}(k) & * & * \\ \bar{\Omega}_{21i}(k) & \bar{\Omega}_{22i}(k) & * \\ \Omega_{31i}(k) & \Omega_{32i}(k) & \Omega_{33i}(k) \end{bmatrix}, \quad \eta(k) = [\bar{x}^{\mathrm{T}}(k) \quad \Psi^{\mathrm{T}}(y_{si}(k)) \quad \omega^{\mathrm{T}}(k)]^{\mathrm{T}},$$

$$\bar{\Omega}_{21i}(k) = \bar{B}_i^{\mathrm{T}}(k)\bar{P}_i(k+1)\bar{A}_i(k), \quad \bar{\Omega}_{22i}(k) = 2\bar{B}_i^{\mathrm{T}}(k)\bar{P}_i(k+1)\bar{B}_i(k). \quad (5.20)$$

Summing up (5.19) on both sides from 0 to N with respect to k, we obtain

$$\mathbb{E}\left\{\sum_{k=0}^{N}\Delta V(\bar{x}(k),r(k))\right\} = \mathbb{E}\left\{\bar{x}^{\mathrm{T}}(N+1)\bar{P}_i(N+1)\bar{x}(N+1)\right\} - \bar{x}^{\mathrm{T}}(0)P_i(0)\bar{x}(0)$$

$$\leq \mathbb{E}\left\{\sum_{k=0}^{N}\eta^{\mathrm{T}}(k)\Lambda_k\eta(k) - \sum_{k=0}^{N}(\bar{z}^{\mathrm{T}}(k)\bar{z}(k) - \gamma^2\omega^{\mathrm{T}}(k)\omega(k))\right\}.$$

Hence, the H_∞ performance index defined in (5.13) is given by

$$\mathbb{E}\left\{\|\bar{z}(k)\|_{[0,N]}^2\right\} - \gamma^2(\mathbb{E}\{\|w(k)\|_{[0,N]}^2\} + e^{\mathrm{T}}(0)Q_i e(0))$$

$$\leq \mathbb{E}\left\{\sum_{k=0}^{N}\eta^{\mathrm{T}}(k)\Lambda_k\eta(k)\right\} - \mathbb{E}\{\bar{x}^{\mathrm{T}}(N+1)\bar{P}_i(N+1)\bar{x}(N+1)\} \qquad (5.21)$$

$$+ \bar{x}^{\mathrm{T}}(0)(P_i(0) - \gamma^2[\,I \quad -I\,]^{\mathrm{T}}Q_i[\,I \quad -I\,])\bar{x}(0).$$

Noting that $\bar{P}_i(N+1) > 0$ and the initial condition $P_i(0) \leq \gamma^2[\,I \quad -I\,]^{\mathrm{T}}Q_i[\,I \quad -I\,]$, we can get (5.13) when the following inequality holds:

$$\eta^{\mathrm{T}}(k)\Lambda_k\eta(k) \leq 0. \qquad (5.22)$$

In terms of the sensor saturation constraint in (5.9), we have

$$\Psi^{\mathrm{T}}(y_{si}(k))(\Psi(y_{si}(k)) - \tilde{C}_i(k)\bar{x}(k)) \leq 0, \qquad (5.23)$$

which can be written in $\eta(k)$ as

$$\eta^{\mathrm{T}}(k)\Phi_k\eta(k) \leq 0, \qquad (5.24)$$

where

$$\Phi_k = \frac{1}{2}\begin{bmatrix} 0 & * & * \\ -\tilde{C}_i(k) & 2I & * \\ 0 & 0 & 0 \end{bmatrix}. \qquad (5.25)$$

Now, it suffices to find a condition such that (5.22) holds subject to the sensor saturation constraints (5.24). By using Lemma 2.4.2, the sufficient condition such that the inequalities (5.24) imply (5.22) is that there exist positive scalars $\tau_i(k)$ such that

$$\Lambda_k - \tau_i(k)\Phi_k \leq 0, \qquad (5.26)$$

and then the rest of the proof follows from the statement of Theorem 5.2.1 immediately. The proof is complete. $\square$

5.3 Design of Robust H_∞ Filters

In this section, given the imperfect TP matrix described in *Case 1* and *Case 2*, we shall discuss the robust H_∞ filter design problem for the discrete time-varying MJSs with randomly occurring nonlinearities and sensor saturation.

Based on the analysis results presented in Section 5.2, we are now ready to solve the addressed filter design problem for systems (5.2) in the following theorem with the unknown TP matrix given in *Case 1*. Before presenting the theorem, let us denote

$$
\bar{P}_{1i}^{(r)}(k) = \sum_{j=1}^{s} \lambda_{ij}^{(r)} P_{1j}(k), \quad \bar{P}_{2i}^{(r)}(k) = \sum_{j=1}^{s} \lambda_{ij}^{(r)} P_{2j}(k), \quad i, j = 1, \ldots, s, \ r = 1, \ldots, \nu,
$$

$$
\Phi_{11i}(k) = \mathrm{diag}\{-\rho_i(k)I, -\rho_i(k)I\}, \quad \tilde{\alpha} = \sqrt{\bar{\alpha}(1-\bar{\alpha})}, \quad \tilde{\beta} = \sqrt{\bar{\beta}(1-\bar{\beta})},
$$

$$
\Phi_{22i}^{(r)}(k) = \mathrm{diag}\left\{ -\bar{P}_{1i}^{(r)}(k+1), -\bar{P}_{2i}^{(r)}(k+1), -\frac{1}{4}\bar{P}_{1i}^{(r)}(k+1), -\frac{1}{4}\bar{P}_{2i}^{(r)}(k+1) \right\},
$$

$$
\Phi_{21i}^{(r)}(k) = \begin{bmatrix} \tilde{\alpha}\bar{P}_{1i}^{(r)}(k+1) & 0 \\ 0 & \tilde{\beta}\bar{N}_i^{(r)}(k) \\ \bar{\alpha}\bar{P}_{1i}^{(r)}(k+1) & 0 \\ 0 & \bar{\beta}\bar{N}_i^{(r)}(k) \end{bmatrix}, \quad \Upsilon_{11i}(k) = \begin{bmatrix} \Xi_{1i}(k) & * \\ \Xi_{2i}(k) & \Xi_{3i}(k) \end{bmatrix},
$$

$$
\Upsilon_{21i}^{(r)}(k) = \begin{bmatrix} \Xi_{4i}^{(r)}(k) & \Xi_{5i}^{(r)}(k) \\ \Xi_{6i}^{(r)}(k) & 0 \end{bmatrix}, \quad \Upsilon_{31i}^{(r)}(k) = \begin{bmatrix} 0 & \Xi_{7i}^{(r)}(k) \\ \Xi_{8i} & 0 \end{bmatrix}, \quad \Upsilon_{32i} = \begin{bmatrix} 0 & 0 \\ \Xi_{9i} & 0 \end{bmatrix},
$$

$$
\Upsilon_{22i}^{(r)}(k) = \mathrm{diag}\{-\bar{P}_{1i}^{(r)}(k+1), -\bar{P}_{2i}^{(r)}(k+1),
$$
$$
-\bar{P}_{1i}^{(r)}(k+1), -\bar{P}_{2i}^{(r)}(k+1), -\bar{P}_{1i}^{(r)}(k+1)\},
$$

$$
\Upsilon_{33i}^{(r)}(k) = \mathrm{diag}\{-\bar{P}_{2i}^{(r)}(k+1), -\bar{P}_{1i}^{(r)}(k+1), -\bar{P}_{2i}^{(r)}(k+1), -\xi_i I, -\xi_i I\},
$$

$$
\Xi_{1i}(k) = \mathrm{diag}\{\Pi_{1i}(k), -P_{2i}(k)\}, \quad \Xi_{3i}(k) = \mathrm{diag}\{-\tau_i(k)I, -\gamma^2 I, -I\},
$$

$$
\Pi_{1i}(k) = -P_{1i}(k) + \rho_i(k)\left(\varepsilon_1(k)E_1^{\mathrm{T}}(k)E_1(k) + \varepsilon_2(k)E_2^{\mathrm{T}}(k)E_2(k)\right),
$$

$$
\Xi_{2i}(k) = \begin{bmatrix} \frac{1}{2}\tau_i(k)KC_i(k) & 0 \\ 0 & 0 \\ L_i(k) & -L_{fi}(k) \end{bmatrix}, \quad \Xi_{6i}^{(r)}(k) = \begin{bmatrix} \bar{P}_{1i}^{(r)}(k+1)A_i(k) & 0 \\ \bar{N}_i^{(r)}(k)K_1C_i(k) & \bar{M}_i^{(r)}(k) \\ 0 & 0 \end{bmatrix},
$$

$$
\Xi_{5i}^{(r)}(k) = \begin{bmatrix} 0 & \bar{P}_{1i}^{(r)}(k+1)D_{1i}(k) & 0 \\ \bar{N}_i^{(r)}(k) & \bar{N}_i^{(r)}(k)D_{2i}(k) & 0 \end{bmatrix}, \quad \Xi_{9i} = \begin{bmatrix} H_{1i}^{\mathrm{T}} & 0 & H_{1i}^{\mathrm{T}} \\ 0 & 0 & 0 \end{bmatrix},
$$

$$
\Xi_{4i}^{(r)}(k) = \begin{bmatrix} \bar{P}_{1i}^{(r)}(k+1)A_i(k) & 0 \\ \bar{N}_i^{(r)}(k)K_1C_i(k) & \bar{M}_i^{(r)}(k) \end{bmatrix}, \quad \Xi_{8i} = [\xi_i N_i \quad 0],
$$

$$
\Xi_{7i}^{(r)}(k) = \begin{bmatrix} \bar{N}_i^{(r)}(k) & 0 & 0 \\ 0 & \bar{P}_{1i}^{(r)}(k+1)D_{1i}(k) & 0 \\ 0 & \bar{N}_i^{(r)}(k)D_{2i}(k) & 0 \\ 0 & 0 & H_{2i}^{\mathrm{T}} \end{bmatrix}.
$$

Theorem 5.3.1   *Consider system (5.2) with unknown TP matrix described in Case 1. Let $\gamma > 0$ be a given disturbance attenuation level. For given positive-definite matrices $Q_i > 0$ ($i \in S$), if there exist families of positive-definite matrices $\{P_{1i}(k)\}_{0 \leq k \leq N+1}$, $\{P_{2i}(k)\}_{0 \leq k \leq N+1}$ ($i = 1, 2, \ldots, s$), families of positive scalars $\{\rho_i(k)\}_{0 \leq k \leq N}$, $\{\tau_i(k)\}_{0 \leq k \leq N}$ ($i = 1, 2, \ldots, s$), and $\{\xi_i\}_{i=1,2,\ldots,s}$, and families of real-valued matrices $\{\bar{M}_i^{(r)}(k)\}_{0 \leq k \leq N}$, $\{\bar{N}_i^{(r)}(k)\}_{0 \leq k \leq N}$, and $\{L_{fi}(k)\}_{0 \leq k \leq N}$ satisfying the following RLMIs:*

$$\begin{bmatrix} \Phi_{11i}(k) & * \\ \Phi_{21i}^{(r)}(k) & \Phi_{22i}^{(r)}(k) \end{bmatrix} < 0, i = 1, 2, \ldots, s, \ r = 1, 2, \ldots, \nu, \tag{5.27}$$

$$\begin{bmatrix} \Upsilon_{11i}(k) & * & * \\ \Upsilon_{21i}^{(r)}(k) & \Upsilon_{22i}^{(r)}(k) & * \\ \Upsilon_{31i}^{(r)}(k) & \Upsilon_{32i} & \Upsilon_{33i}^{(r)}(k) \end{bmatrix} \leq 0, \ i = 1, 2, \ldots, s, \ r = 1, 2, \ldots, \nu, \tag{5.28}$$

with the initial condition

$$\begin{bmatrix} P_{1i}(0) - \gamma^2 Q_i & \gamma^2 Q_i \\ \gamma^2 Q_i & P_{2i}(0) - \gamma^2 Q_i \end{bmatrix} \leq 0, \tag{5.29}$$

where $\bar{P}_{1i}^{(r)}(k)$, $\Phi_{11i}(k)$, $\Phi_{22i}^{(r)}(k)$, $\Phi_{21i}^{(r)}(k)$, $\tilde{\alpha}$, $\Upsilon_{11i}(k)$, $\Upsilon_{21i}^{(r)}(k)$, $\Upsilon_{22i}^{(r)}(k)$, $\Upsilon_{31i}^{(r)}(k)$, Υ_{32i}, $\Upsilon_{33i}^{(r)}(k)$, $\Xi_{1i}(k)$, $\Xi_{3i}(k)$, $\Pi_{1i}(k)$, $\Xi_{2i}(k)$, $\Xi_{4i}^{(r)}(k)$, $\Xi_{5i}^{(r)}(k)$, $\Xi_{6i}^{(r)}(k)$, $\Xi_{7i}^{(r)}(k)$, Ξ_{8i}, and Ξ_{9i} are defined previously, then there exists an nth-order filter of the form (5.10) which ensures the H_∞ performance constraint in (5.13), where $L_{fi}(k)$ is given as part of the RLMI solution and the other two filter parameters are given by

$$A_{fi}(k) = \bar{P}_{2i}^{(r)-1}(k + 1)\bar{M}_i^{(r)}(k), \quad B_{fi}^{(r)}(k) = \bar{P}_{2i}^{(r)-1}(k + 1)\bar{N}_i^{(r)}(k).$$

Proof.   Since the TP matrix $\hat{\Psi} = [\lambda_{ij}]$ belongs to the convex polyhedral set $\mathfrak{R}$, there always exist scalars $\psi_r \geq 0$ ($r = 1, 2, \ldots, \nu$) such that $\hat{\Psi} = \sum_{r=1}^{\nu} \psi_r \hat{\Psi}^{(r)}$, $\sum_{r=1}^{\nu} \psi_r = 1$, where $\hat{\Psi}^{(r)} = [\lambda_{ij}^{(r)}]$ ($r = 1, 2, \ldots, \nu$) are ν vertexes of the polytope. Hence, considering (5.15) and (5.16) in Theorem 5.2.1, we have

$$\begin{bmatrix} -\rho_i(k)I & * & * \\ \hat{G}_i(k) & -\bar{P}_i^{(r)-1}(k + 1) & * \\ G_i(k) & 0 & -\frac{1}{4}\bar{P}_i^{(r)-1}(k + 1) \end{bmatrix} \leq 0, \tag{5.30}$$

$$\begin{bmatrix} \Omega_{11i}^{(r)}(k) & * & * \\ \Omega_{22i}^{(r)}(k) & \Omega_{22i}^{(r)}(k) & * \\ \Omega_{31i}^{(r)}(k) & \Omega_{32i}^{(r)}(k) & \Omega_{33i}^{(r)}(k) \end{bmatrix} \leq 0, \tag{5.31}$$

where

$$\Omega_{11i}^{(r)}(k) = 2\bar{A}_i^T(k)\bar{P}_i^{(r)}(k + 1)\bar{A}_i(k) - P_i(k) + \bar{L}_i^T(k)\bar{L}_i(k) + \rho_i(k)\bar{E}(k),$$

$$\Omega_{21i}^{(r)}(k) = \bar{B}_i^T(k)\bar{P}_i^{(r)}(k + 1)\bar{A}_i(k) + \frac{1}{2}\tau_i(k)\tilde{C}_i(k),$$

$$\Omega_{22i}^{(r)}(k) = 2\bar{B}_i^{\mathrm{T}}(k)\bar{P}_i^{(r)}(k+1)\bar{B}_i(k) - \tau_i(k)I,$$

$$\Omega_{31i}^{(r)}(k) = \bar{D}_i^{\mathrm{T}}(k)\bar{P}_i^{(r)}(k+1)\bar{A}_i(k), \ \Omega_{32i}^{(r)}(k) = \bar{D}_i^{\mathrm{T}}(k)\bar{P}_i^{(r)}(k+1)\bar{B}_i(k),$$

$$\Omega_{33i}^{(r)}(k) = 2\bar{D}_i^{\mathrm{T}}(k)\bar{P}_i^{(r)}(k+1)\bar{D}_i(k) - \gamma^2 I, \ \bar{P}_i^{(r)}(k) = \sum_{j=1}^{s}\lambda_{ij}^{(r)}P_j(k), i, j = 1, \dots, s.$$

Note that $P_i(k) = \mathrm{diag}\{P_{1i}(k), P_{2i}(k)\}$, where $P_{1i}(k) \in \mathbb{R}^{n \times n}$ and $P_{2i}(k) \in \mathbb{R}^{n \times n}$. Noticing (5.30) and (5.31), by using the Schur complement in Lemma 3.2.1, the S-procedure in Lemma 3.2.2, and some algebraic manipulations, we can obtain (5.27) and (5.28), and this completes the proof of the theorem. $\qquad\square$

Now, let us show that it is straightforward to specialize the main results of Theorem 5.3.1 to the case when the TP matrix $\hat{\Psi} = [\lambda_{ij}]$ is known exactly. The following result for robust H_∞ filtering is easily accessible.

Corollary 5.3.2 *Consider the uncertain discrete-time Markovian system (5.2) with known TP matrix $\hat{\Psi}$. Let $\gamma > 0$ be a given disturbance attenuation level. For given positive-definite matrices $Q_i > 0$ ($i \in S$), if there exist families of positive-definite matrices $\{P_{1i}(k)\}_{0 \le k \le N+1}$, $\{P_{2i}(k)\}_{0 \le k \le N+1}$ ($i = 1, 2, \dots, s$), families of positive scalars $\{\rho_i(k)\}_{0 \le k \le N}$, $\{\tau_i(k)\}_{0 \le k \le N}$ ($i = 1, 2, \dots, s$), $\{\xi_i\}_{i=1,2,\dots,s}$, and families of real-valued matrices $\{\bar{M}_i(k)\}_{0 \le k \le N}$, $\{\bar{N}_i(k)\}_{0 \le k \le N}$, and $\{L_{fi}(k)\}_{0 \le k \le N}$ satisfying the RLMIs*

$$\Phi_i(k) = \begin{bmatrix} \Phi_{11i}(k) & * \\ \Phi_{21i}(k) & \Phi_{22i}(k) \end{bmatrix} < 0, \ i = 1, 2, \dots, s, \tag{5.32}$$

$$\Upsilon_i(k) = \begin{bmatrix} \Upsilon_{11i}(k) & * & * \\ \Upsilon_{21i}(k) & \Upsilon_{22i}(k) & * \\ \Upsilon_{31i}(k) & \Upsilon_{32i} & \Upsilon_{33i}(k) \end{bmatrix} \le 0, \ i = 1, 2, \dots, s, \tag{5.33}$$

with the initial condition

$$\begin{bmatrix} P_{1i}(0) - \gamma^2 Q_i & \gamma^2 Q_i \\ \gamma^2 Q_i & P_{2i}(0) - \gamma^2 Q_i \end{bmatrix} \le 0, \tag{5.34}$$

then there exists an nth-order filter of the form (5.10) which ensures the H_∞ performance constraint in (5.13), where $L_{fi}(k)$ is given as part of the RLMI solution and the other two admissible filter parameters are given by

$$A_{fi}(k) = \bar{P}_{2i}^{-1}(k+1)\bar{M}_i(k), \quad B_{fi}(k) = \bar{P}_{2i}^{-1}(k+1)\bar{N}_i(k). \tag{5.35}$$

Theorem 5.3.1 provides a design scheme for a time-varying filter in the presence of an unknown TP matrix in *Case 1*. Now we are going to consider the similar problem with unknown TP matrix in *Case 2*, and the following theorem is established along a similar line.

Before stating the theorem, let us denote

$$
\hat{\Phi}_{21i}(k) = \begin{bmatrix} \bar{h}\hat{P}_{1i}(k+1) & 0 \\ 0 & \sqrt{\bar{\beta}(1-\bar{\beta})}\bar{N}_i(k) \\ \bar{\alpha}\hat{P}_{1i}(k+1) & 0 \\ 0 & \bar{\beta}\bar{N}_i(k) \end{bmatrix}, \quad \hat{\Upsilon}_{31i}(k) = \begin{bmatrix} 0 & \hat{\Xi}_{7i}(k), \\ \Xi_{8i} & 0 \end{bmatrix},
$$

$$
\hat{\Phi}_{22i}(k) = \mathrm{diag}\left\{-\hat{P}_{1i}(k+1), -\hat{P}_{2i}(k+1), -\frac{1}{4}\hat{P}_{1i}(k+1), -\frac{1}{4}\hat{P}_{2i}(k+1)\right\},
$$

$$
\hat{\Upsilon}_{21i}(k) = \begin{bmatrix} \hat{\Xi}_{4i}(k) & \hat{\Xi}_{5i}(k) \\ \hat{\Xi}_{6i}(k) & 0 \end{bmatrix}, \quad \hat{\Xi}_{4i}(k) = \begin{bmatrix} \hat{P}_{1i}(k+1)A_i(k) & 0 \\ \bar{N}_i(k)K_1C_i(k) & \bar{M}_i(k) \end{bmatrix},
$$

$$
\hat{\Upsilon}_{22i}(k) = \mathrm{diag}\{-\hat{P}_{1i}(k+1), -\hat{P}_{2i}(k+1), -\hat{P}_{1i}(k+1), -\hat{P}_{2i}(k+1), -\hat{P}_{1i}(k+1)\},
$$

$$
\hat{\Upsilon}_{33i}(k) = \mathrm{diag}\{-\hat{P}_{2i}(k+1), -\hat{P}_{1i}(k+1), -\hat{P}_{2i}(k+1), -\xi_i I, -\xi_i I\}, \quad \bar{h} = \sqrt{\bar{\alpha}(1-\bar{\alpha})},
$$

$$
\hat{\Xi}_{5i}(k) = \begin{bmatrix} 0 & \hat{P}_{1i}(k+1)D_{1i}(k) & 0 \\ \bar{N}_i(k) & \bar{N}_i(k)D_{2i}(k) & 0 \end{bmatrix}, \quad \hat{\Xi}_{6i}(k) = \begin{bmatrix} cc\hat{P}_{1i}(k+1)A_i(k) & 0 \\ \bar{N}_i(k)K_1C_i(k) & \bar{M}_i(k) \\ 0 & 0 \end{bmatrix}.
$$

Theorem 5.3.3 *Consider system (5.2) with unknown TP matrix described in Case 2. Let $\gamma > 0$ be a given disturbance attenuation level. For given positive-definite matrices $Q_i > 0$ ($i \in S$), assume that there exist families of positive-definite matrices $\{P_{1i}(k)\}_{0 \le k \le N+1}$, $\{P_{2i}(k)\}_{0 \le k \le N+1}$ ($i = 1, 2, \ldots, s$), families of positive scalars $\{\rho_i(k)\}_{0 \le k \le N}$, $\{\tau_i(k)\}_{0 \le k \le N}$ ($i = 1, 2, \ldots, s$), $\{\xi_i\}_{i=1,2,\ldots,s}$, and families of real-valued matrices $\{\bar{M}_i(k)\}_{0 \le k \le N}$, $\{\bar{N}_i(k)\}_{0 \le k \le N}$, and $\{L_{fi}(k)\}_{0 \le k \le N}$ satisfying the following RLMIs:*

$$
\hat{\Phi}_i(k) = \begin{bmatrix} \Phi_{11i}(k) & * \\ \hat{\Phi}_{21i}(k) & \hat{\Phi}_{22i}(k) \end{bmatrix} < 0, \tag{5.36}
$$

$$
\hat{\Upsilon}_i(k) = \begin{bmatrix} \Upsilon_{11i}(k) & * & * \\ \hat{\Upsilon}_{21i}(k) & \hat{\Upsilon}_{22i}(k) & * \\ \hat{\Upsilon}_{31i}(k) & \Upsilon_{32i} & \hat{\Upsilon}_{33i}(k) \end{bmatrix} \le 0,
$$

$$
i = 1, 2, \ldots, s, r = 1, 2, \ldots, \nu, \tag{5.37}
$$

with the initial condition

$$
\begin{bmatrix} P_{1i}(0) - \gamma^2 Q_i & \gamma^2 Q_i \\ \gamma^2 Q_i & P_{2i}(0) - \gamma^2 Q_i \end{bmatrix} \le 0, \tag{5.38}
$$

where $\hat{\Phi}_{21i}(k)$, $\hat{\Phi}_{22i}(k)$, $\hat{\Upsilon}_{21i}(k)$, $\hat{\Upsilon}_{22i}(k)$, $\hat{\Upsilon}_{31i}(k)$, $\hat{\Upsilon}_{33i}(k)$, $\hat{\Xi}_{4i}(k)$, $\hat{\Xi}_{5i}(k)$, and $\hat{\Xi}_{6i}(k)$ are defined previously and $\Phi_{11i}(k)$, $\Upsilon_{11i}(k)$, and Υ_{32i} are the same as defined in Theorem 5.3.1, and if $\hat{\Psi}_k^i = \emptyset$, we take

$$
\hat{P}_{1i}(k) = P_{1i}(k), \quad \hat{P}_{2i}(k) = P_{2i}(k), \quad i = 1, \ldots, s,
$$

otherwise

$$\hat{P}_{1i}(k) = \left(\sum_{j \in \hat{\Psi}_k^i} \lambda_{ij} \right)^{-1} \sum_{j \in \hat{\Psi}_k^i} \lambda_{ij} P_{1j}(k), \quad \hat{P}_{1i}(k) = P_{1i}(k), \ j \in \hat{\Psi}_{uk}^i,$$

$$\hat{P}_{2i}(k) = \left(\sum_{j \in \hat{\Psi}_k^i} \lambda_{ij} \right)^{-1} \sum_{j \in \hat{\Psi}_k^i} \lambda_{ij} P_{2j}(k), \quad \hat{P}_{2i}(k) = P_{2i}(k), \ j \in \hat{\Psi}_{uk}^i.$$

Then, there exists an nth-order filter of the form (5.10) which ensures the H_∞ performance constraint in (5.13), where $L_{\mathrm{fi}}(k)$ is given as part of the RLMI solution and the other two filter parameters are given by

$$A_{\mathrm{fi}}(k) = \hat{P}_{2i}^{-1}(k+1)\bar{M}_i(k), \quad B_{\mathrm{fi}}(k) = \hat{P}_{2i}^{-1}(k+1)\bar{N}_i(k). \tag{5.39}$$

Proof. Denote

$$\zeta_{ki} = \left(\sum_{j \in \hat{\Psi}_k^i} \lambda_{ij} \right)^{-1}.$$

It is clear that, for any $i \in S$, (5.32) and (5.33) can be rewritten as follows

$$\Phi_i(k) = \sum_{j \in \hat{\Psi}_k^i} \lambda_{ij} \hat{\Phi}_i(k) \left| \begin{array}{l} \hat{P}_{1i}(k) = \zeta_{ki} \sum_{j \in \hat{\Psi}_k^i} \lambda_{ij} P_{1j}(k), \\ \hat{P}_{2i}(k) = \zeta_{ki} \sum_{j \in \hat{\Psi}_k^i} \lambda_{ij} P_{2j}(k) \end{array} \right.$$

$$+ \sum_{j \in \hat{\Psi}_{uk}^i} \lambda_{ij} \hat{\Phi}_i(k) \left| \begin{array}{l} \hat{P}_{1i}(k) = P_{1i}(k), \\ \hat{P}_{2i}(k) = P_{2i}(k) \end{array} \right. . \tag{5.40}$$

$$\Upsilon_i(k) = \sum_{j \in \hat{\Psi}_k^i} \lambda_{ij} \hat{\Upsilon}_i(k) \left| \begin{array}{l} \hat{P}_{1i}(k) = \zeta_{ki} \sum_{j \in \hat{\Psi}_k^i} \lambda_{ij} P_{1j}(k), \\ \hat{P}_{2i}(k) = \zeta_{ki} \sum_{j \in \hat{\Psi}_k^i} \lambda_{ij} P_{2j}(k) \end{array} \right.$$

$$+ \sum_{j \in \hat{\Psi}_{uk}^i} \lambda_{ij} \hat{\Upsilon}_i(k) \left| \begin{array}{l} \hat{P}_{1i}(k) = P_{1i}(k), \\ \hat{P}_{2i}(k) = P_{2i}(k) \end{array} \right. . \tag{5.41}$$

Therefore, if

$$\hat{\Phi}_i(k) \left| \begin{array}{l} \hat{P}_{1i}(k) = \zeta_{ki} \sum_{j \in \hat{\Psi}_k^i} \lambda_{ij} P_{1j}(k), \\ \hat{P}_{2i}(k) = \zeta_{ki} \sum_{j \in \hat{\Psi}_k^i} \lambda_{ij} P_{2j}(k) \end{array} \right. < 0,$$

$$\hat{\Phi}_i(k) \left| \begin{array}{l} \hat{P}_{1i}(k) = P_{1i}(k), \\ \hat{P}_{2i}(k) = P_{2i}(k) \end{array} \right. < 0, \quad j \in \hat{\Psi}_{uk}^i, \tag{5.42}$$

$$\hat{\Upsilon}_i(k) \quad \left| \begin{array}{l} \hat{P}_{1i}(k) = \zeta_{ki} \sum_{j \in \hat{\Psi}^i_k} \lambda_{ij} P_{1j}(k), \\ \hat{P}_{2i}(k) = \zeta_{ki} \sum_{j \in \hat{\Psi}^i_k} \lambda_{ij} P_{2j}(k) \end{array} \right. < 0,$$

$$\hat{\Upsilon}_i(k) \quad \left| \begin{array}{l} \hat{P}_{1i}(k) = P_{1i}(k), \\ \hat{P}_{2i}(k) = P_{2i}(k) \end{array} \right. < 0, \quad j \in \hat{\Psi}^i_{uk}, \tag{5.43}$$

we have $\Phi_i(k) < 0$ and $\Upsilon_i(k) < 0$ for any $i \in S$. This completes the proof. $\qquad\square$

Remark 5.2 *Theorem 5.3.3 provides feasible solutions to the filter design problem for time-varying MJS (5.2) under partially unknown TPs. Note that if $\hat{\Psi}^i_{uk} = \emptyset$ holds for any $i \in S$ (i.e., all the TPs are accessible), the corresponding results in (5.36) and (5.37) reduce to (5.32) and (5.33). Similarly, when $\hat{\Psi}^i_k = \emptyset$ holds for any $i \in S$ (i.e., all the TPs are inaccessible), Theorem 5.3.3 is still valid at the cost of the incremental conservatism. More specifically, the more known entries there are in the TP matrix, the less conservatism of the results we would have.*

Based on Theorem 5.3.1 and Theorem 5.3.3, we suggest the following Robust H_∞ filter design (RHFD) algorithm involving RLMIs conditions.

Algorithm RHFD

Step 1. Given the H_∞ performance index γ, positive-definite matrices Q_i ($i = 1, 2, \ldots, s$) and the state initial conditions $x(0)$ and $\hat{x}(0)$. Select the initial values for matrices $\{P_{1i}(0)\}$ and $\{P_{2i}(0)\}$ which satisfy the condition (5.29) and set $k = 0$.

Step 2. Obtain the positive matrices $P_{1i}(k + 1)$, $P_{2i}(k + 1)$ and matrices $\bar{M}_i(k)$, $\bar{N}_i(k)$, and $L_{fi}(k)$ for the sampling instant k by solving the RLMIs (5.27) and (5.28) or (5.36) and (5.37), respectively, with known parameters $P_{1i}(k)$ and $P_{2i}(k)$.

Step 3. Derive the other two filter parameter matrices $A_{fi}(k)$ and $B_{fi}(k)$ by solving (5.39), and set $k = k + 1$.

Step 4. If $k < N$, then go to Step 2, otherwise go to Step 5.

Step 5. Stop.

Remark 5.3 *In Theorem 5.3.1 and Theorem 5.3.3, the robust H_∞ finite-horizon filter is designed by solving a series of RLMIs, as outlined in Algorithm RHFD, where both the current measurement and the previous state estimation are employed to estimate the current state. Such a recursive filtering process is particularly useful for real-time implementation such as online tracking of highly maneuvering targets. On the other hand, the Algorithm RHFD can be easily adjusted to the algorithm of minimizing the H_∞ performance index γ subject to RLMIs (5.27) and (5.28) or (5.36) and (5.37). In this case, the linear search algorithm can be employed to minimize the index γ so as to enhance the filter performance. Also, in the future, it would be interesting to explore the possibility of learning the unknown dynamics (e.g., transition matrix and unknown components) when it is identifiable and consider the adaptive filtering problems. Various learning algorithms could be applied here that include regret minimization approaches for finding out the minimization problem in the literature (e.g., switching-type policies).*

5.4 Fault Detection with Sensor Saturations and Randomly Varying Nonlinearities

5.4.1 Problem Formulation

Given the unknown TP matrices described in *Case 2*, we consider, on a probability space $(\Omega, \mathcal{F}, \text{Prob})$, the following class of Markovian jump discrete systems with RVNs and sensor saturation:

$$
\begin{cases}
x(k+1) = A(r(k))x(k) + \alpha(k)g(r(k), x(k)) + (1 - \alpha(k))h(r(k), x(k)) \\
\qquad + D_1(r(k))w(k) + G(r(k))f(k), \\
y(k) = \sigma(C(r(k))x(k)) + D_2(r(k))w(k) + E(r(k))f(k),
\end{cases}
\tag{5.44}
$$

where $x(k) \in \mathbb{R}^{n_x}$ represents the state vector, $y(k) \in \mathbb{R}^{n_y}$ is the process output, $w(k) \in \mathbb{R}^{n_w}$ is the disturbance input which belongs to $l_2[0, \infty)$, $g(\cdot)$ and $h(\cdot)$ are nonlinear vector functions. $f(k) \in \mathbb{R}^l$ is the fault to be detected. For fixed-system mode, $A(r(k))$, $D_1(r(k))$, $G(r(k))$, $C(r(k))$, $D_2(r(k))$, and $E(r(k))$ are constant matrices with appropriate dimensions.

The stochastic variable $\alpha(k)$ is a Bernoulli-distributed white-noise sequence taking values on 0 and 1 with

$$
\text{Prob}\{\alpha(k) = 1\} = \bar{\alpha}, \quad \text{Prob}\{\alpha(k) = 0\} = 1 - \bar{\alpha}.
$$

In this section, we assume that Markov chain $r(k)$ is independent of the stochastic variable $\alpha(k)$.

The nonlinear functions $g(r(k), x(k))$ and $h(r(k), x(k))$ are assumed to satisfy $g(r(k), 0) = 0$, $h(r(k), 0) = 0$, and

$$
\|g(r(k), x(k) + \delta(k)) - g(r(k), x(k))\| \leq \|B_1(r(k))\delta(k)\|,
$$
$$
\|h(r(k), x(k) + \delta(k)) - h(r(k), x(k))\| \leq \|B_2(r(k))\delta(k)\|,
\tag{5.45}
$$

where, for fixed-system mode, $B_1(r(k))$ and $B_2(r(k))$ are known matrices and $\delta(k)$ is a vector. The saturation function $\sigma: \mathbb{R}^{n_y} \rightarrow \mathbb{R}^{n_y}$ is defined as

$$
\sigma(v) = \begin{bmatrix} \sigma_1^{\mathrm{T}}(v_1) & \sigma_2^{\mathrm{T}}(v_2) & \cdots & \sigma_{n_y}^{\mathrm{T}}(v_{n_y}) \end{bmatrix}^{\mathrm{T}},
\tag{5.46}
$$

with $\sigma_i(v_i) = \text{sign}(v_i) \min\{v_{i,\max}, |v_i|\}$, where $v_{i,\max}$ is the ith element of the vector $v_{\max}$, the saturation level.

Assuming that there exist two diagonal matrices L_1 and L_2 such that $0 \leq L_1 < I \leq L_2$, then the saturation function $\sigma(C(r(k))x(k))$ in (5.44) can be decomposed into a linear and a nonlinear part as

$$
\sigma(C(r(k))x(k)) = L_1 C(r(k))x(k) + \Psi(C(r(k))x(k)),
\tag{5.47}
$$

where $\Psi(C(r(k))x(k))$ is a nonlinear vector-valued function satisfying the sector condition which is described in Definition 5.1.1 with $\bar{H}_1 = 0$, $\bar{H}_2 = L$, and can be described as follows:

$$\Psi^{\mathrm{T}}(C(r(k))x(k))(\Psi(C(r(k))x(k)) - LC(r(k))x(k)) \le 0, \qquad (5.48)$$

where $L = L_2 - L_1$.

Note that the set S comprises various operation modes of the system in (5.44), and the Markov chain $\{r(k)\}$ takes values in the finite set $S = \{1, 2, \ldots, s\}$. For presentation convenience, for each possible $r(k) = i$ $(i \in S)$, a matrix $N(r(k))$ and a function $l(r(k))$ are denoted by N_i and l_i, respectively, where N could be A, D_1, G, C, D_2, E, B_1, and B_2, and l could be g and h.

Let us now work on the system mode $r(k) = i$ $(\forall i \in S)$. As is well known, a typical fault detection system consists of a residual generator and an evaluation of the generated residual. Furthermore, the residual evaluation includes an evaluation function and a threshold. For the purpose of residual generation, we consider a fault detection filter of the following form:

$$\begin{cases} \hat{x}(k+1) = A_i\hat{x}(k) + \bar{\alpha}g_i(\hat{x}(k)) + (1 - \bar{\alpha})h_i(\hat{x}(k)) + K_i[y(k) - C_i\hat{x}(k)], \\ \tilde{r}(k) = M[y(k) - C_i\hat{x}(k)], \end{cases} \qquad (5.49)$$

where $\hat{x}(k) \in \mathbb{R}^{n_x}$ is the state estimate, $\tilde{r}(k) \in \mathbb{R}^l$ is the so-called residual, and K_i and M are appropriately dimensioned filter matrices to be determined. In this chapter, the filter gains K_i $(i \in S)$ are mode dependent and the residual weighting factor M is static.

Letting $e(k) = x(k) - \hat{x}(k)$, it follows immediately that

$$\begin{cases} e(k+1) = (A_i - K_iC_i)e(k) + \bar{\alpha}\tilde{g}_i(e(k)) + (1 - \bar{\alpha})\tilde{h}_i(e(k)) + (\alpha(k) - \bar{\alpha})[g_i(x(k)) \\ \qquad -h_i(x(k))] + (D_{1i} - K_iD_{2i})w(k) + (G_i - K_iE_i)f(k) - K_i\sigma(C_ix(k)) \\ \qquad +K_iC_ix(k), \\ \tilde{r}(k) = M(\sigma(C_ix(k)) + D_{2i}w(k) + E_if(k) - C_ix(k) + C_ie(k)), \end{cases} \qquad (5.50)$$

where $\tilde{g}_i(e(k)) := g_i(x(k)) - g_i(\hat{x}(k))$ and $\tilde{h}_i(e(k)) := h_i(x(k)) - h_i(\hat{x}(k))$.

By augmenting $\eta(k) = [x^{\mathrm{T}}(k) \quad e^{\mathrm{T}}(k)]^{\mathrm{T}}$, the overall fault detection dynamics is governed by the following augmented system:

$$\begin{cases} \eta(k+1) = \mathcal{Y}_i(\eta(k)) + (\alpha(k) - \bar{\alpha})\Lambda_2\mathcal{G}_i(\eta(k)) + \mathcal{D}_{di}w(k) + \mathcal{D}_{fi}f(k) \\ \qquad +\mathcal{K}_{\sigma i}\sigma(C_iH_1\eta(k)), \\ \tilde{r}(k) = M[\sigma(C_iH_1\eta(k)) + \hat{C}_i\eta(k) + D_{2i}w(k) + E_if(k)], \end{cases} \qquad (5.51)$$

where

$$
\mathcal{Y}_i(\eta(k)) = \mathcal{A}_i \eta(k) + \Lambda_1 \mathcal{G}_i(\eta(k)), \quad \mathcal{K}_{\sigma i} = \begin{bmatrix} 0 & -K_i^{\mathrm{T}} \end{bmatrix}^{\mathrm{T}}, \quad H_1 = \begin{bmatrix} I & 0 \end{bmatrix},
$$

$$
\mathcal{A}_i - \begin{bmatrix} A_i & 0 \\ K_i C_i & A_i - K_i C_i \end{bmatrix}, \quad \mathcal{D}_{fi} = \begin{bmatrix} G_i \\ G_i - K_i E_i \end{bmatrix}, \quad \hat{C}_i = \begin{bmatrix} -C_i & C_i \end{bmatrix},
$$

$$
\Lambda_1 = \begin{bmatrix} \bar{\alpha} I & (1-\bar{\alpha}) I & 0 & 0 \\ 0 & 0 & \bar{\alpha} I & (1-\bar{\alpha}) I \end{bmatrix}, \quad \mathcal{D}_{di} = \begin{bmatrix} D_{1i} \\ D_{1i} - K_i D_{2i} \end{bmatrix}, \tag{5.52}
$$

$$
\Lambda_2 = \begin{bmatrix} I & -I & 0 & 0 \\ I & -I & 0 & 0 \end{bmatrix}, \quad \tilde{\mathcal{H}}_i(e(k)) = \begin{bmatrix} \tilde{g}_i^{\mathrm{T}}(e(k)) & \tilde{h}_i^{\mathrm{T}}(e(k)) \end{bmatrix}^{\mathrm{T}},
$$

$$
\mathcal{G}_i(\eta(k)) = \begin{bmatrix} \mathcal{H}_i^{\mathrm{T}}(x(k)) & \tilde{\mathcal{H}}_i^{\mathrm{T}}(e(k)) \end{bmatrix}^{\mathrm{T}}, \quad \mathcal{H}_i(x(k)) = \begin{bmatrix} g_i^{\mathrm{T}}(x(k)) & h_i^{\mathrm{T}}(x(k)) \end{bmatrix}^{\mathrm{T}}.
$$

Moreover, it follows from (5.45), (5.47), and (5.48) that

$$
\|\mathcal{G}_i(\eta(k))\| \le \|\tilde{B}_i \eta(k)\|, \tag{5.53}
$$

$$
\sigma(C_i H_1 \eta(k)) = \tilde{L}_{1i} \eta(k) + \Psi(C_i H_1 \eta(k)), \tag{5.54}
$$

$$
\Psi^{\mathrm{T}}(C_i H_1 \eta(k)) \left(\Psi(C_i H_1 \eta(k)) - \tilde{L}_{2i} \eta(k) \right) \le 0 \tag{5.55}
$$

where

$$
\tilde{B}_i := \begin{bmatrix} B_{1i} & 0 \\ B_{2i} & 0 \\ 0 & B_{1i} \\ 0 & B_{2i} \end{bmatrix}, \quad \tilde{L}_{1i} := \begin{bmatrix} L_1 C_i & 0 \end{bmatrix}, \quad \tilde{L}_{2i} := \begin{bmatrix} L C_i & 0 \end{bmatrix}. \tag{5.56}
$$

Before proceeding further, we introduce the following definition.

Definition 5.4.1   *The fault detection dynamics in (5.51) is said to be stochastically stable in the mean square for any initial conditions $\eta(0)$ and $\theta(0) \in S$ if, when $w(k) = 0$ and $f(k) = 0$, there exists a finite $W(\theta(0)) > 0$ such that*

$$
\mathbb{E} \left\{ \sum_{k=0}^{\infty} \|\eta(k)\|^2 \,\Bigg|\, \eta(0), \theta(0) \right\} < \eta^{\mathrm{T}}(0) W(\theta(0)) \eta(0).
$$

The main purpose of this chapter is to design a fault detection filter of the form (5.49) such that the following requirements are met simultaneously:

(a) The fault detection dynamics (5.51) is stochastically stable.
(b) Under the zero initial condition, we can obtain the following inequality for any nonzero $w(k)$:

$$
\sum_{k=0}^{\infty} \mathbb{E}\{\|\tilde{r}(k)\|^2\} \le \gamma^2 \sum_{k=0}^{\infty} \|w(k)\|^2 \Bigg|_{f(k)=0}, \tag{5.57}
$$

where $\gamma > 0$ is made as small as possible in the feasibility of (5.57) so as to minimize the effect from the exogenous disturbance on the residual.

(c) Under the zero initial condition, we can obtain the following inequality for any nonzero $f(k)$:

$$\sum_{k=0}^{\infty} \mathbb{E}\{\|\tilde{r}(k)\|^2\} \geq \beta^2 \sum_{k=0}^{\infty} \|f(k)\|^2 \Bigg|_{w(k)=0}, \tag{5.58}$$

where $\beta > 0$ is made as large as possible in the feasibility of (5.58) so as to enhance the sensitivity of faults on the residual.

Remark 5.4 *It should be noted that the performance index γ reflects the robustness of residuals against the disturbance in the fault-free case, and the performance index β quantifies the sensitivity of the residuals with respect to the fault in the disturbance-free case. Therefore, in order to achieve a satisfactory trade-off between the robustness against the disturbances and the sensitivity to the faults, the fault detection dynamics (5.51) should be made stochastically stable, where the index*

$$J = \gamma/\beta \tag{5.59}$$

is used to evaluate the overall performance of the designed fault detection filter.

We further adopt a residual evaluation stage including an evaluation function $\bar{J}(\tilde{r})$ and a threshold $\bar{J}_{\text{th}}$ of the following form:

$$\bar{J}(\tilde{r}) = \left\{ \sum_{s=k-\mathcal{L}}^{s=k} \tilde{r}^{\mathrm{T}}(s)\tilde{r}(s) \right\}^{1/2}, \quad \bar{J}_{\text{th}} = \sup_{w \in l_2, f=0} \mathbb{E}\{\bar{J}(\tilde{r})\}, \tag{5.60}$$

where $\mathcal{L}$ denotes the length of the finite evaluating time-horizon. Based on (5.60), the occurrence of faults can be detected by comparing $\bar{J}(\tilde{r})$ with $\bar{J}_{\text{th}}$ according to the following rule:

$$\bar{J}(\tilde{r}) > \bar{J}_{\text{th}} \implies \text{with faults} \implies \text{alarm,}$$
$$\bar{J}(\tilde{r}) \leq \bar{J}_{\text{th}} \implies \text{no faults.}$$

5.4.2 Main Results

By using similar analysis techniques, some main results are listed as follows.

Lemma 5.4.2 We consider the discrete-time MJS (5.44) with known TP matrix $\hat{\Psi}$. Let the filter parameters K_i ($i \in S$), M, and the index $\gamma > 0$ be given. The system (5.51) is stochastically stable and satisfies the constraint (5.57) if there exist a set of matrices $P_i > 0$ ($i \in S$) and positive scalars ε_1 and ε_2 satisfying

$$\hat{\Phi}_i = \begin{bmatrix} \hat{\Phi}_{11} & * \\ \hat{\Phi}_{21} & \hat{\Phi}_{22} \end{bmatrix} \leq 0, \tag{5.61}$$

where

$$\hat{\Phi}_{11} = \begin{bmatrix} \Phi_{11} + \mathcal{M}_i^T \mathcal{M}_i + \varepsilon_1 \tilde{B}_i^T \tilde{B}_i & * \\ \Lambda_1^T \bar{P}_i \bar{A}_i & \Phi_{22} - \varepsilon_1 I \end{bmatrix},$$

$$\hat{\Phi}_{21} = \begin{bmatrix} M^T \mathcal{M}_i + \mathcal{K}_{\sigma i}^T \bar{P}_i \bar{A}_i + \varepsilon_2 \tilde{L}_{2i} & \mathcal{K}_{\sigma i}^T \bar{P}_i \Lambda_1 \\ \mathcal{D}_{di}^T \bar{P}_i \bar{A}_i + \mathcal{D}_{2i}^T \mathcal{M}_i & \mathcal{D}_{di}^T \bar{P}_i \Lambda_1 \end{bmatrix},$$

$$\hat{\Phi}_{22} = \begin{bmatrix} \mathcal{K}_{\sigma i}^T \bar{P}_i \mathcal{K}_{\sigma i} + M^T M - \varepsilon_2 I & * \\ \mathcal{D}_{di}^T \bar{P}_i \mathcal{K}_{\sigma i} + \mathcal{D}_{2i}^T M & \Xi_i \end{bmatrix},$$

$$\bar{P}_i = \sum_{j \in S} \lambda_{ij} P_j, \quad \Phi_{11} = \bar{A}_i^T \bar{P}_i \bar{A}_i - P_i, \quad \mathcal{D}_{2i} = M D_{2i},$$

$$\bar{A}_i = \mathcal{A}_i + \mathcal{K}_{\sigma i} \tilde{L}_{1i}, \quad \Xi_i = \mathcal{D}_{di}^T \bar{P}_i \mathcal{D}_{di} + \mathcal{D}_{2i}^T \mathcal{D}_{2i} - \gamma^2 I,$$

$$\mathcal{M}_i = M(\tilde{L}_{1i} + \hat{C}_i), \quad \Phi_{22} = \Lambda_1^T \bar{P}_i \Lambda_1 + \bar{\alpha}(1 - \bar{\alpha}) \Lambda_2^T \bar{P}_i \Lambda_2.$$

Lemma 5.4.3 We consider the discrete-time MJS (5.44) with known TP matrix $\hat{\Psi}$. Let the filter parameters K_i ($i \in S$), M, and the index $\beta > 0$ be given. For the system (5.51), the constraint (5.58) is met if there exist a set of matrices $P_i > 0$ ($i \in S$) and positive constant scalars ε_1 and ε_2 satisfying

$$\Omega_i = \begin{bmatrix} \hat{\Omega}_{11} & * \\ \hat{\Omega}_{21} & \hat{\Omega}_{22} \end{bmatrix} \le 0, \tag{5.62}$$

where

$$\hat{\Omega}_{11} = \begin{bmatrix} \Phi_{11} - \mathcal{M}_i^T \mathcal{M}_i + \varepsilon_1 \tilde{B}_i^T \tilde{B}_i & * \\ \Lambda_1^T \bar{P}_i \bar{A}_i & \Phi_{22} - \varepsilon_1 I \end{bmatrix},$$

$$\hat{\Omega}_{21} = \begin{bmatrix} -M^T \mathcal{M}_i + \mathcal{K}_{\sigma i}^T \bar{P}_i \bar{A}_i + \varepsilon_2 \tilde{L}_{2i} & \mathcal{K}_{\sigma i}^T \bar{P}_i \Lambda_1 \\ \mathcal{D}_{fi}^T \bar{P}_i \bar{A}_i - E_i^T M^T \mathcal{M}_i & \mathcal{D}_{fi}^T \bar{P}_i \Lambda_1 \end{bmatrix},$$

$$\hat{\Omega}_{22} = \begin{bmatrix} \mathcal{K}_{\sigma i}^T \bar{P}_i \mathcal{K}_{\sigma i} - M^T M - \varepsilon_2 I & * \\ \mathcal{D}_{fi}^T \bar{P}_i \mathcal{K}_{\sigma i} - E_i^T M^T M & \Omega_{33} \end{bmatrix},$$

$$\Omega_{33} = \mathcal{D}_{fi}^T \bar{P}_i \mathcal{D}_{fi} + \beta^2 I - E_i^T M^T M E_i,$$

and the other symbols are the same as defined in Lemma 5.4.2.

Lemma 5.4.4 We consider the discrete-time MJS (5.44) with known TP matrix $\hat{\Psi}$. Let the filter parameters K_i ($i \in S$), M, and the indices $\beta > 0$, $\gamma > 0$ be given. The system (5.51) is stochastically stable while satisfying the constraints (5.57) and (5.58) if there exist a set of matrices $P_i > 0$ ($i \in S$) and positive constant scalars ε_1 and ε_2 such that inequalities (5.61) and (5.62) hold simultaneously.

Next, given the unknown TP matrix described in *Case 2*, we first propose the following performance analysis results with a given fault detection filter (5.49), and then deal with the design problem of the fault detection filter for system (5.44).

Theorem 5.4.5 *Consider the discrete-time MJS (5.44) subject to RVNs, sensor saturation, and incomplete knowledge of TPs. Let the indices $\beta > 0$, $\gamma > 0$ and the fault detection filter parameters K_i ($i \in S$), M be given. The fault detection dynamics (5.51) is stochastically stable while achieving the performance constraints (5.57) and (5.58) if there exist matrices $P_i > 0$ ($i \in S$) and positive constant scalars ε_1 and ε_2 such that the following inequalities hold:*

$$\Pi_{ij} = \begin{bmatrix} \Pi_{11} & * \\ \Pi_{21} & \Pi_{22} \end{bmatrix} \leq 0, \tag{5.63}$$

$$\bar{\Pi}_{ij} = \begin{bmatrix} \bar{\Pi}_{11} & * \\ \bar{\Pi}_{21} & \bar{\Pi}_{22} \end{bmatrix} \leq 0, \tag{5.64}$$

where, if $\lambda_{\mathcal{K}}^i = 0$, Q_j is defined to be $Q_j = P_j$ ($j \in S_{UK}^i$), otherwise

$$\begin{cases} Q_j = \dfrac{1}{\lambda_{\mathcal{K}}^i} P_{\mathcal{K}}^i = \dfrac{1}{\lambda_{\mathcal{K}}^i} \sum_{j \in S_{\mathcal{K}}^i} \lambda_{ij} P_j, & \forall j \in S_{\mathcal{K}}^i \\ Q_j = P_j, & \forall j \in S_{UK}^i \end{cases}$$

and

$$\Pi_{11} = \begin{bmatrix} \bar{\Phi}_{11} + \mathcal{M}_i^{\mathrm{T}} \mathcal{M}_i + \varepsilon_1 \bar{B}_i^{\mathrm{T}} \bar{B}_i & * \\ \Lambda_1^{\mathrm{T}} Q_j \bar{A}_i & \bar{\Phi}_{22} - \varepsilon_1 I \end{bmatrix},$$

$$\Pi_{21} = \begin{bmatrix} M^{\mathrm{T}} \mathcal{M}_i + \mathcal{K}_{\sigma i}^{\mathrm{T}} Q_j \bar{A}_i + \varepsilon_2 \tilde{L}_{2i} & \mathcal{K}_{\sigma i}^{\mathrm{T}} Q_j \Lambda_1 \\ \mathcal{D}_{di}^{\mathrm{T}} Q_j \bar{A}_i + \mathcal{D}_{2i}^{\mathrm{T}} \mathcal{M}_i & \mathcal{D}_{di}^{\mathrm{T}} Q_j \Lambda_1 \end{bmatrix},$$

$$\Pi_{22} = \begin{bmatrix} \mathcal{K}_{\sigma i}^{\mathrm{T}} Q_j \mathcal{K}_{\sigma i} + M^{\mathrm{T}} T M - \varepsilon_2 I & * \\ \mathcal{D}_{di}^{\mathrm{T}} Q_j \mathcal{K}_{\sigma i} + \mathcal{D}_{2i}^{\mathrm{T}} M & \bar{\Phi}_{33} \end{bmatrix},$$

$$\bar{\Pi}_{11} = \begin{bmatrix} \bar{\Phi}_{11} - \mathcal{M}_i^{\mathrm{T}} \mathcal{M}_i + \varepsilon_1 \tilde{B}_i^{\mathrm{T}} \tilde{B}_i & * \\ \Lambda_1^{\mathrm{T}} Q_j \bar{A}_i & \bar{\Phi}_{22} - \varepsilon_1 I \end{bmatrix},$$

$$\bar{\Pi}_{21} = \begin{bmatrix} -M^{\mathrm{T}} \mathcal{M}_i + \mathcal{K}_{\sigma i}^{\mathrm{T}} Q_j \bar{A}_i + \varepsilon_2 \tilde{L}_{2i} & \mathcal{K}_{\sigma i}^{\mathrm{T}} Q_j \Lambda_1 \\ \mathcal{D}_{fi}^{\mathrm{T}} Q_j \bar{A}_i - E_i^{\mathrm{T}} M^{\mathrm{T}} \mathcal{M}_i & \mathcal{D}_{fi}^{\mathrm{T}} Q_j \Lambda_1 \end{bmatrix},$$

$$\bar{\Pi}_{22} = \begin{bmatrix} \mathcal{K}_{\sigma i}^{\mathrm{T}} Q_j \mathcal{K}_{\sigma i} - M^{\mathrm{T}} M - \varepsilon_2 I & * \\ \mathcal{D}_{fi}^{\mathrm{T}} Q_j \mathcal{K}_{\sigma i} - E_i^{\mathrm{T}} M^{\mathrm{T}} M & \bar{\Omega}_{33} \end{bmatrix},$$

$$\bar{\Phi}_{11} = \bar{A}_i^{\mathrm{T}} Q_j \bar{A}_i - P_i, \quad \bar{\Phi}_{22} = \Lambda_1^{\mathrm{T}} Q_j \Lambda_1 + \bar{\alpha}(1 - \bar{\alpha}) \Lambda_2^{\mathrm{T}} Q_j \Lambda_2,$$

$$\bar{\Phi}_{33} = \mathcal{D}_{di}^{\mathrm{T}} Q_j \mathcal{D}_{di} + \mathcal{D}_{2i}^{\mathrm{T}} \mathcal{D}_{2i} - \gamma^2 I, \quad \bar{\Omega}_{33} = \mathcal{D}_{fi}^{\mathrm{T}} Q_j \mathcal{D}_{fi} + \beta^2 I - E_i^{\mathrm{T}} M^{\mathrm{T}} M E_i.$$

Based on the analysis results with a given fault detection filter, we are now ready to solve the filter design problem for system (5.51) in the following theorem with the incomplete knowledge of TPs.

Theorem 5.4.6 *Consider system (5.44) with the unknown TP matrix described in* Case 2. *Let $\beta > 0$, $\gamma > 0$ be the given indices. The fault detection dynamics (5.51) is stochastically* *stable while achieving the performance constraints (5.57) and (5.58) if there exist matrices $P_i > 0$, N_{ij} $(i, j \in S)$, and $\bar{M}$ and positive constant scalars ε_1 and ε_2 such that the following LMIs hold:*

$$
\begin{bmatrix}
\bar{\Upsilon}_{11} & * & * \\
\bar{\Upsilon}_{21} & \bar{\Upsilon}_{22} & * \\
\bar{\Upsilon}_{31} & \bar{\Upsilon}_{32} & \bar{\Upsilon}_{33}
\end{bmatrix} \leq 0,
\tag{5.65}
$$

$$
\begin{bmatrix}
\hat{\Upsilon}_{11} & * & * \\
\hat{\Upsilon}_{21} & \hat{\Upsilon}_{22} & * \\
\hat{\Upsilon}_{31} & \hat{\Upsilon}_{32} & \hat{\Upsilon}_{33}
\end{bmatrix} \leq 0,
\tag{5.66}
$$

where

$$
\bar{\Upsilon}_{11} = \operatorname{diag}\{\Upsilon_{11} + \tilde{\Upsilon}_{11}, -\varepsilon_1 I\}, \quad
\bar{\Upsilon}_{21} = \begin{bmatrix} \bar{M}(\tilde{L}_{1i} + \hat{C}_i) + \varepsilon_2 \tilde{L}_{2i} & 0 \\ D_{2i}^{\mathrm{T}} \bar{M}(\tilde{L}_{1i} + \hat{C}_i) & 0 \end{bmatrix},
$$

$$
\bar{\Upsilon}_{22} = \begin{bmatrix} -\varepsilon_2 I + \bar{M} & * \\ D_{2i}^{\mathrm{T}} \bar{M} & D_{2i}^{\mathrm{T}} \bar{M} D_{2i} - \gamma^2 I \end{bmatrix}, \quad
\bar{\Upsilon}_{32} = \begin{bmatrix} N_{ij} & Q_j \hat{D}_{1i} + N_{ij} D_{2i} \\ 0 & 0 \end{bmatrix},
$$

$$
\bar{\Upsilon}_{31} = \begin{bmatrix} Q_j \mathcal{A}_{0i} + N_{ij}(\hat{C}_i + \tilde{L}_{1i}) & Q_j \Lambda_1 \\ 0 & \sqrt{\bar{\alpha}(1-\bar{\alpha})} Q_j \Lambda_2 \end{bmatrix}, \quad
\Upsilon_{11} = \varepsilon_1 \tilde{B}_i^{\mathrm{T}} \tilde{B}_i - P_i,
$$

$$
\bar{\Upsilon}_{33} = \operatorname{diag}\{-Q_j, -Q_j\}, \quad
\hat{\Upsilon}_{11} = \operatorname{diag}\{\Upsilon_{11} - \tilde{\Upsilon}_{11}, -\varepsilon_1 I\}, \quad
\hat{G}_i = I_2 \otimes G_i,
$$

$$
\hat{\Upsilon}_{21} = \begin{bmatrix} -\bar{M}(\tilde{L}_{1i} + \hat{C}_i) + \varepsilon_2 \tilde{L}_{2i} & 0 \\ -E_i^{\mathrm{T}} \bar{M}(\tilde{L}_{1i} + \hat{C}_i) & 0 \end{bmatrix}, \quad
\hat{\Upsilon}_{22} = \begin{bmatrix} -\varepsilon_2 I - \bar{M} & * \\ -E_i^{\mathrm{T}} \bar{M} & -E_i^{\mathrm{T}} \bar{M} E_i + \beta^2 I \end{bmatrix},
$$

$$
\hat{\Upsilon}_{32} = \begin{bmatrix} N_{ij} & Q_j \hat{G}_i + N_{ij} E_i \\ 0 & 0 \end{bmatrix}, \quad
\tilde{\Upsilon}_{11} = (\tilde{L}_{1i} + \hat{C}_i)^{\mathrm{T}} \bar{M}(\tilde{L}_{1i} + \hat{C}_i),
$$

$$
\bar{H} = \begin{bmatrix} 0 & -I \end{bmatrix}^{\mathrm{T}}, \quad
\hat{D}_{1i} = I_2 \otimes D_{1i}, \quad
\mathcal{A}_{0i} = \operatorname{diag}\{A_i, A_i\},
\tag{5.67}
$$

and the other parameters have been defined in Theorem 5.4.5. Furthermore, if $(P_i, N_{ij}, \bar{M}, \varepsilon_1, \varepsilon_2)$ is a feasible solution of (5.65) and (5.66), then the fault detection filter parameters K_i and M can be obtained by means of the matrices N_{ij} and $\bar{M}$, respectively, where M is a factorization of $\bar{M}$ (i.e., $\bar{M} = M^{\mathrm{T}} M$) and

$$
K_i = (\bar{H}^{\mathrm{T}} Q_j \bar{H})^{-1} \bar{H}^{\mathrm{T}} N_{ij}.
$$

Remark 5.5 *Theorem 5.4.6 provides a solution to the fault detection filter design problem for the discrete MJS (5.44) under partially unknown TPs. Obviously, in the spirit of fault detection, the index $\gamma > 0$ should be made as small as possible subject to (5.65) so as to minimize the effect from the exogenous disturbance on the residual, while the index $\beta > 0$ should be made as large as possible subject to (5.66) in order to maximize the sensitivity of faults on the residual. Based on such a principle, we will propose an algorithm that locally optimizes the gains of the fault detection filters.*

To achieve both the satisfactory robustness against disturbances and the satisfactory sensitivity to faults, we suggest the following locally Optimized Fault Detection Filter Design (OFDFD) algorithm.

Algorithm OFDFD

Step 1. Obtain γ_{min} (the minimum of γ) and β_{max} (the maximum of β) by solving (5.65) and (5.66), respectively, in Theorem 5.4.6.

Step 2. If, with γ and β replaced by γ_{min} and β_{max}, respectively, (5.65) and (5.66) are feasible for Theorem 5.4.6, we can obtain the locally optimized parameters K_i and M for the desired fault detection filter and exit. Otherwise, go to Step 3.

Step 3. Increase γ_{min} by μ and decrease β_{max} by μ where $\mu > 0$ is a sufficiently small scalar, and then solve (5.65) and (5.66) with the updated γ_{min} and β_{max}. Repeat such a procedure until (5.65) and (5.66) are feasible, and therefore obtain the locally optimized filter parameters $\{K_i, M\}$ and the index $J_{min} = \gamma_{min}/\beta_{max}$.

Step 4. Stop.

Remark 5.6 *Based on the proposed Algorithm OFDFD, the main results in Theorem 5.4.6 can be applied to solve the fault detection problem for a wide class of MJSs involving sensor saturations and RVNs that result typically from networked environments. Algorithm OFDFD is developed to check the existence of the desired fault detection filter gains, and the explicit expression of such filter gains is characterized in terms of the solution to a set of LMIs that can be effectively solved by algorithms such as the interior-point method.*

Remark 5.7 *The system (5.44) under consideration is quite comprehensive and reflects partially known mode TPs, RVNs, and sensor saturations. Furthermore, two energy norm indices are used for the fault detection problem: one in order to account for the restraint of disturbance and the other the sensitivity of faults. Note that the main results established contain all the information of the general systems addressed, including the physical parameters, the TPs, occurrence probabilities of the RVNs, and the amplitudes of the sensor saturations.*

5.5   Illustrative Examples

In this section, some simulation examples are presented to demonstrate the theory presented in this chapter.

5.5.1   Example 1

In this example, we consider *Case 1* for robust H_∞ filtering for MJSs with randomly occurring nonlinearities and sensor saturation.

Consider *Case 1* where the TP matrix of the Markov process is unknown but it resides in a polytope with the following two vertices:

$$\hat{\Psi}^{(1)} = \begin{bmatrix} 0.5 & 0.5 \\ 0.3 & 0.7 \end{bmatrix}, \quad \hat{\Psi}^{(2)} = \begin{bmatrix} 0.6 & 0.4 \\ 0.5 & 0.5 \end{bmatrix}.$$

Suppose that the system involves two modes, and the system data are given as follows:

Mode 1

$$A_1(k) = \begin{bmatrix} 0.2 & 0.2\sin(k) \\ 1.1\sin(5k) & 0.5 \end{bmatrix}, \quad D_{11}(k) = \begin{bmatrix} 0.1\sin(3k) \\ -0.3 \end{bmatrix}, \quad H_{11} = \begin{bmatrix} 0.1 \\ 0.3 \end{bmatrix},$$

$$D_{21}(k) = -0.3\sin(3k), \; H_{21} = 0.2, \; L_1(k) = \begin{bmatrix} 0.3\sin(2k) & 0.7 \end{bmatrix},$$

$$C_1(k) = \begin{bmatrix} 0.9 & 0.5\sin(5k) \end{bmatrix}, \quad |F_1(k)| \leq 1, \quad N_1 = \begin{bmatrix} 0 & 0.5 \end{bmatrix}.$$

$\sigma(y_{s1}(k))$ is a saturation function described as follows:

$$\sigma(y_{s1}(k)) = \begin{cases} \sigma(y_{s1}(k)) = y_{s1}(k), & \text{if } -V_{ys1j,max} \leq y_{s1}(k) \leq V_{ys1j,max}; \\ \sigma(y_{s1}(k)) = V_{ys1j,max}, & \text{if } y_{s1}(k) > V_{ys1j,max}; \\ \sigma(y_{s1}(k)) = -V_{ys1j,max}, & \text{if } y_{s1}(k) < -V_{ys1j,max}. \end{cases}$$

Mode 2

$$A_2(k) = \begin{bmatrix} 0.3\sin(k) & 0.1 \\ 1.3 & 0.5\sin(5k) \end{bmatrix}, \quad D_{12}(k) = \begin{bmatrix} 0.1 \\ 0.4\sin(3k) \end{bmatrix}, \quad H_{12} = \begin{bmatrix} 0.2 \\ 0.1 \end{bmatrix},$$

$$D_{22}(k) = -0.2\sin(4k), \quad H_{22} = 0.1, \quad L_2(k) = \begin{bmatrix} 0.4\sin(2k) & 0.2 \end{bmatrix},$$

$$C_2(k) = \begin{bmatrix} 1.3 & 0.2\sin(k) \end{bmatrix}, \quad |F_2(k)| \leq 1, \quad N_2 = \begin{bmatrix} 0 & 0.5 \end{bmatrix}.$$

$$\sigma(y_{s2}(k)) = \begin{cases} \sigma(y_{s2}(k)) = y_{s2}(k), & \text{if } -V_{ys2j,max} \leq y_{s2}(k) \leq V_{ys2j,max}; \\ \sigma(y_{s2}(k)) = V_{ys2j,max}, & \text{if } y_{s2}(k) > V_{ys2j,max}; \\ \sigma(y_{s2}(k)) = -V_{ys2j,max}, & \text{if } y_{s2}(k) < -V_{ys2j,max}; \end{cases}$$

and the nonlinear functions $f(k, x(k))$ and $g(k, x(k))$ are selected as

$$f(k, x(k)) = \begin{bmatrix} \frac{0.2x_1(k)}{2x_2^2(k)+1} & 0.1x_1(k)\sin(x_2(k)) \end{bmatrix}^{\mathrm{T}}, \quad g(k, x(k)) = 0.2x_1(k)\sin(x_2(k)).$$

It is easy to see that the constraint (5.4) can be met with $\varepsilon_1(k) = \varepsilon_2(k) = 1$ and $E_1(k) = E_2(k) = \mathrm{diag}\{0.2, 0.15\}$. In this example, the saturation values are taken as $V_{ys11} = V_{ys21} = 0.06$ and $K = 0.2$, $K_1 = 0.8$. The state initial value is $x(0) = [0.2 \quad -0.5]^{\mathrm{T}}$, $\hat{x}(0) = [-0.2 \quad -0.16]^{\mathrm{T}}$. The exogenous disturbance input $w(k)$ is supposed to be a random noise uniformly distributed over $[-0.5, 0.5]$ and the probabilities are assumed to be $\bar{\alpha} = \bar{\beta} = 0.9$. Set $\gamma = 0.5$ and let $Q_1 = Q_2 = \mathrm{diag}\{1, 1\}$. Choose the parameters' initial values to satisfy (5.29).

Consider the real TP matrix as

$$\hat{\Psi} = \begin{bmatrix} 0.56 & 0.44 \\ 0.42 & 0.58 \end{bmatrix},$$

which means that $\psi_1 = 0.4$ and $\psi_2 = 0.6$ in (5.1). According to the RHFD algorithm, the RLMIs in Theorem 5.3.1 can be solved recursively subject to given initial conditions and prespecified performance indices.

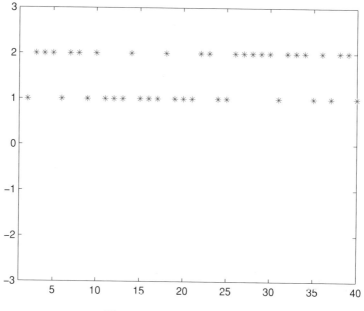

Figure 5.1   Random mode $r(k)$

The simulation results are shown in Figures 5.1–5.6, where Figure 5.1 plots one of the possible realizations of the Markovian jumping mode $r(k)$. Under this mode sequence, the corresponding output $z(k)$ and its estimation $\hat{z}(k)$ are shown in Figure 5.2, whereas the estimation error $\bar{z}(k)$ is depicted in Figure 5.3. The actual states $x_1(k)$, $x_2(k)$ and their estimates $\hat{x}_1(k)$, $\hat{x}_2(k)$ are given in Figure 5.4 and Figure 5.5, respectively. Figure 5.6 shows the sensor output. Note that the sensor outputs is saturated. The simulation confirms that the filter design performs very well.

5.5.2   Example 2

In this example, we consider *Case 2* for robust H_∞ filtering for MJSs with randomly occurring nonlinearities and sensor saturation.

Consider *Case 2* where some elements in the TP matrix of the Markov process are unknown and the possible three cases for $\hat{\Psi}$ are given as follows:

$$\hat{\Psi}_1 = \begin{bmatrix} 0.6 & 0.4 \\ 0.5 & 0.5 \end{bmatrix}, \quad \hat{\Psi}_2 = \begin{bmatrix} 0.6 & 0.4 \\ ? & ? \end{bmatrix}, \quad \hat{\Psi}_3 = \begin{bmatrix} ? & ? \\ ? & ? \end{bmatrix},$$

where $\hat{\Psi}_1$ (respectively, $\hat{\Psi}_2$ and $\hat{\Psi}_3$) shows that the elements in the TP matrix are completely known (respectively, partially known and completely unknown), and the other parameters of the discrete stochastic nonlinear time-varying system (5.2) are the same as in Example 1 (Section 5.5.1). Similarly, according to the RHFD algorithm, the RLMIs in Theorem 5.3.3 can be solved recursively subject to given initial conditions and prespecified performance indices. The corresponding simulation results for the estimation error in these three cases are given

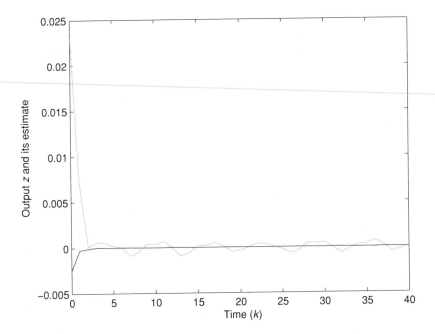

Figure 5.2   Output $z(k)$ and its estimate

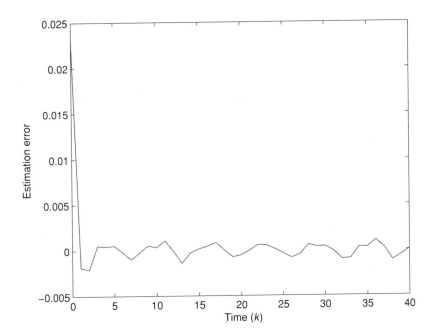

Figure 5.3   Estimation error

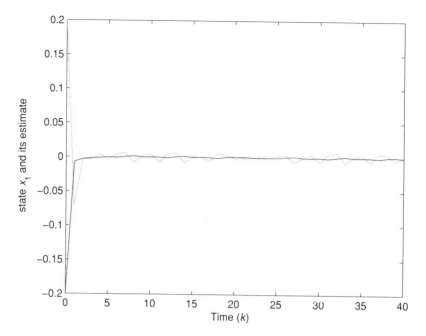

Figure 5.4   The state $x_1(k)$ and its estimate

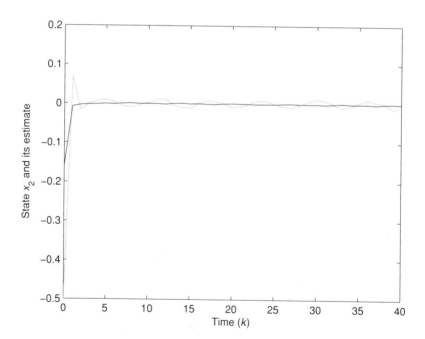

Figure 5.5   The state $x_2(k)$ and its estimate

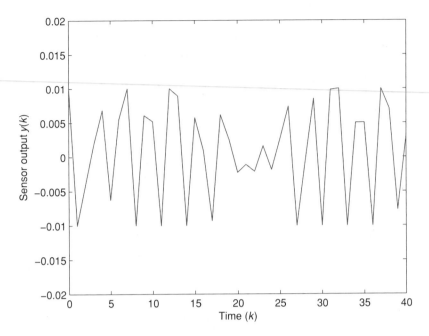

Figure 5.6 Sensor output

in Figure 5.7, Figure 5.9, and Figure 5.11, respectively. The corresponding simulation results for the states $x_1(k)$ and their estimates $\hat{x}_1(k)$ in the same three cases are also presented in Figure 5.8, Figure 5.10, and Figure 5.12, respectively. Again, it can be seen that the more known entries in the TP matrix we have, the less conservatism of the condition there would be.

5.5.3 Example 3

In this example, we consider fault detection for MJSs with sensor saturations and RVNs.

For the class of discrete-time MJSs (5.44) with RVNs and sensor saturations, let us consider the following three cases for the TP matrix $\hat{\Psi}$ of the Markov process:

$$\hat{\Psi}_1 = \begin{bmatrix} 0.3 & 0.7 \\ 0.4 & 0.6 \end{bmatrix}, \quad \hat{\Psi}_2 = \begin{bmatrix} ? & ? \\ 0.4 & 0.6 \end{bmatrix}, \quad \hat{\Psi}_3 = \begin{bmatrix} ? & ? \\ ? & ? \end{bmatrix}.$$

Apparently, the matrix $\hat{\Psi}_1$ (respectively, $\hat{\Psi}_2$ and $\hat{\Psi}_3$) means that the TPs are completely known (respectively, partially known and completely unknown).

Assume that the system involves two modes and the other system data are given as follows:

$$A_1 = \begin{bmatrix} -0.6 & 0.4 \\ 0.3 & 0.5 \end{bmatrix}, \quad A_2 = \begin{bmatrix} 0.3 & 0.5 \\ 0.4 & 0.5 \end{bmatrix}, \quad D_{11} = \begin{bmatrix} -0.1 \\ 0.7 \end{bmatrix},$$

$$D_{12} = \begin{bmatrix} 0.1 \\ -0.5 \end{bmatrix}, \quad G_1 = G_2 = \begin{bmatrix} 1 \\ -1 \end{bmatrix}, \quad C_1 = [0 \quad 0.5],$$

$$C_2 = [0.2 \quad 0.2], \quad D_{21} = D_{22} = 0.4, \quad E_1 = 1, \quad E_2 = 2.2.$$

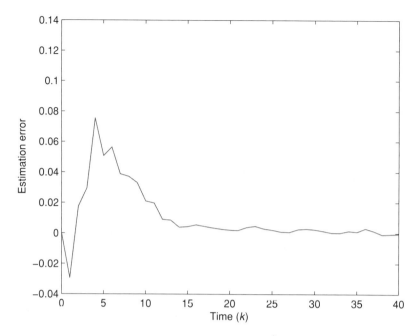

Figure 5.7   Estimation error of $\hat{\Psi}_1$ case

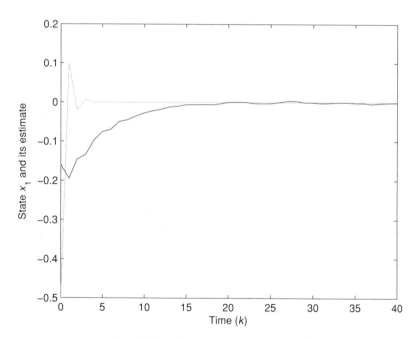

Figure 5.8 $x_1(k)$ and its estimate of $\hat{\Psi}_1$ case

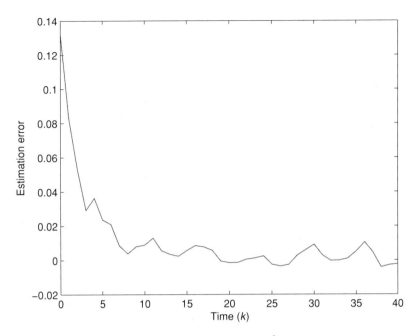

Figure 5.9   Estimation error of $\hat{\Psi}_1$ case

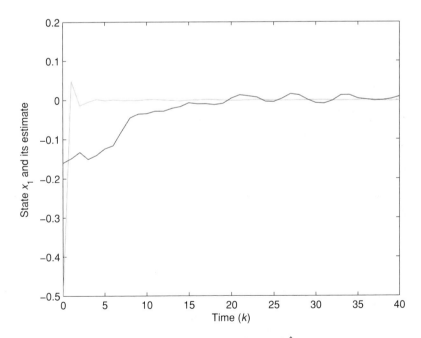

Figure 5.10 $x_1(k)$ and its estimate of $\hat{\Psi}_2$ case

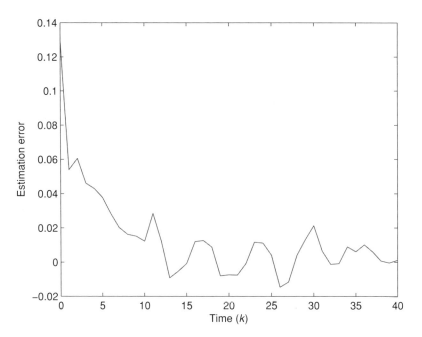

Figure 5.11   Estimation error of $\hat{\Psi}_3$ case

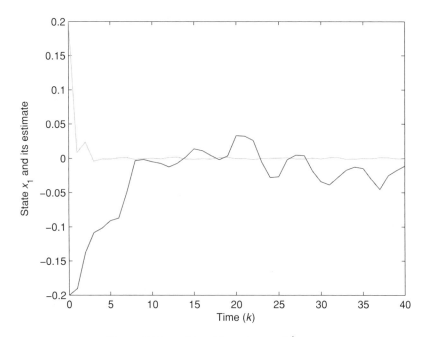

Figure 5.12 $x_1(k)$ and its estimate of $\hat{\Psi}_3$ case

Table 5.1   The optimal indices and filter gains for different cases

TP matrix	$J_{\min}$	K_1	K_2	M
$\hat{\Psi}_1$ (completely known)	0.8992	$\begin{bmatrix} 0.6387 \\ 0.1695 \end{bmatrix}$	$\begin{bmatrix} 0.0058 \\ -0.1199 \end{bmatrix}$	0.0547
$\hat{\Psi}_2$ (partially known)	1.2983	$\begin{bmatrix} 0.5643 \\ 0.3226 \end{bmatrix}$	$\begin{bmatrix} 0.1708 \\ -0.5382 \end{bmatrix}$	0.4362
$\hat{\Psi}_3$ (completely unknown)	1.6180	$\begin{bmatrix} 0.1628 \\ 0.0608 \end{bmatrix}$	$\begin{bmatrix} 0.1166 \\ -0.0127 \end{bmatrix}$	0.3308

Furthermore, let $\bar{\alpha} = \mathbb{E}\{\alpha(k)\} = 0.9$ and suppose that the RVNs are given by

$$g_1(x(k)) = g_2(x(k)) = \begin{bmatrix} 0.05x_1(k) - \tanh(0.05x_1(k)) & 0.2x_2(k) \end{bmatrix}^{\mathrm{T}},$$

$$h_1(x(k)) = h_2(x(k)) = \begin{bmatrix} -0.1x_1(k) & \tanh(0.1x_1(k)) \end{bmatrix}^{\mathrm{T}}.$$

It can be readily seen that (5.45) is satisfied with $B_{11} = B_{12} = \text{diag}\{0.1, 0.2\}$ and $B_{21} = B_{22} = \text{diag}\{0.1, 0.1\}$.

The saturation functions $\sigma(C_i x(k))$ $(i = 1, 2)$ are described as follows:

$$\sigma(C_i x(k)) = \begin{cases} C_i x(k), & \text{if } -v_{C_i x(k),\max} \le C_i x(k) \le v_{C_i x(k),\max}; \\ v_{C_i x(k),\max}, & \text{if } C_i x(k) > v_{C_i x(k),\max}; \\ -v_{C_i x(k),\max}, & \text{if } C_i x(k)(k) < -v_{C_i x(k),\max}. \end{cases}$$

The saturation values are taken as $v_{C_1 x(k),\max} = v_{C_2 x(k),\max} = 0.5$, and $L = 0.3, L_1 = 0.7$.

With the above parameters, the fault detection filter design problem can be solved by using Algorithm OFDFD. For the three different cases of TP matrices, the locally optimized index $J_{\min}$ and the corresponding filter gains are summarized in Table 5.1, where it can be seen that the more known knowledge we have in the TP matrix, the better the fault detection performance the filter can achieve.

For the simulation purpose, we consider the initial value $x(0) = [0.2 \quad -0.5]^{\mathrm{T}}$ and $\hat{x}(0) = [0 \quad 0]^{\mathrm{T}}$ with $k = 0, 1, \ldots, 300$. The exogenous disturbance input is $w(k) = 10^{-4} \sin(5k)v(k)$, where $v(k)$ is a uniformly distributed noise over $[-0.5, 0.5]$. The fault signal $f(k)$ is given by

$$f(k) = \begin{cases} 1, & 100 \le k \le 200, \\ 0, & \text{else.} \end{cases}$$

To demonstrate the mode switches, we take the TP matrix $\hat{\Psi}_1$ as an example and let $\theta(0) = 2$. The stochastic jumps of $\theta(k)$ between the two modes at certain time steps are plotted in Figure 5.13. Accordingly, Figures 5.14–5.16 show the sensor outputs in the different cases of the TPs. Note that the sensor outputs are saturated. Figures 5.17–5.19 present the generated residual signals $\tilde{r}(k)$ in the three cases, and their evolution functions $\bar{J}(\tilde{r}) = \{\sum_{l=0}^{k} \tilde{r}^{\mathrm{T}}(l)\tilde{r}(l)\}^{1/2}$ for both the faulty case and fault-free case are shown in Figures 5.20–5.22, respectively. The selected thresholds $\bar{J}_{\mathrm{th}} = \sup_{f=0} \mathbb{E}\{\sum_{k=0}^{300} \tilde{r}^{\mathrm{T}}(k)\tilde{r}(k)\}^{1/2}$ are obtained in all cases which are listed in Table 5.2. Also, the time steps required for successfully detecting the faults are

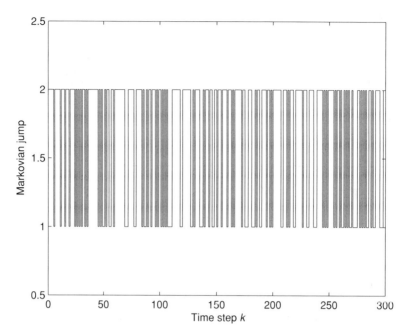

Figure 5.13   Modes evolution

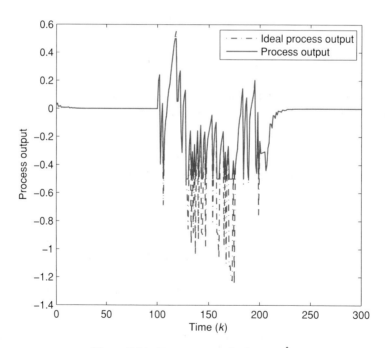

Figure 5.14   Process output in the case $\hat{\Psi}_1$

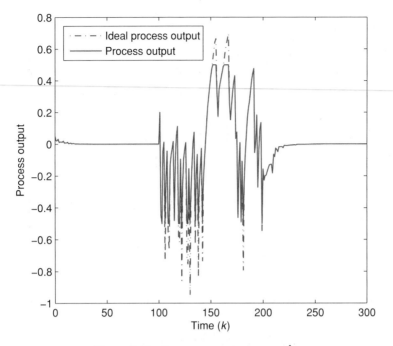

Figure 5.15 Process output in the case $\hat{\Psi}_2$

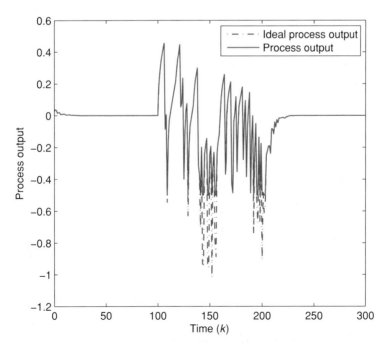

Figure 5.16 Process output in the case $\hat{\Psi}_3$

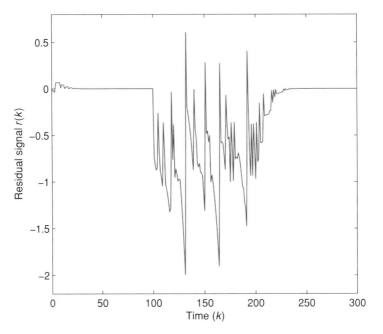

Figure 5.17   Residual $\tilde{r}(k)$ in the case $\hat{\Psi}_1$

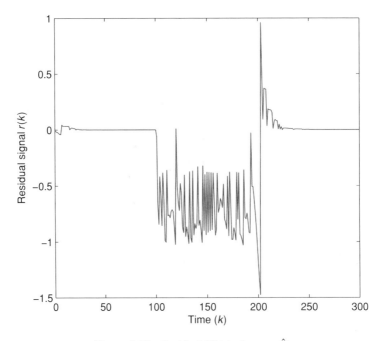

Figure 5.18   Residual $\tilde{r}(k)$ in the case $\hat{\Psi}_2$

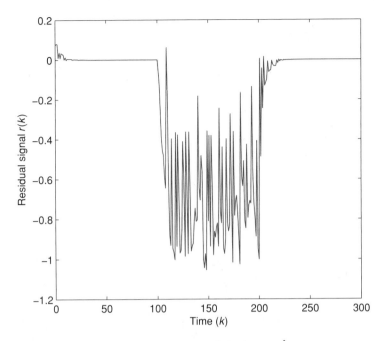

Figure 5.19 Residual $\tilde{r}(k)$ in the case $\hat{\Psi}_3$

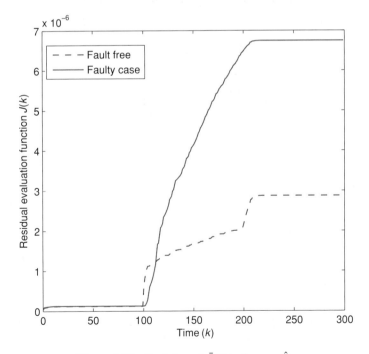

Figure 5.20 Evolution of $\bar{J}(\tilde{r})$ in the case $\hat{\Psi}_1$

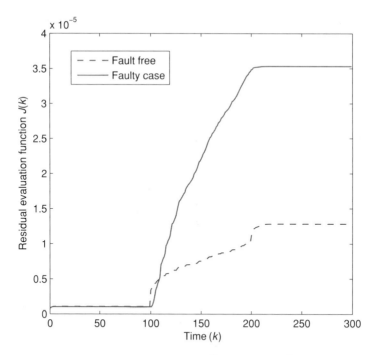

Figure 5.21   Evolution of $\bar{J}(\tilde{r})$ in the case $\hat{\Psi}_2$

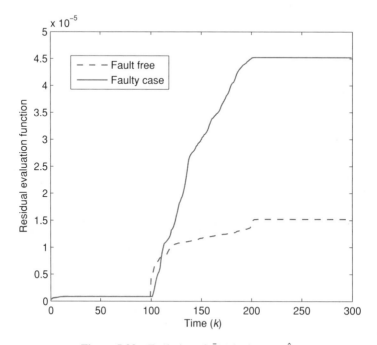

Figure 5.22   Evolution of $\bar{J}(\tilde{r})$ in the case $\hat{\Psi}_3$

Table 5.2 Thresholds and time steps of fault detection for different cases

TP matrix	$\hat{\Psi}_1$ (completely known)	$\hat{\Psi}_2$ (partially known)	$\hat{\Psi}_3$ (completely unknown)
Thresholds	1.2113×10^{-6}	0.4823×10^{-5}	0.8116×10^{-5}
Time steps	111	112	117

calculated and outlined in Table 5.2. Obviously, the more knowledge about the TPs we have, the faster the fault detection process will be.

5.5.4 Example 4

Following [173, 174], we consider a single-link robot arm whose dynamic equations are given as follows:

$$\ddot{\theta}(t) = -\frac{M_i g l}{J_i} \sin(\theta(t)) - \frac{D_i}{J_i}\dot{\theta}(t) + \frac{1}{J_i}u(t),$$

where $\theta(t)$ is the angle position of the arm, $u(t)$ is the control input (manipulated variable), g is the acceleration due to gravity, and l is the length of the arm. What makes the robot arm an MJS is that its mass of the payload M_i, inertia J_i, and damping D_i form a discrete state, and are changing depending on the angle $\theta(t)$. This can physically happen if the robot works under different environmental conditions and with changing payloads.

Letting $x_1(t) = \theta(t)$ and $x_2(t) = \dot{\theta}(t)$, considering that the robot works in a network environment, the system model includes RVNs and the process output is subject to saturation and sensor fault. By discretizing the plant with a sampling period of 0.05 s, we obtain the following discrete-time model:

$$x(k + 1) = A_i x(k) + \hat{B}_i u(k) + \alpha(k)g_i(x(k)) + (1 - \alpha(k))h_i(x(k)),$$
$$y(k) = \sigma(C_i x(k)) + D_{2i} w(k) + E_i f(k),$$

where

$$A_i = \begin{bmatrix} 1 & 0.05 \\ 0 & 1 - \frac{0.05D_i}{J_i} \end{bmatrix}, \quad \hat{B}_i = \begin{bmatrix} 0 \\ \frac{a_1}{J_i} \end{bmatrix}, \quad g_i(x(k)) = \begin{bmatrix} 0 \\ \frac{-0.05M_i g l}{J_i} \sin(x_1(k)) \end{bmatrix},$$

$$h_i(x(k)) = \begin{bmatrix} 0 \\ \frac{-0.05M_i g l}{J_i} \tanh(0.2x_1(k)) \end{bmatrix},$$

and the parameters M_i, J_i, and D_i have two different modes, as shown in Table 5.3. The TP matrix that relates the two operation modes is

$$\hat{\Psi} = \begin{bmatrix} 0.3 & 0.7 \\ ? & ? \end{bmatrix}.$$

Table 5.3   Modes of the parameters M_i, J_i, and D_i

Mode i	Parameter M_i	Parameter J_i	Parameter D_i
1	1	1	2
2	5	5	2

The model parameters are chosen as $g = 9.8$, $l = 0.5$, $a_1 = 0$, and the other parameters are the same as in Example 3 (Section 5.5.3). Similarly, the fault detection filter design problem can be solved by using Algorithm OFDFD, and the solution can be obtained as follows:

$$K_1 = \begin{bmatrix} -0.0194 \\ 0.7746 \end{bmatrix}, \quad K_2 = \begin{bmatrix} -0.0267 \\ 0.1347 \end{bmatrix}, \quad M = 0.5921, \quad J_{\min} = 1.0877.$$

A similar calculation shows that the fault can be detected in 17 time steps after its occurrence.

5.6 Summary

In this chapter, the filtering and fault detection problems have been investigated for discrete-time MJSs with RVNs and sensor saturation. The TP matrices considered include the case with polytopic uncertainties and the case with partially unknown TPs. Also, the cases with completely known or completely unknown TPs have been studied as two special cases. The H_∞ filtering problem was considered first, where the randomly occurring nonlinearities were modeled by the Bernoulli-distributed white sequences with known conditional probabilities. Sufficient conditions have been derived for the filtering augmented system under consideration to satisfy the H_∞ performance constraint. The corresponding robust H_∞ filters have been designed by solving sets of RLMIs. Second, the fault detection problem was investigated for discrete-time MJSs with RVNs and sensor saturation. Two energy norm indices were used for the fault detection problem: one in order to account for the restraint of disturbance and the other the sensitivity of faults. A locally optimized fault detection filter was designed such that (1) the fault detection dynamics is stochastically stable, (2) the effect from the exogenous disturbance on the residual is attenuated with respect to a minimized H_∞-norm, and (3) the sensitivity of the residual to the fault is enhanced in terms of a maximized H_∞-norm. Finally, the results of this chapter have been demonstrated by some simulation examples.

6

Quantized Fault Detection with Mixed Time-Delays and Packet Dropouts

This chapter is concerned with the quantized fault detection problem for two classes of discrete-time nonlinear systems with stochastic mixed time-delays and successive packet dropouts. The mixed time-delays comprise both the multiple discrete time-delays and the infinite distributed delays that occur in a random way. The fault detection problem is first considered for a class of discrete-time systems with randomly occurring nonlinearities, mixed stochastic time-delays, and measurement quantizations. A sequence of stochastic variables is introduced to govern the random occurrences of the nonlinearities, discrete time-delays, and distributed time-delays, where all the stochastic variables are mutually independent but obey the Bernoulli distribution. Moreover, by using similar analysis techniques, the network-based robust fault detection problem is also studied for a class of uncertain discrete-time T–S fuzzy systems with stochastic mixed time-delays and successive packet dropouts. The main purpose is to design a fault detection filter such that the overall fault detection dynamics is exponentially stable in the mean square and, at the same time, the error between the residual signal and the fault signal is made as small as possible. Sufficient conditions are established via intensive stochastic analysis for the existence of the desired fault detection filters, and then the explicit expression of the desired filter gains is derived by means of the feasibility of certain matrix inequalities. Also, the optimal performance index for the fault detection problem addressed can be obtained by solving an auxiliary convex optimization problem. Some illustrative examples are provided to show the usefulness and effectiveness of the proposed design method.

Filtering, Control and Fault Detection with Randomly Occurring Incomplete Information, First Edition.
Hongli Dong, Zidong Wang, and Huijun Gao.
© 2013 John Wiley & Sons, Ltd. Published 2013 by John Wiley & Sons, Ltd.

6.1   Problem Formulation for Fault Detection Filter Design

Consider the following discrete-time systems with randomly occurring nonlinearities and mixed stochastic time-delays:

$$
\begin{cases}
x(k+1) = Ax(k) + A_{d1} \displaystyle\sum_{i=1}^{q} \alpha_i(k)x(k - \tau_i(k)) + \beta(k)A_{d2} \displaystyle\sum_{d=1}^{\infty} \mu_d x(k-d) \\
\qquad\quad + \gamma(k)g(k, x(k)) + D_1 w(k) + Gf(k), \\
y(k) = Cx(k) + D_2 w(k) + Hf(k), \\
x(k) = \psi(k), \forall k \in \mathbb{Z}^-,
\end{cases}
\tag{6.1}
$$

where $x(k) \in \mathbb{R}^n$ represents the state vector, $y(k) \in \mathbb{R}^m$ is the process output, $w(k) \in \mathbb{R}^p$ is the unknown input belonging to $l_2[0, \infty)$, and $f(k) \in \mathbb{R}^l$ is the fault to be detected. $\tau_i(k)$ $(i = 1, 2, \ldots, q)$ denotes the discrete time-delays, while d $(d = 1, 2, \ldots, \infty)$ describes the distributed time-delays, $\psi(k)$ is a given initial sequence, and A, A_{d1}, A_{d2}, D_1, G, C, D_2, and H are all constant matrices with appropriate dimensions.

The nonlinear function $g(k, x(k))$ satisfies the following condition:

$$
\|g(k, x(k))\|^2 \leq \varepsilon(k)\|E(k)x(k)\|^2,
\tag{6.2}
$$

where $\varepsilon(k) > 0$ is a known positive scalar and $E(k)$ is a known constant matrix.

The constants $\mu_d \geq 0$ $(d = 1, 2, \ldots, \infty)$ satisfy the following convergence condition:

$$
\bar{\mu} := \sum_{d=1}^{\infty} \mu_d \leq \sum_{d=1}^{\infty} d\mu_d < +\infty.
\tag{6.3}
$$

The stochastic variables $\alpha_i(k)$ $(i = 1, 2, \ldots, q)$, $\beta(k)$, and $\gamma(k)$ are mutually uncorrelated Bernoulli-distributed white sequences that account for, respectively, the phenomena of randomly occurring discrete time-delays, distributed time-delays, and nonlinearities. A natural assumption on the sequences $\alpha_i(k)$ $(i = 1, 2, \ldots, q)$, $\beta(k)$, and $\gamma(k)$ are made as follows:

$$
\begin{aligned}
&\text{Prob}\{\alpha_i(k) = 1\} = \mathbb{E}\{\alpha_i(k)\} = \bar{\alpha}_i, \quad \text{Prob}\{\alpha_i(k) = 0\} = 1 - \bar{\alpha}_i, \\
&\text{Prob}\{\beta(k) = 1\} = \mathbb{E}\{\beta(k)\} = \bar{\beta}, \quad \text{Prob}\{\beta(k) = 0\} = 1 - \bar{\beta}, \\
&\text{Prob}\{\gamma(k) = 1\} = \mathbb{E}\{\gamma(k)\} = \bar{\gamma}, \quad \text{Prob}\{\gamma(k) = 0\} = 1 - \bar{\gamma},
\end{aligned}
\tag{6.4}
$$

where $\bar{\alpha}_i \in [0, 1]$, $\bar{\beta} \in [0, 1]$, and $\bar{\gamma} \in [0, 1]$ are known constants.

Remark 6.1 *The nonlinearities described by $g(k, x(k))$ could occur in a probabilistic way based on an individual probability distribution specified a priori through statistical tests. The concept of such randomly occurring nonlinearities was put forward by Wang et al. [92] to reflect the stochastic nonlinearities for complex networks. In this chapter, the randomly occurring nonlinearities are addressed for the fault detection problems that are of more practical significance in a networked environment. On the other hand, the term $\sum_{d=1}^{\infty} \mu_d x(k - d)$ in (6.1) represents the so-called infinitely distributed delay in the discrete-time setting, which* *can be regarded as the discretization of the infinite integral form $\int_{-\infty}^{t} k(t-s)x(s)\,ds$ for the continuous-time system. The importance of distributed delays has been widely recognized, but the corresponding results for discrete-time systems have been very few, especially when the fault detection problem becomes a research focus.*

Assumption 6.1   *The communication delays $\tau_i(k)$ $(i = 1, 2, \cdots, q)$ are time-varying and satisfy $d_m \leq \tau_i(k) \leq d_M$, where d_m and d_M are constant positive scalars representing the lower and upper bounds on the communication delays, respectively.*

Remark 6.2   *The description of the communication delays in (6.1) exhibits the following two features: (1) the communication delays are allowed to occur in three fashions – namely, discrete, successive, and distributed; and (2) each possible delay could occur independently according to an individual probability distribution that can be specified a priori through a statistical test.*

In a networked environment, it is quite common that the measurements $y(k)$ of the system are quantized during the signal transmission. Let us denote the quantizer as $h(\cdot) = [h_1(\cdot) \quad h_2(\cdot) \quad \cdots \quad h_m(\cdot)]^{\mathrm{T}}$ which is symmetric; that is, $h_j(-v) = -h_j(v)$, $j = 1, \ldots, m$. The map of the quantization process is

$$\tilde{y}(k) = h(y(k)) = \left[h_1(y^{(1)}(k)) \quad h_2(y^{(2)}(k)) \quad \cdots \quad h_m(y^{(m)}(k)) \right]^{\mathrm{T}}.$$

In this chapter, we are interested in the logarithmic static and time-invariant quantizer. For each $h_j(\cdot)$ $(1 \leq j \leq m)$, the set of quantization levels is described by

$$\mathcal{U}_j = \left\{ \pm \hat{\mu}_i^{(j)}, \hat{\mu}_i^{(j)} = \chi_j^i \hat{\mu}_0^{(j)}, i = 0, \pm 1, \pm 2, \cdots \right\} \cup \{0\}, \quad 0 < \chi_j < 1, \quad \hat{\mu}_0^{(j)} > 0,$$

and each of the quantization level corresponds to a segment such that the quantizer maps the whole segment to this quantization level.

According to Fu and Xie [69], the logarithmic quantizer is given by

$$h_j(y^{(j)}(k)) = \left\{ \begin{array}{ll} \hat{\mu}_i^{(j)}, & \frac{1}{1+\delta_j}\hat{\mu}_i^{(j)} \leq y^{(j)}(k) \leq \frac{1}{1-\delta_j}\hat{\mu}_i^{(j)}, \\ 0, & y^{(j)}(k) = 0, \\ -h_j(-y^{(j)}(k)), & y^{(j)}(k) < 0, \end{array} \right.$$

where $\delta_j = (1 - \chi_j)/(1 + \chi_j)$. It can be easily seen from the above definition that $h_j(y^{(j)}(k)) = (1 + \Delta_k^{(j)})y^{(j)}(k)$ with $|\Delta_k^{(j)}| \leq \delta_j$. According to the transformation discussed above, the quantizing effect can be transformed into the sector-bounded uncertainties.

Defining $\Delta_k = \mathrm{diag}\{\Delta_k^{(1)}, \ldots, \Delta_k^{(m)}\}$, the measurements with quantization effect can be expressed as

$$\begin{aligned} \tilde{y}(k) &= (I + \Delta_k)y(k) \\ &= (I + \Delta_k)Cx(k) + (I + \Delta_k)D_2w(k) + (I + \Delta_k)Hf(k). \end{aligned} \tag{6.5}$$

Consider a full-order fault detection filter of the following structure:

$$\begin{cases} \hat{x}(k+1) = A_F \hat{x}(k) + B_F \bar{y}(k) \\ \quad r(k) = C_F \hat{x}(k) + D_F \bar{y}(k) \end{cases}, \tag{6.6}$$

where $\hat{x}(k) \in \mathbb{R}^n$ represents the filter state vector, $r(k) \in \mathbb{R}^l$ is the so-called residual that is compatible with the fault vector $f(k)$, and A_F, B_F, C_F, and D_F are appropriately dimensioned filter matrices to be determined.

By defining $\bar{\Delta} = \text{diag}\{\delta_1, \ldots, \delta_m\}$, $F_k = \Delta_k \bar{\Delta}^{-1}$, we can obtain an unknown real-valued time-varying matrix satisfying $F_k F_k^T \leq I$. From (6.1), (6.5), and (6.6), we have the overall fault detection dynamics governed by the following system:

$$\begin{cases} \bar{x}(k+1) = (\bar{A} + \Delta\bar{A})\bar{x}(k) + \sum_{i=1}^{q}(\bar{A}_{di} + \tilde{A}_{di})\bar{x}(k - \tau_i(k)) + (\bar{A}_d + \tilde{A}_d) \\ \qquad \times \sum_{d=1}^{\infty} \mu_d \bar{x}(k-d) + (\bar{\gamma} + \tilde{\gamma}(k))Zg(k, x(k)) + (\bar{D} + \Delta\bar{D})v(k), \tag{6.7} \\ \bar{r}(k) = (\bar{C} + \Delta\bar{C})\bar{x}(k) + (\bar{D}_F + \Delta\bar{D}_F)v(k), \end{cases}$$

where

$$\bar{x}(k) = [x^T(k) \quad \hat{x}^T(k)]^T, \quad \bar{r}(k) = r(k) - f(k), \quad v(k) = [w^T(k) \quad f^T(k)]^T,$$

$$\bar{A} = \begin{bmatrix} A & 0 \\ B_F C & A_F \end{bmatrix}, \quad \bar{A}_{di} = \begin{bmatrix} \bar{\alpha}_i A_{d1} & 0 \\ 0 & 0 \end{bmatrix}, \quad \tilde{A}_{di} = \begin{bmatrix} \tilde{\alpha}_i(k) A_{d1} & 0 \\ 0 & 0 \end{bmatrix},$$

$$\bar{A}_d = \begin{bmatrix} \bar{\beta} A_{d2} & 0 \\ 0 & 0 \end{bmatrix}, \quad \tilde{A}_d = \begin{bmatrix} \tilde{\beta}(k) A_{d2} & 0 \\ 0 & 0 \end{bmatrix}, \quad \bar{D} = \begin{bmatrix} D_1 & G \\ B_F D_2 & B_F H \end{bmatrix},$$

$$Z = [I \quad 0]^T, \quad \bar{C} = [D_F C \quad C_F], \quad \bar{D}_F = [D_F D_2 \quad D_F H - I],$$

$$\Delta\bar{A} = H_F F_k E_C, \quad \Delta\bar{D} = H_F F_k E_D, \quad \Delta\bar{C} = D_F F_k E_C, \quad \Delta\bar{D}_F = D_F F_k E_D,$$

$$H_F = [0 \quad B_F^T]^T, \quad E_C = [\bar{\Delta}C \quad 0], \quad E_D = [\bar{\Delta}D_2 \quad \bar{\Delta}H],$$

with $\tilde{\alpha}_i(k) = \alpha_i(k) - \bar{\alpha}_i$, $\tilde{\beta}(k) = \beta(k) - \bar{\beta}$, and $\tilde{\gamma}(k) = \gamma(k) - \bar{\gamma}$. It is clear that $\mathbb{E}\{\tilde{\alpha}_i(k)\} = 0$, $\mathbb{E}\{\tilde{\alpha}_i^2(k)\} = \bar{\alpha}_i(1 - \bar{\alpha}_i)$, $\mathbb{E}\{\tilde{\beta}(k)\} = 0$, $\mathbb{E}\{\tilde{\beta}^2(k)\} = \bar{\beta}(1 - \bar{\beta})$, $\mathbb{E}\{\tilde{\gamma}(k)\} = 0$, and $\mathbb{E}\{\tilde{\gamma}^2(k)\} = \bar{\gamma}(1 - \bar{\gamma})$.

Definition 6.1.1 *[25] The fault detection dynamics in (6.7) is said to be exponentially stable in the mean square if, in case of $v(k) = 0$ and for any initial conditions, there exist constants $\delta > 0$ and $0 < \kappa < 1$ such that*

$$\mathbb{E}\{\|\bar{x}(k)\|^2\} \leq \delta\kappa^k \sup_{i \in \mathbb{Z}^-} \mathbb{E}\{\|\psi(i)\|^2\}, \quad \forall k \geq 0.$$

Our aim in this chapter is to design a filter of the form (6.6) that makes the error between residual and fault signal as small as possible. By means of definition 6.1.1, the aim of this chapter can be restated as finding the filter parameters A_F, B_F, C_F, and D_F such that the following two requirements are satisfied simultaneously:

(R1) The overall fault detection dynamics (6.7) is exponentially stable in the mean square.

(R2) Under zero initial condition, the residual error $\bar{r}(k)$ satisfies

$$\sum_{k=0}^{\infty} \mathbb{E}\{\|\bar{r}(k)\|^2\} \leq \gamma^2 \sum_{k=0}^{\infty} \mathbb{E}\{\|v(k)\|^2\} \tag{6.8}$$

for all nonzero $v(k)$, where $\gamma > 0$ is made as small as possible in the feasibility of (6.8).

We further adopt a residual evaluation stage including an evaluation function $J(k)$ and a threshold J_{th} of the following form:

$$J(k) = \left\{ \sum_{h=0}^{k} r^T(h) r(h) \right\}^{1/2}, \quad J_{th} = \sup_{w_k \in l_2, f_k = 0} \mathbb{E}\{J(L)\}, \tag{6.9}$$

where L denotes the maximum time step of the evaluation function. Based on (6.9), the occurrence of faults can be detected by comparing $J(k)$ with J_{th} according to the following rule:

$$J(k) > J_{th} \Longrightarrow \text{with faults} \Longrightarrow \text{alarm},$$

$$J(k) \leq J_{th} \Longrightarrow \text{no faults}.$$

6.2 Main Results

In this section, let us investigate the both the analysis and synthesis problems for the fault detection filter design of system (6.1) in the presence of measurement quantization (6.5). The following lemmas will be used in deriving our main results.

Lemma 6.2.1 [149] Let $x \in \mathbb{R}^n$, $y \in \mathbb{R}^n$ and matrix $Q > 0$. Then, we have $x^T Q y + y^T Q x \leq x^T Q x + y^T Q y$.

Lemma 6.2.2 [167] Let $\mathcal{M} \in \mathbb{R}^{n \times n}$ be a positive semi-definite matrix, $x_i \in \mathbb{R}^n$, and constant $a_i > 0$ ($i = 1, 2, \ldots$). If the series concerned is convergent, then we have

$$\left(\sum_{i=1}^{\infty} a_i x_i \right)^T \mathcal{M} \left(\sum_{i=1}^{\infty} a_i x_i \right) \leq \left(\sum_{i=1}^{\infty} a_i \right) \sum_{i=1}^{\infty} a_i x_i \mathcal{M} x_i. \tag{6.10}$$

For presentation convenience, we first discuss the *nominal* system of (6.7) (i.e., without the parameter uncertainties $\Delta\bar{A}$, $\Delta\bar{D}$, $\Delta\bar{C}$, and $\Delta\bar{D}_F$) and will eventually extend our main results to the more general case. In the following theorem, a sufficient condition is presented for the residual dynamics (6.7) to be exponentially stable with (6.8) satisfied under zero initial conditions.

Theorem 6.2.3   *Consider the nominal system of (6.7) with given filter parameters and a prescribed H_∞ index $\gamma > 0$. The fault detection dynamics is exponentially stable in the mean square and satisfies (6.8) if there exist matrices $P > 0$, $Q_j > 0$ $(j = 1, 2, \ldots, q)$, $Q > 0$, and positive constant scalar ρ satisfying*

$$\Phi = \begin{bmatrix} \Omega_{11} + \bar{C}^T \bar{C} & * & * & * \\ \hat{Z}^T P \bar{A} & \Omega_{22} & * & * \\ \bar{A}_d^T P \bar{A} & \bar{A}_d^T P \hat{Z} & \Omega_{33} & * \\ \bar{D}^T P \bar{A} + \bar{D}_F^T \bar{C} & \bar{D}^T P \hat{Z} & \bar{D}^T P \bar{A}_d & \Omega_{44} \end{bmatrix} < 0, \tag{6.11}$$

$$Z^T P Z \leq \rho I, \tag{6.12}$$

where

$$\Omega_{11} = 2\bar{A}^T P \bar{A} + \rho \bar{E}(k) + \bar{\mu} Q + \sum_{j=1}^{q} (d_M - d_m + 1) Q_j - P,$$

$$\Omega_{22} = 2\hat{Z}^T P \hat{Z} + \text{diag}\{-Q_1 + \tilde{A}_1, -Q_2 + \tilde{A}_2, \ldots, -Q_q + \tilde{A}_q\},$$

$$\Omega_{33} = 2\bar{A}_d^T P \bar{A}_d + \bar{\beta}(1 - \bar{\beta}) \hat{A}_{d2}^T P \hat{A}_{d2} - \frac{1}{\bar{\mu}} Q, \quad \Omega_{44} = 2\bar{D}^T P \bar{D} + \bar{D}_F^T \bar{D}_F - \gamma^2 I,$$

$$\tilde{A}_i = \bar{\alpha}_i (1 - \bar{\alpha}_i) \hat{A}_{d1}^T P \hat{A}_{d1} (i = 1, 2, \ldots, q),$$

$$\bar{E}(k) = \text{diag}\{(4\bar{\gamma}^2 + \bar{\gamma}) \varepsilon(k) E^T(k) E(k), 0\}, \quad \hat{Z} = [\bar{A}_{d1} \quad \bar{A}_{d2} \quad \cdots \quad \bar{A}_{dq}],$$

$$\hat{A}_{d1} = \text{diag}\{A_{d1}, 0\}, \quad \hat{A}_{d2} = \text{diag}\{A_{d2}, 0\}.$$

Proof.   Choose the following Lyapunov functional for system (6.7):

$$V(k) = \sum_{i=1}^{4} V_i(k), \tag{6.13}$$

where

$$V_1(k) = \bar{x}^T(k) P \bar{x}(k), \quad V_2(k) = \sum_{j=1}^{q} \sum_{i=k-\tau_j(k)}^{k-1} \bar{x}^T(i) Q_j \bar{x}(i),$$

$$V_3(k) = \sum_{j=1}^{q} \sum_{m=-d_M+1}^{-d_m} \sum_{i=k+m}^{k-1} \bar{x}^T(i) Q_j \bar{x}(i), \quad V_4(k) = \sum_{d=1}^{\infty} \mu_d \sum_{\tau=k-d}^{k-1} \bar{x}^T(\tau) Q \bar{x}(\tau),$$

with $P > 0$, $Q > 0$, $Q_j > 0$ $(j = 1, 2, \ldots, q)$ being matrices to be determined.

Notice that

$$\mathbb{E}\left\{\tilde{A}_{di}^{\mathrm{T}} P \tilde{A}_{di}\right\} = \bar{\alpha}_i (1 - \bar{\alpha}_i) \hat{A}_{d1}^{\mathrm{T}} P \hat{A}_{d1}, \tag{6.14}$$

$$\mathbb{E}\{\tilde{A}_{d}^{\mathrm{T}} P \tilde{A}_{d}\} = \bar{\beta}(1 - \bar{\beta}) \hat{A}_{d2}^{\mathrm{T}} P \hat{A}_{d2}. \tag{6.15}$$

According to Lemma 6.2.1, we have

$$2\bar{\gamma}\bar{x}^{\mathrm{T}}(k)\bar{A}^{\mathrm{T}} P Z g(k, x(k))$$
$$\leq \bar{x}^{\mathrm{T}}(k)\bar{A}^{\mathrm{T}} P \bar{A}\bar{x}(k) + \bar{\gamma}^2 g^{\mathrm{T}}(k, x(k)) Z^{\mathrm{T}} P Z g(k, x(k)), \tag{6.16}$$

$$2\bar{\gamma}g^{\mathrm{T}}(k, x(k)) Z^{\mathrm{T}} P \bar{D} v(k)$$
$$\leq \bar{\gamma}^2 g^{\mathrm{T}}(k, x(k)) Z^{\mathrm{T}} P Z g(k, x(k)) + v^{\mathrm{T}}(k)\bar{D}^{\mathrm{T}} P \bar{D} v(k), \tag{6.17}$$

$$2\bar{\gamma}\left(\sum_{i=1}^{q} \bar{A}_{di}\bar{x}(k - \tau_i(k))\right)^{\mathrm{T}} P Z g(k, x(k))$$

$$\leq \left(\sum_{i=1}^{q} \bar{A}_{di}\bar{x}(k - \tau_i(k))\right)^{\mathrm{T}} P \left(\sum_{i=1}^{q} \bar{A}_{di}\bar{x}(k - \tau_i(k))\right)$$
$$+ \bar{\gamma}^2 g^{\mathrm{T}}(k, x(k)) Z^{\mathrm{T}} P Z g(k, x(k)), \tag{6.18}$$

$$2\bar{\gamma}\left(\bar{A}_d \sum_{d=1}^{\infty} \mu_d \bar{x}(k - d)\right)^{\mathrm{T}} P Z g(k, x(k))$$

$$\leq \left(\bar{A}_d \sum_{d=1}^{\infty} \mu_d \bar{x}(k - d)\right)^{\mathrm{T}} P \left(\bar{A}_d \sum_{d=1}^{\infty} \mu_d \bar{x}(k - d)\right)$$
$$+ \bar{\gamma}^2 g^{\mathrm{T}}(k, x(k)) Z^{\mathrm{T}} P Z g(k, x(k)). \tag{6.19}$$

Also, it follows from (6.2) that

$$g^{\mathrm{T}}(k, x(k))(4\bar{\gamma}^2 + \bar{\gamma}) Z^{\mathrm{T}} P Z g(k, x(k)) \leq x^{\mathrm{T}}(k)(4\bar{\gamma}^2 + \bar{\gamma})\rho\varepsilon(k) E^{\mathrm{T}}(k) E(k) x(k)$$
$$= \bar{x}^{\mathrm{T}}(k)\rho \bar{E}(k)\bar{x}(k). \tag{6.20}$$

Then, along the trajectory of system (6.7), we have from (6.14)–(6.20) that

$$\mathbb{E}\{\Delta V_1(k)\}$$
$$= \mathbb{E}\{\bar{x}^{\mathrm{T}}(k + 1) P \bar{x}(k + 1) - \bar{x}^{\mathrm{T}}(k) P \bar{x}(k)\}$$
$$\leq \mathbb{E}\left\{\bar{x}^{\mathrm{T}}(k)(2\bar{A}^{\mathrm{T}} P \bar{A} - P + \rho \bar{E}(k))\bar{x}(k) + 2\bar{x}^{\mathrm{T}}(k)\bar{A}^{\mathrm{T}} P \left(\sum_{i=1}^{q} \bar{A}_{di}\bar{x}(k - \tau_i(k))\right)\right.$$

$$+2\bar{x}^{\mathrm{T}}(k)\bar{A}^{\mathrm{T}}P\bar{A}_d\left(\sum_{d=1}^{\infty}\mu_d\bar{x}(k-d)\right)+2\bar{x}^{\mathrm{T}}(k)\bar{A}^{\mathrm{T}}P\bar{D}v(k)$$

$$+2\left(\sum_{i=1}^{q}\bar{A}_{di}\bar{x}(k-\tau_i(k))\right)^{\mathrm{T}}P\left(\sum_{i=1}^{q}\bar{A}_{di}\bar{x}(k-\tau_i(k))\right)$$

$$+\sum_{i=1}^{q}\bar{x}^{\mathrm{T}}(k-\tau_i(k))\tilde{A}_{di}^{\mathrm{T}}P\tilde{A}_{di}\bar{x}(k-\tau_i(k))+2\left(\sum_{i=1}^{q}\bar{A}_{di}\bar{x}(k-\tau_i(k))\right)^{\mathrm{T}}P\bar{D}v(k)$$

$$+2\left(\sum_{i=1}^{q}\bar{A}_{di}\bar{x}(k-\tau_i(k))\right)^{\mathrm{T}}P\bar{A}_d\left(\sum_{d=1}^{\infty}\mu_d\bar{x}(k-d)\right)+2v^{\mathrm{T}}(k)\bar{D}^{\mathrm{T}}P\bar{D}v(k)$$

$$+2\left(\bar{A}_d\sum_{d=1}^{\infty}\mu_d\bar{x}(k-d)\right)^{\mathrm{T}}P\left(\bar{A}_d\sum_{d=1}^{\infty}\mu_d\bar{x}(k-d)\right)+2\left(\bar{A}_d\sum_{d=1}^{\infty}\mu_d\bar{x}(k-d)\right)^{\mathrm{T}}$$

$$\times P\bar{D}v(k)+\left(\tilde{A}_d\sum_{d=1}^{\infty}\mu_d\bar{x}(k-d)\right)^{\mathrm{T}}P\left(\tilde{A}_d\sum_{d=1}^{\infty}\mu_d\bar{x}(k-d)\right)\Bigg\}.\qquad(6.21)$$

Next, it can be derived that

$$\mathbb{E}\{\Delta V_2(k)\}\leq\mathbb{E}\left\{\sum_{j=1}^{q}\left(\bar{x}^{\mathrm{T}}(k)Q_j\bar{x}(k)-\bar{x}^{\mathrm{T}}(k-\tau_j(k))Q_j\bar{x}(k-\tau_j(k))\right.\right.$$

$$\left.\left.+\sum_{i=k-d_M+1}^{k-d_m}\bar{x}^{\mathrm{T}}(i)Q_j\bar{x}(i)\right)\right\},$$

$$\mathbb{E}\{\Delta V_3(k)\}=\mathbb{E}\left\{\sum_{j=1}^{q}\left((d_M-d_m)\bar{x}^{\mathrm{T}}(k)Q_j\bar{x}(k)-\sum_{i=k-d_M+1}^{k-d_m}\bar{x}^{\mathrm{T}}(i)Q_j\bar{x}(i)\right)\right\},$$

$$\mathbb{E}\{\Delta V_4(k)\}=\mathbb{E}\left\{\bar{\mu}\bar{x}^{\mathrm{T}}(k)Q\bar{x}(k)-\sum_{d=1}^{\infty}\mu_d\bar{x}^{\mathrm{T}}(k-d)Q\bar{x}(k-d)\right\}.\qquad(6.22)$$

From Lemma 6.2.2, it can be easily seen that

$$-\sum_{d=1}^{\infty}\mu_d\bar{x}^{\mathrm{T}}(k-d)Q\bar{x}(k-d)\leq-\frac{1}{\bar{\mu}}\left(\sum_{d=1}^{\infty}\mu_d\bar{x}(k-d)\right)^{\mathrm{T}}Q\left(\sum_{d=1}^{\infty}\mu_d\bar{x}(k-d)\right),\quad(6.23)$$

where $\bar{\mu}$ is defined in (6.3). For notational convenience, we denote the following matrix variables:

$$\xi(k) := \begin{bmatrix} \bar{x}^{\mathrm{T}}(k) & \bar{x}^{\mathrm{T}}(k - \tau_1(k)) & \cdots & \bar{x}^{\mathrm{T}}(k - \tau_q(k)) & \sum_{d=1}^{\infty} \mu_d \bar{x}^{\mathrm{T}}(k - d) & v^{\mathrm{T}}(k) \end{bmatrix}^{\mathrm{T}},$$
$$\zeta(k) := \begin{bmatrix} \bar{x}^{\mathrm{T}}(k) & \bar{x}^{\mathrm{T}}(k - \tau_1(k)) & \cdots & \bar{x}^{\mathrm{T}}(k - \tau_q(k)) & \sum_{d=1}^{\infty} \mu_d \bar{x}^{\mathrm{T}}(k - d) \end{bmatrix}^{\mathrm{T}}.$$

We are now ready to prove the exponential stability of the system (6.7) with $v(k) = 0$. Obviously, the combination of (6.21)–(6.23) results in

$$\mathbb{E}\{\Delta V(k)\} \leq \mathbb{E}\{\zeta^{\mathrm{T}}(k)\Omega\zeta(k)\}, \tag{6.24}$$

where

$$\Omega = \begin{bmatrix} \Omega_{11} & * & * \\ \hat{Z}^{\mathrm{T}}P\bar{A} & \Omega_{22} & * \\ \bar{A}_d^{\mathrm{T}}P\bar{A} & \bar{A}_d^{\mathrm{T}}P\hat{Z} & \Omega_{33} \end{bmatrix}.$$

It follows immediately from Theorem 6.2.3 that $\Omega < 0$. Furthermore, along the same line of the proof for Theorem 1 of Wang et al. [25], the exponential stability of system (6.7) can be confirmed in the mean-square sense.

Let us now move to the proof of the H_∞ performance for the system (6.7). To do so, we assume zero initial condition and consider the following index:

$$J_N = \mathbb{E}\sum_{k=0}^{\infty}[\bar{r}^{\mathrm{T}}(k)\bar{r}(k) - \gamma^2 v^{\mathrm{T}}(k)v(k)]$$

$$= \mathbb{E}\sum_{k=0}^{\infty}[\bar{r}^{\mathrm{T}}(k)\bar{r}(k) - \gamma^2 v^{\mathrm{T}}(k)v(k) + \Delta V(k)] - \mathbb{E}V(k+1)$$

$$\leq \mathbb{E}\sum_{k=0}^{\infty}[\bar{r}^{\mathrm{T}}(k)\bar{r}(k) - \gamma^2 v^{\mathrm{T}}(k)v(k) + \Delta V(k)] = \xi^{\mathrm{T}}(k)\Phi\xi(k).$$

According to Theorem 6.2.3, we have $J_N \leq 0$ and therefore (6.8), which completes the proof of Theorem 6.2.3. □

Having established the analysis results, we are in a position to deal with the filter design problem. In the following theorem, sufficient conditions are provided for the existence of the desired fault detection filters.

Theorem 6.2.4 *Consider the nominal system of (6.7) and let $\gamma > 0$ be a given scalar. A desired full-order fault detection filter of the form (6.6) exists if there exist positive-definite matrices P, Q, Q_j $(j = 1, 2, \ldots, q)$, positive constant scalar ρ, and matrices X and K*

satisfying

$$\Lambda = \begin{bmatrix} \hat{\Lambda}_{11} & * & * & * & * \\ 0 & -\gamma^2 I & * & * & * \\ \hat{\Lambda}_{31} & P\hat{D}_0 + X\hat{R}_2 & -P & * & * \\ \hat{\Lambda}_{41} & 0 & 0 & -\bar{P} & * \\ \hat{\Lambda}_{51} & \hat{\Lambda}_{52} & 0 & 0 & -\hat{P} \end{bmatrix} < 0, \tag{6.25}$$

$$Z^{\mathrm{T}} P Z \le \rho I, \tag{6.26}$$

where

$$\hat{\Lambda}_{11} = \mathrm{diag}\{\Lambda_{11}, \Lambda_{22}, \Lambda_{33}\}, \quad \hat{\Lambda}_{31} = [\, P\hat{A}_0 + X\hat{R}_1 \quad P\hat{Z} \quad P\bar{A}_d \,],$$

$$\hat{\Lambda}_{41} = \mathrm{diag}\{P\hat{A}_0 + X\hat{R}_1, P\hat{Z}, P\bar{A}_d\}, \quad \hat{E}_2 = [\,0_{m \times n} \quad I_{m \times m}\,]^{\mathrm{T}},$$

$$\Lambda_{11} = \rho \bar{E}(k) + \bar{\mu} Q + \sum_{j=1}^{q} (d_M - d_m + 1) Q_j - P,$$

$$\Lambda_{22} = \mathrm{diag}\{-Q_1 + \tilde{A}_1, \ldots, -Q_q + \tilde{A}_q\}, \quad \Lambda_{33} = \bar{\beta}(1 - \bar{\beta})\hat{A}_{d2}^{\mathrm{T}} P \hat{A}_{d2} - \frac{1}{\bar{\mu}} Q,$$

$$\hat{\Lambda}_{51} = \begin{bmatrix} 0 & 0 & 0 \\ K\hat{R}_1 & 0 & 0 \end{bmatrix}, \quad \hat{\Lambda}_{52} = \begin{bmatrix} P\hat{D}_0 + X\hat{R}_2 \\ K\hat{R}_2 - \hat{E}_1^{\mathrm{T}} \end{bmatrix}, \quad \hat{E} = \begin{bmatrix} 0_{n \times n} \\ I_{n \times n} \end{bmatrix},$$

$$\bar{P} = \mathrm{diag}\{P, P, P\}, \quad \hat{P} = \mathrm{diag}\{P, I\}, \quad \hat{E}_1 = [\,0_{l \times p} \quad I_{l \times l}\,]^{\mathrm{T}},$$

$$\hat{A}_0 = \begin{bmatrix} A & 0 \\ 0 & 0 \end{bmatrix}, \quad \hat{D}_0 = \begin{bmatrix} D_1 & G \\ 0 & 0 \end{bmatrix}, \quad \hat{R}_1 = \begin{bmatrix} 0 & I \\ C & 0 \end{bmatrix}, \quad \hat{R}_2 = \begin{bmatrix} 0 & 0 \\ D_2 & H \end{bmatrix}.$$

Furthermore, if (P, Q, Q_j, X, K, ρ) is a feasible solution of (6.25) and (6.26), then the fault detection filter parameters in the form of (6.6) are given as follows:

$$[A_{\mathrm{F}} \quad B_{\mathrm{F}}] = [\hat{E}^{\mathrm{T}} P \hat{E}]^{-1} \hat{E}^{\mathrm{T}} X, \quad [C_{\mathrm{F}} \quad D_{\mathrm{F}}] = K.$$

Proof. In order to avoid partitioning the positive-definite matrices P, Q, and Q_j, we rewrite the parameters in Theorem 6.2.3 in the following form:

$$\begin{aligned} \bar{A} &= \hat{A}_0 + \hat{E} K_1 \hat{R}_1, \quad \bar{D} = \hat{D}_0 + \hat{E} K_1 \hat{R}_2, \\ \bar{C} &= K \hat{R}_1, \quad \bar{D}_{\mathrm{F}} = K \hat{R}_2 - \hat{E}_1^{\mathrm{T}}, \quad H_{\mathrm{F}} = \hat{E} K_1 \hat{E}_2, \end{aligned} \tag{6.27}$$

where $K_1 = [A_F \quad B_F]$. Noticing (6.27) and using the Schur complement lemma, (6.11) can be rewritten as

$$
\begin{bmatrix}
\hat{\Lambda}_{11} & * & * & * & * \\
0 & -\gamma^2 I & * & * & * \\
\check{\Lambda}_{31} & \hat{D}_0 + \hat{E} K_1 \hat{R}_2 & -P^{-1} & * & * \\
\check{\Lambda}_{41} & 0 & 0 & -\bar{P}^{-1} & * \\
\hat{\Lambda}_{51} & \check{\Lambda}_{52} & 0 & 0 & -\hat{P}^{-1}
\end{bmatrix} < 0,
\tag{6.28}
$$

where

$$
\check{\Lambda}_{31} = [\hat{A}_0 + \hat{E} K_1 \hat{R}_1 \quad \hat{Z} \quad \bar{A}_d], \quad \check{\Lambda}_{41} = \text{diag}\{\hat{A}_0 + \hat{E} K_1 \hat{R}_1, \hat{Z}, \bar{A}_d\},
$$

$$
\check{\Lambda}_{52} = \begin{bmatrix} \hat{D}_0 + \hat{E} K_1 \hat{R}_2 \\ K \hat{R}_2 - \hat{E}_1^{\mathrm{T}} \end{bmatrix}.
$$

Pre- and post-multiplying the inequality (6.28) by $\text{diag}\{I, I, P, \bar{P}, \hat{P}\}$ and letting $X = P \hat{E} K_1$, we can obtain (6.25) readily, and the proof is then complete. $\qquad\square$

So far, we have obtained the main results for nominal systems, and now we show how the results can be extended to the general case where the parameter uncertainties are included.

Theorem 6.2.5 *Consider the uncertain system (6.7) and let $\gamma > 0$ be a given scalar. A desirable full-order fault detection filter of the form (6.6) exists if there exist positive-definite matrices P, Q, and Q_j ($j = 1, 2, \ldots, q$), positive constant scalars ρ and φ, and matrices X and K satisfying*

$$
\Psi = \begin{bmatrix}
\hat{\Lambda}_{11} & * & * & * & * & * & * \\
0 & -\gamma^2 I & * & * & * & * & * \\
\hat{\Lambda}_{31} & P \hat{D}_0 + X \hat{R}_2 & -P & * & * & * & * \\
\hat{\Lambda}_{41} & 0 & 0 & -\bar{P} & * & * & * \\
\hat{\Lambda}_{51} & \hat{\Lambda}_{52} & 0 & 0 & -\hat{P} & * & * \\
0 & 0 & \bar{X} & \hat{X} & \bar{K} & -\varphi I & * \\
\bar{E}_C & \bar{E}_D & 0 & 0 & 0 & 0 & -\varphi I
\end{bmatrix} < 0,
\tag{6.29}
$$

$$
Z^{\mathrm{T}} P Z \leq \rho I,
\tag{6.30}
$$

where

$$
\bar{E}_C = \begin{bmatrix} \varphi E_C & 0 & 0 \\ 0 & 0 & 0 \end{bmatrix}, \quad \hat{X} = \begin{bmatrix} \hat{E}_2^{\mathrm{T}} X^{\mathrm{T}} & 0 & 0 \\ 0 & 0 & 0 \end{bmatrix}, \quad \bar{K} = \begin{bmatrix} 0 & \hat{E}_2^{\mathrm{T}} K^{\mathrm{T}} \\ \hat{E}_2^{\mathrm{T}} X^{\mathrm{T}} & \hat{E}_2^{\mathrm{T}} K^{\mathrm{T}} \end{bmatrix},
$$

$$
\bar{E}_D = [0 \quad \varphi E_D^{\mathrm{T}}]^{\mathrm{T}}, \quad \bar{X} = [X \hat{E}_2 \quad X \hat{E}_2]^{\mathrm{T}},
$$

with $\hat{\Lambda}_{11}$, $\hat{\Lambda}_{31}$, $\hat{\Lambda}_{41}$, $\hat{\Lambda}_{51}$, $\hat{\Lambda}_{52}$, $\bar{P}$, and $\hat{P}$ defined in Theorem 6.2.4. Furthermore, if $(P, Q, Q_j, X, K, \rho, \varphi)$ is a feasible solution of (6.29) and (6.30), then the fault detection *filter parameters in the form of (6.6) are given as follows:*

$$[A_F \quad B_F] = [\hat{E}^T P \hat{E}]^{-1} \hat{E}^T X, \tag{6.31}$$

$$[C_F \quad D_F] = K. \tag{6.32}$$

Proof. In (6.25), let us replace $\bar{A}$, $\bar{C}$, $\bar{D}$, and $\bar{D}_F$ with $\bar{A} + \Delta\bar{A}$, $\bar{C} + \Delta\bar{C}$, $\bar{D} + \Delta\bar{D}$, and $\bar{D}_F + \Delta\bar{D}_F$, respectively, where $\Delta\bar{A} = \hat{E} K_1 \hat{E}_2 F_k E_C$, $\Delta\bar{D} = \hat{E} K_1 \hat{E}_2 F_k E_D$, $\Delta\bar{C} = K \hat{E}_2 F_k E_C$, and $\Delta\bar{D}_F = K \hat{E}_2 F_k E_D$. Then, rewrite (6.25) in terms of the S-procedure as $\Lambda + M F_k N + N^T F_k^T M^T < 0$ with

$$M = \begin{bmatrix} 0 & 0 & 0 & 0 & \hat{E}_2^T X^T & \hat{E}_2^T X^T & 0 & 0 & 0 & \hat{E}_2^T K^T \\ 0 & 0 & 0 & 0 & \hat{E}_2^T X^T & 0 & 0 & 0 & \hat{E}_2^T X^T & \hat{E}_2^T K^T \end{bmatrix}^T,$$

$$N = \begin{bmatrix} E_C & 0 & 0 & 0 & 0 & 0 & 0 & 0 & 0 & 0 \\ 0 & 0 & 0 & E_D & 0 & 0 & 0 & 0 & 0 & 0 \end{bmatrix}.$$

From the Schur complement lemma and the S-procedure, (6.29) can be easily obtained, which ends the proof. □

Remark 6.3 *In Theorem 6.2.5, sufficient conditions are presented that ensure the residual dynamics to be exponential stable in the mean square with a guaranteed performance index γ. It is shown that the feasibility of the fault detection filter design problem can be readily checked by the solvability of inequalities (6.29) and (6.30). Among these feasible solutions, the optimal performance index γ^* can be found by solving the following convex optimization problem: minimize γ subject to (6.29) and (6.30) over matrix variables P, Q, Q_j ($j = 1, 2, \ldots, q$), X, and K and scalars ρ and φ.*

6.3 Fuzzy-Model-Based Robust Fault Detection

6.3.1 Problem Formulation

In this section, we consider the fault detection problem for a class of uncertain discrete-time fuzzy systems with stochastic mixed time-delays and successive packet dropouts in NCSs, where the framework is shown in Figure 6.1. The sensors are connected to the fault detection filter via a network which is subject to possible successive packet dropouts.

The Physical Plant

Consider a discrete-time nonlinear system with stochastic mixed time-delays which can be represented by the following T–S fuzzy dynamic model:

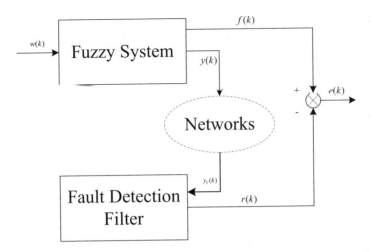

Figure 6.1   The framework of the fuzzy fault detection filter design over network environments

△ **Plant Rule i:** IF $\theta_1(k)$ is M_{i1} and $\theta_2(k)$ is M_{i2} and ... and $\theta_p(k)$ is M_{ip} THEN

$$x(k+1) = A_i(k)x(k) + A_{d1i} \sum_{m=1}^{h} \alpha_m(k)x(k - \tau_m(k))$$

$$+\beta(k)A_{d2i} \sum_{d=1}^{\infty} \mu_d x(k - d) + D_{1i}w(k) + G_i f(k),$$

$$y(k) = C_i x(k) + D_{2i}w(k),$$

$$x(k) = \psi(k), \ \forall k \in \mathbb{Z}^-, \ i = 1, \ldots, r, \tag{6.33}$$

where M_{ij} is the fuzzy set, r is the number of IF–THEN rules, and $\theta(k) = [\theta_1(k), \theta_2(k), \ldots, \theta_p(k)]$ is the premise variable vector. $x(k) \in \mathbb{R}^n$ represents the state vector; $y(k) \in \mathbb{R}^m$ is the process output; $w(k) \in \mathbb{R}^q$ is the unknown disturbance input; $f(k) \in \mathbb{R}^l$ is the fault to be detected; $w(k)$ and $f(k)$ belong to $l_2[0, \infty)$; $\tau_m(k)$ denotes the discrete communication delay that occurs according to the stochastic variable $\alpha(k)$; d describes the distributed time delay; $A_i(k) = A_i + \Delta A_i(k)$ and $(A_i, A_{d1i}, A_{d2i}, D_{1i}, G_i, C_i, \text{and } D_{2i})$ are known constant matrices of compatible dimensions; $\psi(k), \ k \in \mathbb{Z}^-$, are given random initial conditions satisfying $\sup_{k \in \mathbb{Z}^-} \mathbb{E}\{\|\psi(k)\|^2\} < \infty$.

The real-valued matrix $\Delta A_i(k)$ represents the norm-bounded parameter uncertainty of the following structure:

$$\Delta A_i(k) = H_{ai} F(k) E_a, \ i = 1, \ldots, r, \tag{6.34}$$

where H_{ai} and E_a are known constant matrices of appropriate dimensions, and $F(k)$ is an unknown matrix function satisfying

$$F^T(k)F(k) \le I. \tag{6.35}$$

The parameter uncertainty $\Delta A_i(k)$ is said to be admissible if both (6.34) and (6.35) hold.

The variable $\tau_m(k)$ denotes the time-varying delay satisfying

$$d_{\min} \leq \tau_m(k) \leq d_{\max},$$

where $d_{\min}$ and $d_{\max}$ are constant positive integers representing the lower and upper bounds on the communication delay, respectively. The constants $\mu_d \geq 0$ ($d = 1, 2, \ldots, \infty$) satisfy the following convergence conditions:

$$\bar{\mu} := \sum_{d=1}^{\infty} \mu_d < +\infty. \tag{6.36}$$

To account for the phenomena of randomly occurring discrete time-delays and distributed time-delays, we introduce the following stochastic variables: $\alpha_m(k) \in \mathbb{R}$ ($m = 1, 2, \ldots, h$) and $\beta(k) \in \mathbb{R}$, which are mutually independent Bernoulli-distributed white sequences and also independent of the premise variables $\theta(k)$. A natural assumption on $\alpha_m(k)$ and $\beta(k)$ is as follows:

$$\text{Prob}\{\alpha_m(k) = 1\} = \mathbb{E}\{\alpha_m(k)\} = \bar{\alpha}_m, \quad \text{Prob}\{\alpha_m(k) = 0\} = 1 - \bar{\alpha}_m,$$
$$\text{Prob}\{\beta(k) = 1\} = \mathbb{E}\{\beta(k)\} = \bar{\beta}, \quad \text{Prob}\{\beta(k) = 0\} = 1 - \bar{\beta}.$$

By using a center average defuzzifier, product interference and a singleton fuzzifier, the global dynamics of the T–S fuzzy systems (6.33) can be inferred as follows:

$$x(k+1) = \sum_{i=1}^{r} h_i(\theta(k)) \left[A_i(k)x(k) + A_{d1i} \sum_{m=1}^{h} \alpha_m(k)x(k - \tau_m(k)) + \beta(k)A_{d2i} \right.$$
$$\left. \times \sum_{d=1}^{\infty} \mu_d x(k - d) + D_{1i}w(k) + G_i f(k) \right],$$

$$y(k) = \sum_{i=1}^{r} h_i(\theta(k))[C_i x(k) + D_{2i}w(k)],$$

$$x(k) = \psi(k), \ \forall k \in \mathbb{Z}^-, \tag{6.37}$$

where the fuzzy basis functions are given by

$$h_i(\theta(k)) = \frac{\vartheta_i(\theta(k))}{\sum\limits_{i=1}^{r} \vartheta_i(\theta(k))},$$

with $\vartheta_i(\theta(k)) = \prod_{j=1}^{p} M_{ij}(\theta_j(k))$. Where $\vartheta_i(\theta(k)) \geq 0$, $i = 1, 2, \ldots, r$, $\sum_{i=1}^{r} \vartheta_i(\theta(k)) > 0$, and $M_{ij}(\theta_j(k))$ represents the grade of membership of $\theta_j(k)$ in M_{ij}. Hence, we have

$$h_i(\theta(k)) \geq 0, i = 1, 2, \ldots, r, \quad \sum_{i=1}^{r} h_i(\theta(k)) = 1.$$

In what follows, we write $h_i := h_i(\theta(k))$ for brevity.

Communication Channel with Packet Dropouts

In this section, we assume that an unreliable network medium is present between the physical plant and the fault detection filter, and the packet dropout phenomenon constitutes another focus of our present research. The signal received by the fault detection filter can be described by

$$y_f(k) = \hat{\gamma}(k)y(k) + (1 - \hat{\gamma}(k))y_f(k-1), \tag{6.38}$$

where $y_f(k) \in \mathbb{R}^m$ is the *actual* measurement signal of $y(k)$ and $\hat{\gamma}(k) \in \mathbb{R}$ is a binary distributed random variable with $\text{Prob}\{\hat{\gamma}(k) = 1\} = \mathbb{E}\{\hat{\gamma}(k)\} = \bar{\gamma}$ and $\text{Prob}\{\hat{\gamma}(k) = 0\} = 1 - \bar{\gamma}$. In this section, we assume that the premise variables $\theta(k)$ do not depend on the stochastic variables $\alpha(k)$, $\beta(k)$, and $\hat{\gamma}(k)$. Also, all the stochastic variables are assumed to be mutually independent Bernoulli-distributed white sequences.

Remark 6.4 *The dropout model (6.38) was introduced by Sahebsara et al. [29] to describe successive packet dropouts. For example, if $\hat{\gamma}(k) = 1$, then we have $y(k) = y_f(k)$, which means that there is no packet dropout; if $\hat{\gamma}(k) = 0$ but $\hat{\gamma}(k-1) = 1$, then we have $y(k) = y_f(k-1)$, which means that the measured output at time point k is missing but one at time point $k-1$ has been received. As shown by Sun et al. [19], it is easy to further confirm that (6.38) can be a model for multiple consecutive packet dropouts where the latest measurement received in the buffers will be utilized if the current measurement is lost during packet transmissions. Such a scheme is certainly more realistic than the one setting the measurement signals to zero when the current measurements are lost [26, 25].*

Fuzzy Fault Detection Filter

As discussed previously, the key step of fault detection schemes is the construction of a dynamic system called a fault detection observer/filter, in which the residual signal is generated in order to decide whether a fault has occurred or not [82].

In this section, for the physical plant represented by (6.33) and (6.37), we adopt a fuzzy fault detection filter whose model is described as follows:

△ **Filter Rule i:** IF $\theta_1(k)$ is M_{i1} and $\theta_2(k)$ is M_{i2} and $\ldots$ and $\theta_p(k)$ is M_{ip} THEN

$$\begin{aligned}\hat{x}(k+1) &= A_{fi}\hat{x}(k) + B_{fi}y_f(k), \\ r(k) &= C_{fi}\hat{x}(k) + D_{fi}y_f(k),\end{aligned} \tag{6.39}$$

where $\hat{x}(k) \in \mathbb{R}^n$ represents the filter state vector, $r(k) \in \mathbb{R}^l$ is the so-called residual that is compatible with the fault vector $f(k)$, and A_{fi}, B_{fi}, C_{fi}, and D_{fi} are appropriately dimensioned filter matrices to be determined. Then, the overall fuzzy fault detection filter can be represented in the following form:

$$\hat{x}(k+1) = \sum_{i=1}^{r} h_i [A_{fi} \hat{x}(k) + B_{fi} y_f(k)],$$

$$r(k) = \sum_{i=1}^{r} h_i [C_{fi} \hat{x}(k) + D_{fi} y_f(k)]. \tag{6.40}$$

Our aim is to design a fault detection filter of the form in (6.39) that makes the error between residual signal $r(k)$ and fault signal $f(k)$ as small as possible. From (6.37), (6.38) and (6.40), we have the overall fault detection dynamics governed by the following system:

$$\eta(k+1) = \sum_{i=1}^{r} \sum_{j=1}^{r} h_i h_j \left[(\bar{A}_{ij}(k) + \tilde{\gamma}(k)\hat{A}_{ij})\eta(k) + \sum_{m=1}^{h} (\bar{A}_{d1mi} + \tilde{\alpha}_m(k)\hat{A}_{d1i}) \right.$$

$$\times \eta(k - \tau_m(k)) + (\bar{A}_{d2i} + \tilde{\beta}(k)\hat{A}_{d2i}) \sum_{d=1}^{\infty} \mu_d \eta(k-d)$$

$$\left. + (\bar{B}_{ij} + \tilde{\gamma}(k)\hat{B}_{ij})v(k) \right], \tag{6.41}$$

$$e(k) = \sum_{i=1}^{r} \sum_{j=1}^{r} h_i h_j [(\bar{C}_{ij} + \tilde{\gamma}(k)\hat{C}_{ij})\eta(k) + (\bar{D}_{ij} + \tilde{\gamma}(k)\hat{D}_{ij})v(k)],$$

where

$$\eta(k) = \begin{bmatrix} x^{\mathrm{T}}(k) & \hat{x}^{\mathrm{T}}(k) & y_f^{\mathrm{T}}(k-1) \end{bmatrix}^{\mathrm{T}}, \quad \bar{C}_{ij} = \begin{bmatrix} \bar{\gamma} D_{fj} C_i & C_{fj} & (1-\bar{\gamma})D_{fj} \end{bmatrix},$$

$$\bar{A}_{d1mi} = \mathrm{diag}\{\bar{\alpha}_m A_{d1i}, 0, 0\}, \quad \hat{A}_{d1i} = \mathrm{diag}\{A_{d1i}, 0, 0\}, \quad v(k) = \begin{bmatrix} w^{\mathrm{T}}(k) & f^{\mathrm{T}}(k) \end{bmatrix}^{\mathrm{T}},$$

$$\bar{D}_{ij} = \begin{bmatrix} \bar{\gamma} D_{fj} D_{2i} & -I \end{bmatrix}, \quad \bar{A}_{d2i} = \mathrm{diag}\{\bar{\beta} A_{d2i}, 0, 0\}, \quad \hat{D}_{ij} = \begin{bmatrix} D_{fj} D_{2i} & 0 \end{bmatrix},$$

$$\hat{A}_{d2i} = \mathrm{diag}\{A_{d2i}, 0, 0\}, \quad e(k) = r(k) - f(k), \quad \hat{C}_{ij} = \begin{bmatrix} D_{fj} C_i & 0 & -D_{fj} \end{bmatrix},$$

$$\bar{A}_{ij}(k) = \begin{bmatrix} A_i(k) & 0 & 0 \\ \bar{\gamma} B_{fj} C_i & A_{fj} & (1-\bar{\gamma})B_{fj} \\ \bar{\gamma} C_i & 0 & (1-\bar{\gamma})I \end{bmatrix}, \quad \hat{A}_{ij} = \begin{bmatrix} 0 & 0 & 0 \\ B_{fj} C_i & 0 & -B_{fj} \\ C_i & 0 & -I \end{bmatrix},$$

$$\bar{B}_{ij} = \begin{bmatrix} D_{1i} & G_i \\ \bar{\gamma} B_{fj} D_{2i} & 0 \\ \bar{\gamma} D_{2i} & 0 \end{bmatrix}, \quad \hat{B}_{ij} = \begin{bmatrix} 0 & 0 \\ B_{fj} D_{2i} & 0 \\ D_{2i} & 0 \end{bmatrix} \tag{6.42}$$

with $\tilde{\alpha}_m(k) = \alpha_m(k) - \bar{\alpha}_m$, $\tilde{\beta}(k) = \beta(k) - \bar{\beta}$, and $\tilde{\gamma}(k) = \hat{\gamma}(k) - \bar{\gamma}$. It is clear that $\mathbb{E}\{\tilde{\alpha}_m(k)\} = 0$, $\mathbb{E}\{\tilde{\beta}(k)\} = 0$, $\mathbb{E}\{\tilde{\gamma}(k)\} = 0$ and $\mathbb{E}\{\tilde{\alpha}_m^2(k)\} = \bar{\alpha}_m(1 - \bar{\alpha}_m)$, $\mathbb{E}\{\tilde{\beta}^2(k)\} = \bar{\beta}(1 - \bar{\beta})$, $\mathbb{E}\{\tilde{\gamma}^2(k)\} = \bar{\gamma}(1 - \bar{\gamma})$.

Definition 6.3.1   *With system (6.41) and every initial condition ψ, the fault detection dynamics in (6.41) is said to be exponentially mean-square stable if, in the case of $v(k) = 0$, there exist constants $\delta > 0$ and $0 < \kappa < 1$ such that*

$$\mathbb{E}\{\|\eta(k)\|^2\} \le \delta \kappa^k \sup_{i \in \mathbb{Z}^-} \mathbb{E}\{\|\psi(i)\|^2\}, \ \forall k \ge 0.$$

To this end, the fault detection problem to be addressed in this chapter can be described by the following two steps:

Step 1. *Generate a residual signal.* For system (6.33), design a fuzzy fault detection filter in the form of (6.39) to generate a residual signal $r(k)$. Furthermore, the filter is designed so that the overall fault detection system (6.41) is exponentially mean-square stable with the following H_∞ performance constraint under zero initial condition:

$$\sum_{k=0}^{\infty} \mathbb{E}\{\|e(k)\|^2\} \le \gamma^2 \sum_{k=0}^{\infty} \|v(k)\|^2, \tag{6.43}$$

where $v(k) \ne 0$, and $\gamma > 0$ is made as small as possible in the feasibility of (6.43).

Step 2. *Set up a fault detection measure.* We adopt a residual evaluation stage including an evaluation function $J(k)$ and a threshold J_{th} of the following form:

$$J(k) = \left\{ \sum_{k=s-\mathcal{L}}^{k=s} r^T(k) r(k) \right\}^{1/2}, \quad J_{th} = \sup_{w \in l_2, f=0} \mathbb{E}\{J(k)\}, \tag{6.44}$$

where $\mathcal{L}$ denotes the length of the finite evaluating time-horizon. Based on (6.44), the occurrence of faults can be detected by comparing $J(k)$ with J_{th} according to the following rule:

$$J(k) > J_{th} \Longrightarrow \text{with faults} \Longrightarrow \text{alarm,}$$

$$J(k) \le J_{th} \Longrightarrow \text{no faults.}$$

6.3.2   Main Results

For convenience of presentation, we first discuss the nominal system of (6.41) (that is, without parameter uncertainty ΔA_i) and will eventually extend our main results to the general case. We have the following analysis result that serves as a theoretical basis for the subsequent design problem.

Theorem 6.3.2   *Consider the nominal fuzzy system of (6.33) with given filter parameters and a prescribed H_∞ performance $\gamma > 0$. The nominal fuzzy fault detection system in (6.41) is exponentially mean-square stable with a disturbance attenuation level γ if there exist matrices $P > 0$, $Q_k > 0$ $(k = 1, 2, \ldots, h)$, and $R > 0$ satisfying*

$$\Psi_{ii}^{\mathrm{T}} \check{P} \Psi_{ii} + \hat{\Psi}_{ii}^{\mathrm{T}} \check{P} \hat{\Psi}_{ii} + \bar{P}_{ii} < 0, \tag{6.45}$$

$$(\Psi_{ij} + \Psi_{ji})^{\mathrm{T}} \check{P} (\Psi_{ij} + \Psi_{ji}) + (\hat{\Psi}_{ij} + \hat{\Psi}_{ji})^{\mathrm{T}} \check{P} (\hat{\Psi}_{ij} + \hat{\Psi}_{ji}) + 2(\bar{P}_{ij} + \bar{P}_{ji}) < 0, \tag{6.46}$$

where

$$\check{A}_{ij} = \begin{bmatrix} g\hat{A}_{ij} & 0 & 0 & g\hat{B}_{ij} \end{bmatrix}^{\mathrm{T}}, \quad \hat{\Psi}_{ij} = \begin{bmatrix} \check{A}_{ij}^{\mathrm{T}} & \check{C}_{ij}^{\mathrm{T}} \end{bmatrix}^{\mathrm{T}}, \quad g^2 = \bar{\gamma}(1 - \bar{\gamma}),$$

$$\bar{A}_{ij} = \begin{bmatrix} \bar{A}_{ij} & \hat{Z}_{1mi} & \bar{A}_{d2i} & \bar{B}_{ij} \end{bmatrix}, \quad C_{ij} = \begin{bmatrix} \bar{C}_{ij} & 0 & 0 & \bar{D}_{ij} \end{bmatrix}, \quad \Psi_{ij} = \begin{bmatrix} \bar{A}_{ij}^{\mathrm{T}} & C_{ij}^{\mathrm{T}} \end{bmatrix}^{\mathrm{T}},$$

$$\check{C}_{ij} = \begin{bmatrix} g\hat{C}_{ij} & 0 & 0 & g\hat{D}_{ij} \end{bmatrix}^{\mathrm{T}}, \quad \check{P} = \mathrm{diag}\{P, I\}, \quad \hat{P}_{ij} = \mathrm{diag}\{\bar{Q}_k, \mathcal{F}_{ij}, \check{A}_{d2ij}\},$$

$$\bar{P}_{ij} = \mathrm{diag}\{\hat{P}_{ij}, -\gamma^2 I\}, \quad \tilde{P} = I_h \otimes P, \quad \hat{Z}_{1mi} = \begin{bmatrix} \bar{A}_{d11i} & \cdots & \bar{A}_{d1hi} \end{bmatrix},$$

$$\bar{Q}_k = \sum_{k=1}^{h}(d_{\max} - d_{\min} + 1)Q_k + \bar{\mu}R - P, \quad \check{A}_{d2ij} = g_\beta^2 \hat{A}_{d2i}^{\mathrm{T}} P \hat{A}_{d2j} - \frac{1}{\bar{\mu}}R,$$

$$\mathcal{F}_{ij} = \check{A}_{d1i}^{\mathrm{T}} \tilde{P} \check{A}_{d1j} - \hat{Q}, \quad \hat{Q} = \mathrm{diag}\{Q_1, \ldots, Q_h\}, \quad \check{A}_{d1i} = \hat{g}_m \otimes \hat{A}_{d1i},$$

$$\hat{g}_m^2 = \mathrm{diag}\{\bar{\alpha}_1(1 - \bar{\alpha}_1), \ldots, \bar{\alpha}_h(1 - \bar{\alpha}_h)\}, \quad g_\beta^2 = \bar{\beta}(1 - \bar{\beta}). \tag{6.47}$$

Having established the analysis results, we are now in a position to deal with the fuzzy fault detection filter design problem.

Theorem 6.3.3   *Consider the nominal fuzzy system of (6.41) and let $\gamma > 0$ be a given scalar. A desired full-order fault detection filter of the form (6.39) exists if there exist matrices $P > 0$, $R > 0$, and $Q_k > 0$ $(k = 1, 2, \ldots, h)$ and matrices X_i and K_i satisfying*

$$\Omega_1 = \begin{bmatrix} \bar{P}_{ii} & * \\ \Gamma_{ii} & -\check{\mathcal{P}} \end{bmatrix} < 0, \quad (i = 1, 2, \ldots, r), \tag{6.48}$$

$$\Omega_2 = \begin{bmatrix} 2(\bar{P}_{ij} + \bar{P}_{ij}) & * \\ \Gamma_{ij} + \Gamma_{ji} & -\check{\mathcal{P}} \end{bmatrix} < 0, \quad (1 \leq i < j \leq r), \tag{6.49}$$

where

$$\Gamma_{11ij} = \begin{bmatrix} P\hat{A}_{0i} + X_i \hat{R}_{1j} & P\hat{Z}_{1mi} \\ K_i \hat{R}_{1j} & 0 \end{bmatrix}, \quad \Gamma_{12ij} = \begin{bmatrix} P\bar{A}_{d2i} & P\bar{B}_{0i} + \bar{\gamma} X_i \hat{R}_{2j} \\ 0 & \hat{E}_0 + \bar{\gamma} K_i \hat{R}_{2j} \end{bmatrix},$$

$$\Gamma_{21ij} = \begin{bmatrix} gP\hat{A}_{1i} + gX_i \hat{R}_{4j} & 0 \\ gK_i \hat{R}_{4j} & 0 \end{bmatrix}, \quad \Gamma_{22ij} = \begin{bmatrix} 0 & gP\hat{D}_{0i} + gX_i \hat{R}_{2j} \\ 0 & gK_i \hat{R}_{2j} \end{bmatrix}, \quad \check{\mathcal{P}} = I_2 \otimes \check{P},$$

$$\hat{R}_{1j} = \begin{bmatrix} 0 & I & 0 \\ \bar{\gamma}C_j & 0 & (1-\bar{\gamma})I \end{bmatrix}, \quad \Gamma_{ij} = \begin{bmatrix} \Gamma_{11ij} & \Gamma_{12ij} \\ \Gamma_{21ij} & \Gamma_{22ij} \end{bmatrix}, \quad \hat{R}_{4j} = \begin{bmatrix} 0 & 0 & 0 \\ C_j & 0 & -I \end{bmatrix},$$

$$\hat{A}_{0i} = \begin{bmatrix} A_i & 0 & 0 \\ 0 & 0 & 0 \\ \bar{\gamma}C_i & 0 & (1-\bar{\gamma})I \end{bmatrix}, \quad \hat{A}_{1i} = \begin{bmatrix} 0 & 0 & 0 \\ 0 & 0 & 0 \\ C_i & 0 & -I \end{bmatrix}, \quad \hat{D}_{0i} = \begin{bmatrix} 0 & 0 \\ 0 & 0 \\ D_{2i} & 0 \end{bmatrix},$$

$$\bar{B}_{0i} = \begin{bmatrix} D_{1i} & G_i \\ 0 & 0 \\ \bar{\gamma}D_{2i} & 0 \end{bmatrix}, \quad \hat{R}_{2j} = \begin{bmatrix} 0 & 0 \\ D_{2j} & 0 \end{bmatrix}, \quad \hat{E}_0 = \begin{bmatrix} 0 & -I \end{bmatrix}, \quad \hat{E} = \begin{bmatrix} 0 & I & 0 \end{bmatrix}^{\mathrm{T}},$$

and $\bar{P}_{ii}$ and $\check{P}$ are defined in Theorem 6.2.2. Furthermore, if (P, R, Q_k, X_i, K_i) is a feasible solution of (6.48) and (6.49), then the fault detection filter parameters in the form of (6.39) are given as follows:

$$\begin{bmatrix} A_{\mathrm{f}i} & B_{\mathrm{f}i} \end{bmatrix} = (\hat{E}^{\mathrm{T}} P \hat{E})^{-1} \hat{E}^{\mathrm{T}} X_i, \quad \begin{bmatrix} C_{\mathrm{f}i} & D_{\mathrm{f}i} \end{bmatrix} = K_i. \tag{6.50}$$

In the following, the results obtained for nominal systems will be extended to fuzzy system with uncertainty described in (6.33).

Theorem 6.3.4   *Consider the uncertain fuzzy fault detection system (6.41) and let $\gamma > 0$ be a given scalar. A desired full-order fault detection filter of the form (6.39) exists if there exist matrices $P > 0$, $R > 0$, $Q_k > 0$ $(k = 1, 2, \ldots, h)$, matrices X_i and K_i, and positive constant scalars $\varepsilon_{ij} > 0$ satisfying*

$$\begin{bmatrix} \Omega_1 & * & * \\ \bar{H}_{ai}^{\mathrm{T}} & -\varepsilon_{ii}I & * \\ \varepsilon_{ii}\bar{E}_a & 0 & -\varepsilon_{ii}I \end{bmatrix} < 0, \quad (i = 1, 2, \ldots, r), \tag{6.51}$$

$$\begin{bmatrix} \Omega_2 & * & * \\ \bar{H}_{ai}^{\mathrm{T}} + \bar{H}_{aj}^{\mathrm{T}} & -\varepsilon_{ij}I & * \\ \varepsilon_{ij}\bar{E}_a & 0 & -\varepsilon_{ij}I \end{bmatrix} < 0, \quad (1 \le i < j \le r), \tag{6.52}$$

where

$$\bar{H}_{ai} = \begin{bmatrix} 0 & 0 & 0 | & 0 & |\hat{H}_{ai}^{\mathrm{T}} & 0| & 0 & 0 \end{bmatrix}^{\mathrm{T}}, \quad \overline{E}_a = \begin{bmatrix} \widehat{E}_a & 0 & 0| & 0 & |0 & 0| & 0 & 0 \end{bmatrix},$$

$$\hat{H}_{ai} = \begin{bmatrix} H_{ai}^{\mathrm{T}} & 0 & 0 \end{bmatrix}^{\mathrm{T}}, \quad \widehat{E}_a = \begin{bmatrix} E_a & 0 & 0 \end{bmatrix}, \tag{6.53}$$

and Ω_1 and Ω_2 are defined in Theorem 6.2.3. Moreover, if $(P, R, Q_k, X_i, K_i, \varepsilon_{ij})$ is a feasible solution of (6.51) and (6.52), then the fault detection filter parameters in the form of (6.39) are given as follows:

$$[A_{\mathrm{f}i} \quad B_{\mathrm{f}i}] = (\hat{E}^{\mathrm{T}} P \hat{E})^{-1} \hat{E}^{\mathrm{T}} X_i, \quad [C_{\mathrm{f}i} \quad D_{\mathrm{f}i}] = K_i. \tag{6.54}$$

Remark 6.5 *In Theorem 6.2.4, the fuzzy fault detection filter is designed such that the overall fault detection dynamics is exponentially stable in the mean square and, at the same time, the error between the residual signal and the fault signal is made as small as possible. Sufficient conditions are first established for the existence of the desired fuzzy fault detection filters, and then the corresponding solvability conditions for the desired filter gains are established. Also, the optimal performance index for the robust fuzzy fault detection problem addressed can be obtained by solving an auxiliary convex optimization problem. Note that the sufficient conditions involve the occurrence probabilities of the discrete time-delays, distributed time-delays, and packet dropouts, thereby reflecting the nature of the randomly occurring phenomena.*

6.4 Illustrative Examples

In this section, some simulation examples are presented to demonstrate the theory presented in this chapter.

6.4.1 Example 1

This example considers the design of quantized fault detection filters with randomly occurring nonlinearities and mixed tim-delays.

In this section, we aim to demonstrate the effectiveness and applicability of the proposed method. Following Gao *et al.* [26], we consider the networked fault detection problem for an industrial continuous-stirred tank reactor system, where chemical species A reacts to form species B. Figure 6.2 illustrates the physical structure of the system. Assuming that

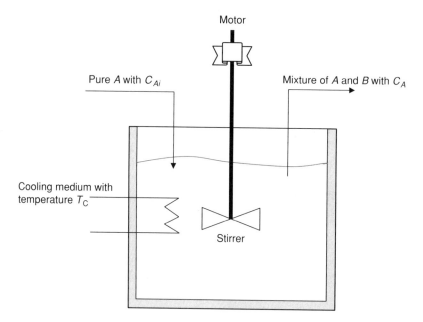

Figure 6.2   A continuous-stirred tank reactor model

the network-induced delays and randomly occurring nonlinearities exist in this system, a discrete-space model is obtained as

$$x(k+1) = Ax(k) + Bu(k) + A_{d1}\sum_{i=1}^{2}\alpha_i(k)x(k - \tau_i(k)) + \gamma(k)g(k, x(k)) + D_1w(k),$$

$$y(k) = Cx(k) + D_2w(k),$$

where the state variables are chosen as $x_1 = C_A$ and $x_2 = T_C$, the input variables are chosen as $u_1 = T$ and $u_2 = C_{Ai}$, and C_A, T_C, T, C_{Ai} are, respectively, the output concentration of chemical species A, the cooling medium temperature, the reaction temperature, and the input concentration of a key reactant A. Our purpose is to detect the fault appearing on the cooling medium temperature T_C. Therefore, the above system can be represented in the form of (6.1) with matrices given by

$$A = \begin{bmatrix} 0.9719 & -0.0013 \\ -0.0340 & 0.8628 \end{bmatrix}, \quad A_{d1} = \begin{bmatrix} 0.14 & 0.2 \\ 0 & 0.2 \end{bmatrix}, \quad D_1 = \begin{bmatrix} 0.1 & 0 \\ 0 & 0.3 \end{bmatrix}, \quad A_{d2} = 0,$$

$$C = [1 \quad 0.1], \quad D_2 = [0 \quad 0.1], \quad G = [-0.0839 \quad 0.0761]^{\mathrm{T}}, \quad B = 0, \quad H = 0.$$

Let the time-varying communication delays satisfy $1 \le \tau_i(k) \le 3$ ($i = 1, 2$) and assume that $\bar{\alpha}_1 = \mathbb{E}\{\alpha_1(k)\} = 0.9$, $\bar{\alpha}_2 = \mathbb{E}\{\alpha_1(k)\} = 0.7$, and $\bar{\gamma} = \mathbb{E}\{\gamma(k)\} = 0.8$. The nonlinear function $g(k, x(k))$ is selected as $g(k, x(k)) = 0.5x_1(k)\sin(x_2(k))$. It is easy to see that the constraint (6.2) is met with $\varepsilon(k) = 1$ and $E(k) = \mathrm{diag}\{0.2, 0.15\}$. For the measurement quantization, the parameters of the logarithmic quantizer are set as $\hat{\mu}_0 = 2$ and $\chi = 0.8$. Then, the fault detection filter parameters can be obtained from Theorem 6.2.5 as follows:

$$A_F = \begin{bmatrix} -0.3276 & 0.2003 \\ -0.2621 & -0.1353 \end{bmatrix}, \quad B_F = \begin{bmatrix} -0.0057 \\ -0.0027 \end{bmatrix}, \quad C_F = [-0.2984 \quad -0.0015],$$

$$D_F = 0.0063,$$

and the optimal performance index given in (6.8) is $\gamma^* = 1.0007$.

It is worth noting that the optimal performance index γ^* obtained will change as the values of $\bar{\alpha}_1$, $\bar{\alpha}_2$, and $\bar{\gamma}$ change. Letting $\bar{\alpha}_1 = 0.9$, for different combinations of $\bar{\alpha}_2$ and $\bar{\gamma}$, the corresponding optimal performance indices γ^* are shown in Table 6.1. It can be concluded from Table 6.1 that the optimal trade-off between the robustness and sensitivity is affected not only by the randomly occurring communication time-delays but also by the randomly occurring nonlinearities.

Table 6.1 Optimal performance index

γ^*	$\bar{\alpha}_2 = 0.9$	$\bar{\alpha}_2 = 0.7$	$\bar{\alpha}_2 = 0.5$
$\bar{\gamma} = 0.8$	1.0004	1.0007	1.0011
$\bar{\gamma} = 0.6$	1.0010	1.0012	1.0106

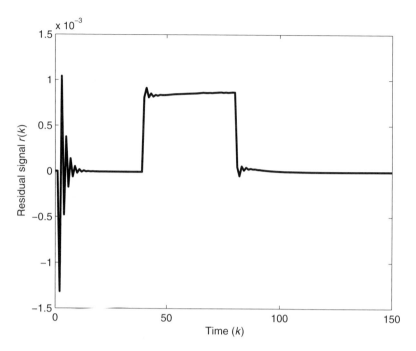

Figure 6.3   Residual signal without $w(k)$

To further illustrate the effectiveness of the fault detection filter design, for $k = 0, 1, \ldots, 150$, let the fault signal $f(k)$ be given as:

$$f(k) = \begin{cases} 1, & 40 \leq k \leq 80, \\ 0, & \text{else.} \end{cases} \tag{6.55}$$

First, in the case that the external disturbance is $w(k) = 0$, the residual response $r(k)$ and evolution of residual evaluation function $J(k)$ are shown in Figure 6.3 and Figure 6.4, respectively, which indicate that the filter design can detect the fault effectively when it occurs.

Next, assume that the disturbance is given by

$$w(k) = \begin{cases} [\,\text{rand}[0, 1] \quad 1.2 \quad \text{rand}[0, 1]\,]^{\mathrm{T}}, & 0 \leq k \leq 50, \\ 0, & \text{else,} \end{cases} \tag{6.56}$$

where the "rand" function generates arrays of random numbers whose elements are uniformly distributed on the interval [0, 1]. The residual response $r(k)$ and evolution of residual evaluation function $J(k)$ are shown in Figure 6.5 and Figure 6.6, respectively. Selecting a threshold as $J_{\text{th}} = \sup_{f=0} \mathbb{E}\{\sum_{s=0}^{150} r'(s) r(s)\}^{1/2}$, after 200 runs of the simulations, we get an average value of $J_{\text{th}} = 0.0031$. From Figure 6.6, we can see that $0.0026 = J(45) < J_{\text{th}} < J(46) = 0.0034$, which means that the fault can be detected in six time steps after its occurrence. Therefore, it

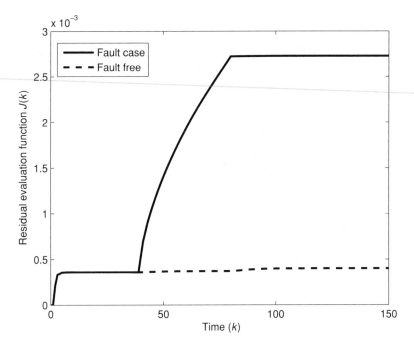

Figure 6.4 Evolution of $J(k)$ without $w(k)$

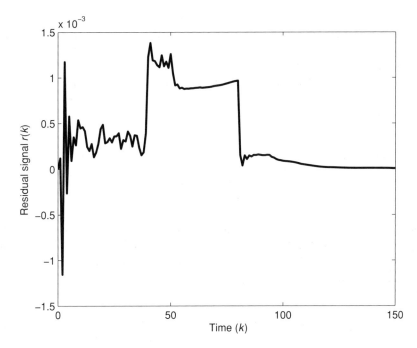

Figure 6.5 Residual signal with $w(k)$

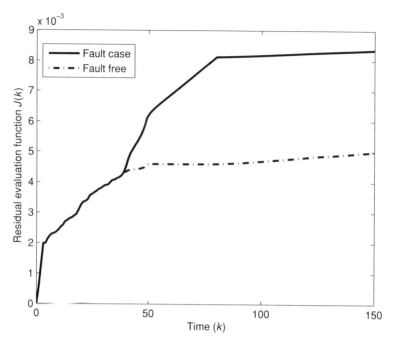

Figure 6.6   Evolution of $J(k)$ with $w(k)$

can be seen that not only can the residual reflect the fault in time, but it can also detect the fault without confusing it with the disturbance $w(k)$.

In summary, all the simulation results have further confirmed our theoretical analysis for the problem of quantized fault detection for networked systems with randomly occurring nonlinearities and mixed time-delays.

6.4.2   Example 2

In this example, we examine fuzzy-model-based robust fault detection with stochastic mixed time-delays and successive packet dropouts.

In this section, we use a nonlinear pendulum to demonstrate the effectiveness and applicability of the proposed method. The pendulum system is modified from Feng [175] by adding one damping term for stability of the system. It is assumed that two components of the system (that is, angle and angular velocity) are randomly perturbed by time delays. The equations of motion of the pendulum are described as follows:

$$\dot{\theta}(t) = \lambda\bar{\theta}(t) + \alpha_1(t)(1-\lambda)\bar{\theta}(t - d(t)) + \alpha_2(t)(1-\lambda)$$
$$\times \theta(t - d(t)) + \beta(t)\int_{-\infty}^{t} \bar{\lambda}(t - s)\theta(s)\,ds,$$

$$\dot{\bar{\theta}}(t) = -\frac{g\sin(\theta(t)) + (b/lm)[\lambda\bar{\theta}(t) + (1-\lambda)\bar{\theta}(t-d(t))]}{\frac{2}{3}l - \frac{a}{2}ml\cos^2(\theta(t))} \tag{6.57}$$

$$-\frac{(aml/4)[\lambda\bar{\theta}(t) + (1-\lambda)\bar{\theta}(t-d(t))]^2\sin(2\theta(t))}{\frac{2}{3}l \quad \frac{a}{2}ml\cos^2(\theta(t))} + w_1(t),$$

$$y(t) = \sin(\theta(t)) + \lambda\bar{\theta}(t) + w_2(t),$$

where θ denotes the angle of the pendulum from the vertical, $\bar{\theta}$ is the angular velocity, $g = 9.8\,\text{m/s}^2$ is the acceleration due to gravity, m is the mass of the pendulum, $a = 1/(m+M)$, M is the mass of the cart, l is the length of the pendulum, b is the damping coefficient of the pendulum around the pivot, and w_1 and w_2 are the disturbances applied to the cart and measurement noise, respectively. In this simulation, the pendulum parameters are chosen as $m = 2\,\text{kg}$, $M = 8\,\text{kg}$, $l = 0.5\,\text{m}$, and $b = 0.7\,\text{Nm/s}$, and the retarded coefficient $\lambda = 0.6$.

Letting $x_1(t) = \theta(t)$, $x_2(t) = \bar{\theta}(t)$, we linearize the plant around the origin $x = (\pm\pi/2)$ and $x = (\pm\pi/3)$, and consider the differences between the linearized local model and the original nonlinear model as the uncertainties. By discretizing the plant with a sampling period 0.05 s, we obtain the following discrete-time T–S fuzzy model:

$$x(k+1) = \sum_{i=1}^{3} h_i(\theta(k))\left[(A_i + \Delta A_i(k))x(k) + A_{d1i}\sum_{m=1}^{h}\alpha_m(k)x(k - \tau_m(k))\right.$$

$$\left. + \beta(k)A_{d2i}\sum_{d=1}^{\infty}\mu_d x(k-d) + D_{1i}w(k)\right],$$

$$y(k) = \sum_{i=1}^{3} h_i(\theta(k))[C_i x(k) + D_{2i}w(k)].$$

The model parameters are given as follows:

$$A_1 = \begin{bmatrix} 1.000 & 0.0450 \\ 0.8558 & 0.7894 \end{bmatrix}, \quad A_2 = \begin{bmatrix} 1.000 & 0.0450 \\ 0.6315 & 0.8018 \end{bmatrix}, \quad A_3 = \begin{bmatrix} 1.000 & 0.0450 \\ -0.4679 & 0.8055 \end{bmatrix},$$

$$A_{d11} = A_{d12} = A_{d13} = \begin{bmatrix} 0.14 & 0.02 \\ 0 & 0.094 \end{bmatrix}, \quad A_{d21} = A_{d22} = A_{d23} = \begin{bmatrix} 0 & 0.12 \\ 0.1 & 0.02 \end{bmatrix},$$

$$D_{11} = D_{12} = D_{13} = \begin{bmatrix} 0 & 1 \end{bmatrix}^{\text{T}}, \quad C_1 = \begin{bmatrix} 0.9949 & 0.9 \end{bmatrix}, \quad C_2 = \begin{bmatrix} 0.8270 & 0.9 \end{bmatrix},$$

$$C_3 = \begin{bmatrix} 0.6366 & 0.9 \end{bmatrix}, \quad H_{a1} = H_{a2} = H_{a3} = \begin{bmatrix} 0.2 & 0.01 \end{bmatrix}^{\text{T}}, \quad E_a = \begin{bmatrix} 0 & 2 \end{bmatrix},$$

$$F(k) = \sin(k), \quad D_{21} = D_{22} = D_{23} = 1,$$

and the membership functions are shown in Figure 6.7. Assume that the time-varying communication delays $\tau_1(k)$ and $\tau_2(k)$ are random variables whose elements are uniformly distributed on the interval $[2, 6]$, and

$$\bar{\alpha}_1 = \mathbb{E}\{\alpha_1(k)\} = 0.8, \quad \bar{\alpha}_2 = \mathbb{E}\{\alpha_2(k)\} = 0.6, \quad \bar{\beta} = \mathbb{E}\{\beta(k)\} = 0.9.$$

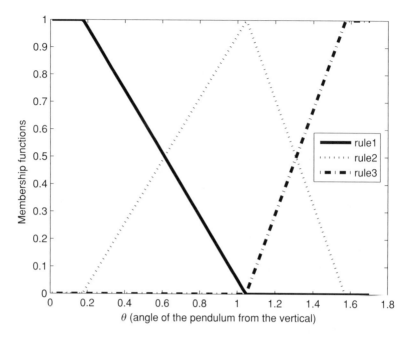

θ (angle of the pendulum from the vertical)

Figure 6.7 Membership functions

Choosing the constants $\mu_d = 2^{-3-d}$, we can easily find that $\bar{\mu} = \sum_{d=1}^{\infty} \mu_d = 2^{-3} < \sum_{d=1}^{\infty} d\mu_d = 2^{-2} < \infty$, which satisfies the convergence condition (6.36).

Assume that there are faults on the angular velocity of the pendulum, with fault matrices given by

$$G_1 = \begin{bmatrix} 0.9887 & -0.0180 \end{bmatrix}^{\mathrm{T}}, \quad G_2 = \begin{bmatrix} 0.9033 & -0.0172 \end{bmatrix}^{\mathrm{T}}, \quad G_3 = \begin{bmatrix} 0.6237 & 0.0180 \end{bmatrix}^{\mathrm{T}}.$$

Let the probability of $\hat{\gamma}(k)$ be given by $\bar{\gamma} = 0.7$. Applying Theorem 6.3.4, we can obtain the desired $\mathcal{H}_{\infty}$ filter parameters as follows:

$$A_{f1} = \begin{bmatrix} -0.3879 & -0.4043 \\ -0.3840 & 0.4032 \end{bmatrix}, \quad A_{f2} = \begin{bmatrix} 0.4279 & -0.4243 \\ -0.4840 & -0.5132 \end{bmatrix},$$

$$A_{f3} = \begin{bmatrix} 0.3868 & 0.4093 \\ 0.5420 & 0.5132 \end{bmatrix}, \quad B_{f1} = \begin{bmatrix} -0.4690 & -0.4690 \end{bmatrix}^{\mathrm{T}},$$

$$B_{f2} = \begin{bmatrix} 0.5679 & 0.4420 \end{bmatrix}^{\mathrm{T}}, \quad B_{f3} = \begin{bmatrix} -0.3868 & -0.3420 \end{bmatrix}^{\mathrm{T}},$$

$$C_{f1} = \begin{bmatrix} -0.7846 & -0.6585 \end{bmatrix}, \quad C_{f2} = \begin{bmatrix} -0.7579 & -0.5664 \end{bmatrix},$$

$$C_{f3} = \begin{bmatrix} -0.5052 & 0.4335 \end{bmatrix}, \quad D_{f1} = -3.5656, \quad D_{f2} = -1.3585, \quad D_{f3} = -0.1792,$$

with the optimized performance index $\gamma^* = 1.1598$.

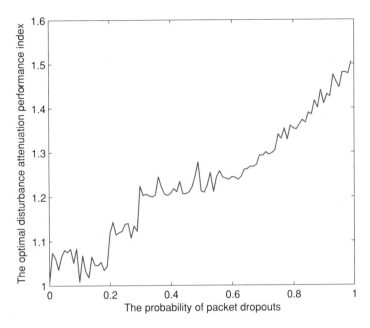

Figure 6.8 The probability of packet dropouts versus the optimal performance γ^*

Now, let us show how the probabilities in the measurement equation (6.38) affect the H_∞ performance of the fault detection filtering process. Figure 6.8 provides the plot of the average optimal disturbance attenuation level γ^* versus the probability of packet dropouts after 100 Monte Carlo simulations. It can be seen clearly that a better performance can be achieved with less missing measurements.

To further illustrate the effectiveness of the fault detection filter design, for $k = 0$, $1, \ldots, 150$, let the fault signal $f(k)$ be given as

$$f(k) = \begin{cases} 1, & 50 \le k \le 100, \\ 0, & \text{else.} \end{cases} \tag{6.58}$$

First, in the case that the initial conditions $\psi(k)$, $\forall k \in \mathbb{Z}^-$, $\psi \in \mathbb{R}^2$ are 200 random state vectors whose elements are uniformly distributed on the interval $[0, 0.1]$, $\tau_1(0) = 3$, $\tau_2(0) = 4$, $x(0) = [\pi/8 \quad 0]^T$, $\hat{x}(0) = [0 \quad 0]^T$, $y_f(-1) = 0$, $T = 20$, and the external disturbance is $w(k) = 0$. The residual signal $r(k)$ and evolution of residual evaluation function $J(k)$ are shown in Figure 6.9 and Figure 6.10, respectively, which indicate that the filter design can detect the fault effectively when it occurs.

Next, assume that the disturbance is given by

$$w(k) = \begin{cases} 0.5 \times \text{rand}[0, 1], & 30 \le k \le 130, \\ 0, & \text{else,} \end{cases} \tag{6.59}$$

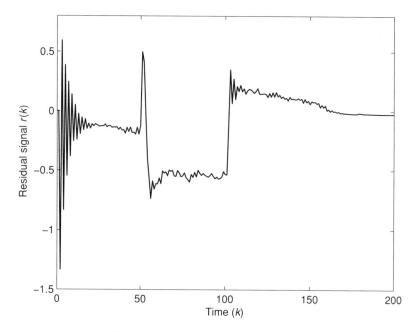

Figure 6.9   Residual signal without $w(k)$

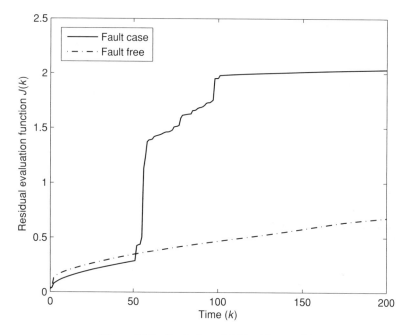

Figure 6.10   Evolution of $J(k)$ without $w(k)$

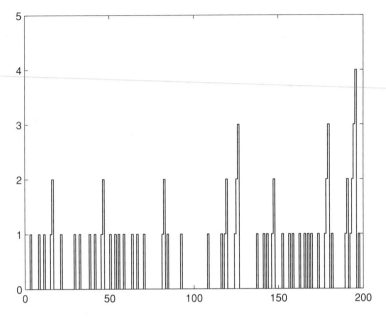

Figure 6.11   The distribution of the packet dropout number under $\bar{\gamma} = 0.7$

where the rand function generates arrays of random numbers whose elements are uniformly distributed on the interval [0, 1].

The rand distribution of successive packet dropout numbers is shown in Figure 6.11. From this we can see that, if the number on the Y-axis is 0, this means that the current measurement output of the physical plant is transmitted to the fault detection filter successfully. Furthermore, when the number is i ($i = 1, 2, \ldots$), this means that we have experienced i successive packet dropouts and the received measurement at the time $k - i$ will be used for the current estimation. The residual signal $r(k)$ and evolution of residual evaluation function $J(k)$ are shown in Figure 6.12 and Figure 6.13, respectively. It can be seen that the residual not only can reflect the fault in time, but also detect the fault without confusing it with the disturbance $w(k)$.

Remark 6.6   *In the simulation, we increase the magnitude of $w(k)$ in (6.59) with the hope of seeing how a larger disturbance would influence the performance of the fault detection filter. For example, we take $w(k)$ as $1 \times$ rand[0, 1] and $2 \times$ rand[0, 1], and then show the corresponding evolutions of residual evaluation function $J(k)$ in Figure 6.14 and Figure 6.15, respectively. For simulation purposes, the threshold is selected as $J_{\text{th}} = \sup_{f=0} \mathbb{E}\{\sum_{k=0}^{200} r^T(k)r(k)\}^{1/2}$ and, accordingly, it can be obtained that $J_{\text{th}} = 1.2643$ in Figure 6.14 after 200 Monte Carlo simulations with no faults. From Figure 6.14, it can be seen that $1.1036 = J(111) < J_{\text{th}} < J(112) = 1.3657$, which means that the fault can be detected in 12 time steps after its occurrence. Similarly, we can conclude from Figure 6.15 that the fault can be detected in 21 time steps after its occurrence. From simulation results, it can be clearly observed that the smaller $w(k)$ we have, the smaller the threshold we obtain and the faster the fault detection will take.*

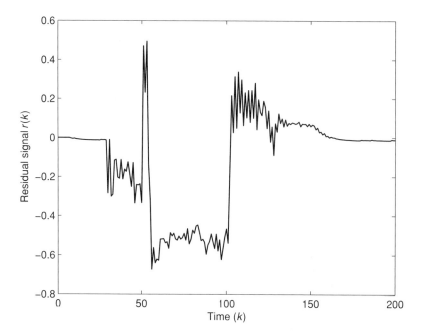

Figure 6.12 Residual signal with $w(k)$

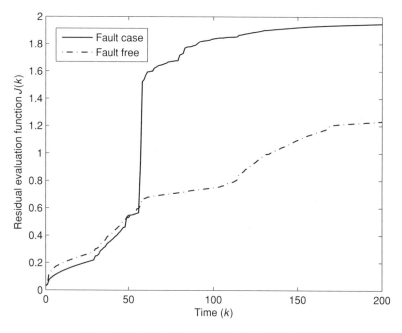

Figure 6.13 Evolution of $J(k)$ with $w(k)$ in (6.59)

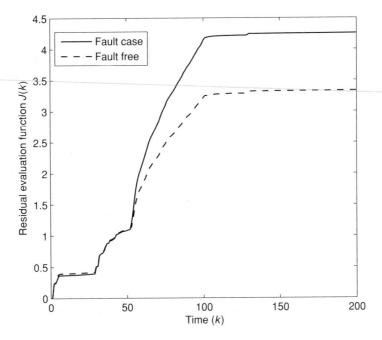

Figure 6.14   Evolution of $J(k)$ with $w(k) = 1 \times \text{rand}[0, 1]$, $30 \leq k \leq 130$

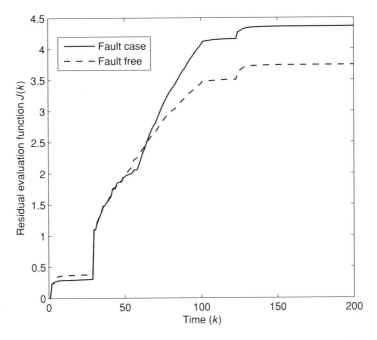

Figure 6.15   Evolution of $J(k)$ with $w(k) = 2 \times \text{rand}[0, 1]$, $30 \leq k \leq 130$

6.5   Summary

In this chapter, the fault detection problems have been dealt with for two class of discrete-time nonlinear systems with randomly occurring mixed time-delays, successive packet dropouts, and measurement quantizations. The mixed time-delays involve both multiple time-varying discrete delays and infinite distributed delays. The successive packet dropouts are modeled by a stochastic variable satisfying the Bernoulli random binary distribution. The fault detection problem was first addressed for a class of discrete-time systems with randomly occurring nonlinearities, mixed stochastic time-delays, and measurement quantizations. A fault detection filter was designed such that the overall fault detection dynamics is exponentially stable in the mean square and, at the same time, the error between the residual signal and the fault signal is made as small as possible. Sufficient conditions have been established via intensive stochastic analysis for the existence of the desired fault detection filters, and then the explicit expression of the desired filter gains has been derived by means of the feasibility of certain matrix inequalities. Also, the optimal performance index for the addressed fault detection problem was obtained by solving an auxiliary convex optimization problem. Moreover, the robust fault detection problem was investigated for a class of uncertain discrete-time T–S fuzzy systems comprising randomly occurring mixed time-delays and successive packet dropouts. Then, some parallel results were also derived by using similar analysis techniques. Two practical examples have been provided to show the usefulness and effectiveness of the proposed design methods.

7

Distributed Filtering over Sensor Networks with Saturations

This chapter is concerned with the distributed H_∞ filtering problem for a class of nonlinear systems with randomly occurring sensor saturations (ROSSs) and successive packet dropouts over sensor networks. The issue of ROSSs is brought up to account for the random nature of sensor saturations in a networked environment of sensors and, accordingly, a novel sensor model is proposed to describe both the ROSSs and successive packet dropouts within a unified framework. Two sets of Bernoulli-distributed white sequences are introduced to govern the random occurrences of the sensor saturations and successive packet dropouts. Through available output measurements from not only the individual sensor but also its neighboring sensors, a sufficient condition is established for the desired distributed filter to ensure that the filtering dynamics is exponentially mean-square stable and the prescribed H_∞ performance constraint is satisfied. The solution of the distributed filter gains is characterized by solving an auxiliary convex optimization problem. Finally, a simulation example is provided to show the effectiveness of the proposed filtering scheme.

7.1 Problem Formulation

Consider the filter configuration with n sensors as shown in Figure 7.1, where each sensor can receive information from both the plant and its neighboring sensors. The information received by sensor i from the plant is transmitted via communication cables which are of limited capacity and, therefore, may suffer from the phenomena of ROSSs and packet dropouts. On the other hand, sensor i can also obtain information from its neighboring sensors according to the topology of the sensor network.

In this chapter, we assume that the n sensor nodes are distributed in space according to a fixed network topology represented by a directed graph $\mathcal{G} = (\mathcal{V}, \mathcal{E}, \mathcal{A})$ of order n with the set of nodes $\mathcal{V} = 1, 2, \dots, n$, the set of edges $\mathcal{E} \in \mathcal{V} \times \mathcal{V}$, and the weighted adjacency matrix $\mathcal{A} = [a_{ij}]$ with nonnegative adjacency element a_{ij}. An edge of $\mathcal{G}$ is denoted by ordered pair (i, j). The adjacency elements associated with the edges of the graph are positive (i.e.,

Filtering, Control and Fault Detection with Randomly Occurring Incomplete Information, First Edition.
Hongli Dong, Zidong Wang, and Huijun Gao.
© 2013 John Wiley & Sons, Ltd. Published 2013 by John Wiley & Sons, Ltd.

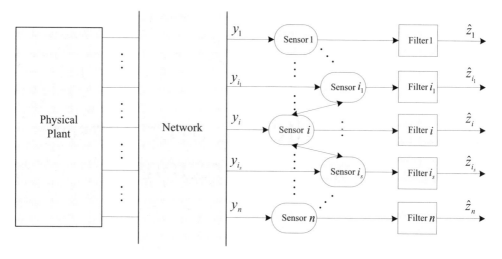

Figure 7.1   The filtering problem over sensor networks

$a_{ij} > 0 \Longleftrightarrow (i, j) \in \mathcal{E}$), which means that sensor i can obtain information from sensor j. Also, we assume that $a_{ii} = 1$ for all $i \in \mathcal{V}$; therefore, (i, i) can be regarded as an additional edge. The set of neighbors of node $i \in \mathcal{V}$ plus the node itself are denoted by $\mathcal{N}_i = \{j \in \mathcal{V} : (i, j) \in \mathcal{E}\}$.

Consider a discrete-time nonlinear system

$$\begin{cases} x(k + 1) = Ax(k) + f(x(k)) + Bw(k), \\ \quad\quad z(k) = Mx(k), \end{cases} \tag{7.1}$$

where $x(k) \in \mathbb{R}^{n_x}$ represents the state vector which cannot be observed directly, $z(k) \in \mathbb{R}^{n_z}$ is the output to be estimated, and $w(k) \in \mathbb{R}^{n_w}$ is the disturbance input belonging to $l_2[0, \infty)$. The nonlinear function $f(x(\cdot))$ satisfies the following condition:

$$\| f(x(\cdot)) \|^2 \leq \| Ex(\cdot) \|^2, \tag{7.2}$$

where E is a known constant matrix.

In this chapter, the n sensors with both saturation and packet dropouts are modeled by

$$\begin{aligned} y_i(k) = &\beta_i(k)\sigma(C_ix(k)) + (1 - \beta_i(k))\gamma_i(k)C_ix(k) + (1 - \beta_i(k))(1 - \gamma_i(k)) \\ &\times y_i(k - 1) + D_iv(k), \quad i = 1, 2, \dots, n, \end{aligned} \tag{7.3}$$

where $y_i(k) \in \mathbb{R}^{n_y}$ is the measurement output measured by sensor i from the plant, $\beta_i(k)$ and $\gamma_i(k)$ $(i = 1, 2, \dots, n)$ are Bernoulli-distributed white sequences taking values on 0 and 1 with

$$\begin{cases} \text{Prob}\{\beta_i(k) = 1\} = \bar{\beta}_i \\ \text{Prob}\{\beta_i(k) = 0\} = 1 - \bar{\beta}_i \end{cases} \text{and} \quad \begin{cases} \text{Prob}\{\gamma_i(k) = 1\} = \bar{\gamma}_i \\ \text{Prob}\{\gamma_i(k) = 0\} = 1 - \bar{\gamma}_i \end{cases},$$

respectively, where $\bar{\beta}_i$, $\bar{\gamma}_i \in [0, 1]$ are known constants. Throughout the chapter, the stochastic variables $\beta_i(k)$ and $\gamma_i(k)$ are independent mutually in all i ($1 \leq i \leq n$).

The saturation function $\sigma(\cdot) : \mathbb{R}^{n_y} \mapsto \mathbb{R}^{n_y}$ is defined as

$$\sigma(\vartheta) = \begin{bmatrix} \sigma_1^{\mathrm{T}}(\vartheta_1) & \sigma_2^{\mathrm{T}}(\vartheta_2) & \cdots & \sigma_{n_y}^{\mathrm{T}}(\vartheta_{n_y}) \end{bmatrix}^{\mathrm{T}}, \tag{7.4}$$

with $\sigma_i(\vartheta_i) = \mathrm{sign}(\vartheta_i) \min \{\vartheta_{i,\max}, |\vartheta_i|\}$, where $\vartheta_{i,\max}$ is the ith element of the vector $\vartheta_{\max}$, the saturation level.

Definition 7.1.1   *[150] A nonlinearity $\Psi : \mathbb{R}^m \mapsto \mathbb{R}^m$ is said to satisfy a sector condition if*

$$(\Psi(\vartheta) - \mathcal{H}_1 \vartheta)^{\mathrm{T}} (\Psi(\vartheta) - \mathcal{H}_2 \vartheta) \leq 0, \quad \forall \vartheta \in \mathbb{R}^{n_y} \tag{7.5}$$

for some real matrices $\mathcal{H}_1, \mathcal{H}_2 \in \mathbb{R}^{n_y \times n_y}$, where $\mathcal{H} = \mathcal{H}_2 - \mathcal{H}_1$ is a symmetric positive-definite matrix. In this case, we say that Ψ belongs to the sector $[\mathcal{H}_1, \mathcal{H}_2]$.

As in Refs [88, 148, 176], assuming that there exist diagonal matrices $\underline{L}_i$ and $\overline{L}_i$ such that $0 \leq \underline{L}_i < I \leq \overline{L}_i$, the saturation function $\sigma(C_i x(k))$ in (7.3) can be decomposed into a linear and a nonlinear part as

$$\sigma(C_i x(k)) = \underline{L}_i C_i x(k) + \Psi(C_i x(k)), \tag{7.6}$$

where $\Psi(C_i x(k))$ is a nonlinear vector-valued function satisfying a sector condition with $\mathcal{H}_1 = 0$ and $\mathcal{H}_2 = L_i$, which can be described as follows:

$$\Psi^{\mathrm{T}}(C_i x(k))(\Psi(C_i x(k)) - L_i C_i x(k)) \leq 0, \tag{7.7}$$

where $L_i = \overline{L}_i - \underline{L}_i$.

Remark 7.1   *The proposed measurement model in (7.3) provides a novel unified framework to account for the phenomena of both ROSSs and successive packet dropouts. The stochastic variable $\beta_i(k)$ characterizes the random nature of sensor saturation and, on the other hand, the stochastic variable $\gamma_i(k)$ describes possible successive packet dropouts. It can be seen clearly from (7.3) that, if $\beta_i(k) = 1$, model (7.3) is reduced to the one with saturation; if $\beta_i(k) = 0$ and $\gamma_i(k) = 1$, model (7.3) specializes to the one with neither saturations nor packet dropouts (i.e., the sensor i works normally); if $\beta_i(k) = 0$ and $\gamma_i(k) = 0$, model (7.3) degenerates to the one with successive packet dropouts (i.e., the latest measurement received for sensor i will be used if the current measurement is lost during transmissions). Therefore, model (7.3) is comprehensive in that it takes into account both the probabilistic sensor saturations and probabilistic successive packet dropouts in an environment of sensor networks.*

As is well known, a key point in designing distributed filters for sensor networks is how to fuse the information available for the filter on the sensor node i both from the sensor i itself and from its neighbors. Keeping such a fact in mind, the following filter structure is adopted in this chapter on sensor node i:

$$\begin{cases} \hat{x}_i(k+1) = \sum_{j\in\mathcal{N}_i} a_{ij} K_{ij}\hat{x}_j(k) + \sum_{j\in\mathcal{N}_i} a_{ij} H_{ij} y_j(k), \\ \hat{z}_i(k) = M\hat{x}_i(k), \end{cases} \tag{7.8}$$

where $\hat{x}_i(k) \in \mathbb{R}^{n_x}$ is the state estimate on sensor node i and $\hat{z}_i(k) \in \mathbb{R}^{n_z}$ is the estimate of $z(k)$ from the filter on sensor node i. Here, matrices K_{ij} and H_{ij} are the filter parameters on node i to be determined. The initial values of filters $\hat{x}_i(0)$ $(i = 1, 2, \ldots, n)$ are assumed to be known vectors.

Remark 7.2 *The filter structure in (7.8) establishes the communications between sensor node i and its neighboring nodes, in which the sensor nodes are distributed over a spatial region. It is worth mentioning that (7.8) represents a quite general filter model structure. To see this, assuming that there is no communication between sensor node i and its neighboring nodes, the filter (7.8) can be reduced to*

$$\hat{x}_i(k+1) = K_{ii}\hat{x}_i(k) + H_{ii} y_i(k), \tag{7.9}$$

which has been widely adopted for filter design in the literature.

For convenience of later analysis, we denote

$$\hat{x}(k) = \begin{bmatrix} \hat{x}_1^\mathrm{T}(k) & \hat{x}_2^\mathrm{T}(k) & \cdots & \hat{x}_n^\mathrm{T}(k) \end{bmatrix}^\mathrm{T}, \quad \bar{x}(k) = \mathbf{1}_n \otimes x(k), \quad \bar{C}_i = (e_i e_i^\mathrm{T}) \otimes C_i,$$

$$\hat{z}(k) = \begin{bmatrix} \hat{z}_1^\mathrm{T}(k) & \hat{z}_2^\mathrm{T}(k) & \cdots & \hat{z}_n^\mathrm{T}(k) \end{bmatrix}^\mathrm{T}, \quad \bar{z}(k) = \mathbf{1}_n \otimes z(k), \quad \bar{B} = \mathbf{1}_n \otimes B,$$

$$\tilde{y}(k) = \begin{bmatrix} y_1^\mathrm{T}(k) & y_2^\mathrm{T}(k) & \cdots & y_n^\mathrm{T}(k) \end{bmatrix}^\mathrm{T}, \quad \bar{A} = I_n \otimes A, \quad \bar{M} = I_n \otimes M,$$

$$\tilde{D} = \begin{bmatrix} D_1^\mathrm{T} & D_2^\mathrm{T} & \cdots & D_n^\mathrm{T} \end{bmatrix}^\mathrm{T}, \quad g(\bar{x}(k)) = \mathbf{1}_n \otimes f(x(k)), \quad \bar{\Lambda}_\gamma = \mathrm{diag}\{\bar{\gamma}_1, \bar{\gamma}_2, \ldots, \bar{\gamma}_n\},$$

$$\tilde{C} = \mathrm{diag}\{C_1, C_2, \ldots, C_n\}, \quad \tilde{L} = \mathrm{diag}\{\underline{L_1}, \underline{L_2}, \ldots, \underline{L_n}\}, \quad \bar{\Lambda}_\beta = \mathrm{diag}\{\bar{\beta}_1, \bar{\beta}_2, \ldots, \bar{\beta}_n\},$$

$$\hat{L} = \mathrm{diag}\{L_1, L_2, \ldots, L_n\}, \quad \bar{E}_n = I_n \otimes E^\mathrm{T} E, \tag{7.10}$$

where

$$\bar{K} = [\bar{K}_{ij}]_{n\times n} \quad \text{with} \quad \bar{K}_{ij} = \begin{cases} a_{ij} K_{ij}, & i = 1, 2, \ldots, n; \quad j \in \mathcal{N}_i \\ 0, & i = 1, 2, \ldots, n; \quad j \notin \mathcal{N}_i \end{cases},$$

$$\bar{H} = [\bar{H}_{ij}]_{n\times n} \quad \text{with} \quad \bar{H}_{ij} = \begin{cases} a_{ij} H_{ij}, & i = 1, 2, \ldots, n; \quad j \in \mathcal{N}_i \\ 0, & i = 1, 2, \ldots, n; \quad j \notin \mathcal{N}_i \end{cases}. \tag{7.11}$$

Obviously, since $a_{ij} = 0$ when $j \notin \mathcal{N}_i$, $\bar{K}$ and $\bar{H}$ are two matrices that can be expressed as

$$\bar{K} \in \mathcal{T}_{n_x \times n_x}, \quad \bar{H} \in \mathcal{T}_{n_x \times n_y}, \tag{7.12}$$

where $\mathcal{T}_{p\times q} = \{\bar{U} = [U_{ij}] \in \mathbb{R}^{np\times nq} \mid U_{ij} \in \mathbb{R}^{p\times q}, U_{ij} = 0 \text{ if } j \notin \mathcal{N}_i\}.$

Letting $\eta(k) = [\,\bar{x}^T(k) \quad \hat{x}^T(k) \quad \tilde{y}^T(k-1)\,]^T$ and $\vec{z}(k) = \bar{z}(k) - \hat{z}(k)$, the following augmented system is obtained that governs the filtering dynamics for the sensor network:

$$\eta(k+1) = \vec{A}\eta(k) + \vec{g}(\vec{H}\eta(k)) + \hat{H}\sigma(\tilde{C}\vec{H}\eta(k)) + \vec{B}\vec{w}(k) + \sum_{i=1}^{n}(\beta_i(k) - \bar{\beta}_i)$$

$$\times \vec{E}_i\sigma(\tilde{C}\vec{H}\eta(k)) + \sum_{i=1}^{n}((1-\beta_i(k))\gamma_i(k) - (1-\bar{\beta}_i)\bar{\gamma}_i)\vec{E}_i\tilde{C}\vec{H}\eta(k)$$

$$+ \sum_{i=1}^{n}((1-\beta_i(k))(1-\gamma_i(k)) - (1-\bar{\beta}_i)(1-\bar{\gamma}_i))\vec{E}_i\vec{F}\eta(k),$$

$$\vec{z}(k) = \vec{M}\eta(k), \tag{7.13}$$

where

$$\vec{A} = \begin{bmatrix} \bar{A} & 0 & 0 \\ \bar{H}(I-\bar{\Lambda}_\beta)\bar{\Lambda}_\gamma\tilde{C} & \bar{K} & \bar{H}(I-\bar{\Lambda}_\beta)(I-\bar{\Lambda}_\gamma) \\ (I-\bar{\Lambda}_\beta)\bar{\Lambda}_\gamma\tilde{C} & 0 & (I-\bar{\Lambda}_\beta)(I-\bar{\Lambda}_\gamma) \end{bmatrix}, \quad \vec{g}(\vec{H}\eta(k)) = \begin{bmatrix} g(\bar{x}(k)) \\ 0 \\ 0 \end{bmatrix},$$

$$\vec{M} = \begin{bmatrix} \bar{M} & -\bar{M} & 0 \end{bmatrix}, \quad E_i = e_i e_i^T, \quad \vec{H} = \begin{bmatrix} I & 0 & 0 \end{bmatrix}, \quad \vec{F} = \begin{bmatrix} 0 & 0 & I \end{bmatrix}, \tag{7.14}$$

$$\vec{B} = \begin{bmatrix} \bar{B} & 0 \\ 0 & \bar{H}\tilde{D} \\ 0 & \tilde{D} \end{bmatrix}, \quad \vec{w}(k) = \begin{bmatrix} w(k) \\ v(k) \end{bmatrix}, \quad \vec{E}_i = \begin{bmatrix} 0 \\ \bar{H}E_i \\ E_i \end{bmatrix}, \quad \hat{H} = \begin{bmatrix} 0 \\ \bar{H}\bar{\Lambda}_\beta \\ \bar{\Lambda}_\beta \end{bmatrix}.$$

Also, from (7.2), (7.6), and (7.7), we have

$$\vec{g}^T(\vec{H}\eta(k))\vec{g}(\vec{H}\eta(k)) \le \eta^T(k)\vec{H}^T\bar{E}_n\vec{H}\eta(k), \tag{7.15}$$

$$\sigma(\tilde{C}\vec{H}\eta(k)) = \tilde{L}\tilde{C}\vec{H}\eta(k) + \Psi(\tilde{C}\vec{H}\eta(k)), \tag{7.16}$$

$$\Psi^T(\tilde{C}\vec{H}\eta(k))\left(\Psi(\tilde{C}\vec{H}\eta(k)) - \hat{L}\tilde{C}\vec{H}\eta(k)\right) \le 0, \tag{7.17}$$

where

$$\sigma(\tilde{C}\vec{H}\eta(k)) := [\,\sigma^T(C_1 x(k)) \quad \sigma^T(C_2 x(k)) \quad \cdots \quad \sigma^T(C_n x(k))\,]^T,$$

$$\Psi(\tilde{C}\vec{H}\eta(k)) := [\,\Psi^T(C_1 x(k)) \quad \Psi^T(C_2 x(k)) \quad \cdots \quad \Psi^T(C_n x(k))\,]^T.$$

Here, the notations σ and Ψ have been slightly abused to denote the vector-valued saturation functions and vector nonlinear functions of different dimensions, respectively.

Before proceeding further, we introduce the following definition.

Definition 7.1.2 *The augmented system in (7.13) is said to be exponentially mean-square stable if, with $\vec{w}(k) = 0$, there exist constants $\delta > 0$ and $0 < \kappa < 1$ such that*

$$\mathbb{E}\{\|\eta(k)\|^2\} \le \delta\kappa^k \mathbb{E}\{\|\eta(0)\|^2\}, \ \forall \ \eta(0) \in \mathbb{R}^n, \ k \in \mathbb{I}^+.$$

Our aim in this chapter is to design a filter of the form in (7.8) on each node i of the sensor network for system (7.1). In other words, we are going to find the filter parameters K_{ij} and H_{ij} such that the following two requirements are satisfied simultaneously:

- *Exponentially mean-square stability.* The zero-solution of the augmented system (7.13) with $\vec{w}(k) = 0$ is exponentially mean-square stable.
- *H_∞ performance.* Under zero initial conditions, for a given disturbance attenuation level $\gamma > 0$ and all nonzero $\vec{w}(k)$, the filtering error $\vec{z}(k)$ from (7.13) satisfies the following condition:

$$\sum_{k=0}^{\infty} \mathbb{E}\{\|\vec{z}(k)\|^2\} \leq \gamma^2 \sum_{k=0}^{\infty} \|\vec{w}(k)\|^2. \tag{7.18}$$

7.2 Main Results

In this section, we investigate both the filter analysis and design problems for the distributed H_∞ filtering of system (7.1) with n sensors whose topology is determined by the given graph $\mathcal{G} = (\mathcal{V}, \mathcal{E}, \mathcal{A})$. The following lemma will be needed in establishing our main results.

Lemma 7.2.1 [143] Let $P = \text{diag}\{P_1, P_2, \ldots, P_n\}$ with $P_i \in \mathbb{R}^{p \times p}$ ($1 \leq i \leq$) being invertible matrices. If $X = PW$ for $W \in \mathbb{R}^{np \times nq}$, then we have $W \in \mathcal{T}_{p \times q} \Longleftrightarrow X \in \mathcal{T}_{p \times q}$.

The following theorem gives a sufficient condition under which the augmented system (7.13) is exponentially mean-square stable in the sense of Definition 7.2.3 with H_∞ performance constraint given in (7.18).

Theorem 7.2.2 *For given filter parameters K_{ij}, H_{ij} and a prescribed H_∞ index $\gamma > 0$, the filtering dynamics in (7.13) is exponentially mean-square stable and also satisfies the H_∞ performance constraint (7.18) if there exist a positive-definite matrix $P > 0$ and positive scalars ε_1 and ε_2 satisfying*

$$\Sigma = \begin{bmatrix} \Sigma_{11} & * & * & * \\ P(\vec{A} + \hat{H}\tilde{L}\tilde{C}\vec{H}) & P - \varepsilon_2 I & * & * \\ \Sigma_{31} + \varepsilon_1 \hat{L}\tilde{C}\vec{H} & \hat{H}^{\mathrm{T}}P & \Sigma_{33} - \varepsilon_1 I & * \\ \vec{B}^{\mathrm{T}}P(\vec{A} + \hat{H}\tilde{L}\tilde{C}\vec{H}) & \vec{B}^{\mathrm{T}}P & \vec{B}^{\mathrm{T}}P\hat{H} & \vec{B}^{\mathrm{T}}P\vec{B} - \gamma^2 I \end{bmatrix} < 0, \tag{7.19}$$

where

$$\Sigma_{11} = \bar{\Upsilon}_{11} + \varepsilon_2 \hat{H}^{\mathrm{T}}\bar{E}_n \hat{H} + \vec{M}^{\mathrm{T}}\vec{M}, \quad \Sigma_{31} = \bar{\Upsilon}_{31} + \hat{H}^{\mathrm{T}}P(\vec{A} + \hat{H}\tilde{L}\tilde{C}\vec{H}),$$

$$\Sigma_{33} = \bar{\Upsilon}_{33} + \hat{H}^{\mathrm{T}}P\hat{H}, \quad \bar{\Upsilon}_{31} = \sum_{i=1}^{n} \varphi_i^2 \vec{E}_i^{\mathrm{T}}P\vec{E}_i \tilde{L}\tilde{C}\vec{H}, \quad \phi_i^2 = \bar{\beta}_i(1 - \bar{\beta}_i)\bar{\gamma}_i,$$

$$\bar{\Upsilon}_{11} = (\vec{A} + \hat{H}\tilde{L}\tilde{C}\vec{H})^{\mathrm{T}}P(\vec{A} + \hat{H}\tilde{L}\tilde{C}\vec{H}) + \sum_{i=1}^{n}(2\bar{\phi}_i^2 + \hat{\phi}_i^2 + \varphi_i^2)\vec{F}^{\mathrm{T}}\vec{E}_i^{\mathrm{T}}P\vec{E}_i\vec{F}$$

$$+ \sum_{i=1}^{n}(\bar{\phi}_i^2 + \phi_i^2 + \varphi_i^2)\hat{H}^{\mathrm{T}}\tilde{C}^{\mathrm{T}}\tilde{L}^{\mathrm{T}}\vec{E}_i^{\mathrm{T}}P\vec{E}_i\tilde{L}\tilde{C}\vec{H} - P$$

$$+ \sum_{i=1}^{n}(2\phi_i^2 + \hat{\phi}_i^2 + \bar{\varphi}_i^2)\hat{H}^{\mathrm{T}}\tilde{C}^{\mathrm{T}}\vec{E}_i^{\mathrm{T}}P\vec{E}_i\tilde{C}\vec{H},$$

$$\bar{\Upsilon}_{33} = \sum_{i=1}^{n}(\bar{\phi}_i^2 + \hat{\phi}_i^2 + \varphi_i^2)\vec{E}_i^{\mathrm{T}}P\vec{E}_i, \quad \bar{\varphi}_i^2 = (1 - \bar{\beta}_i)\bar{\gamma}_i - (1 - \bar{\beta}_i)^2\bar{\gamma}_i^2,$$

$$\varphi_i^2 = \bar{\beta}_i(1 - \bar{\beta}_i), \quad \hat{\phi}_i^2 = (1 - \bar{\beta}_i)(1 - \bar{\gamma}_i) - (1 - \bar{\beta}_i)^2(1 - \bar{\gamma}_i)^2,$$

$$\bar{\phi}_i^2 = \bar{\beta}_i(1 - \bar{\beta}_i)(1 - \bar{\gamma}_i), \quad \hat{\phi}_i^2 = \bar{\gamma}_i(1 - \bar{\beta}_i)^2(1 - \bar{\gamma}_i).$$

Proof. Choose the following Lyapunov function for system (7.13):

$$V(\eta(k)) = \eta^{\mathrm{T}}(k)P\eta(k); \tag{7.20}$$

the difference of the Lyapunov function is described as follows:

$$\Delta V(\eta(k)) = \mathbb{E}\{V(\eta(k+1))|\eta(k)\} - V(\eta(k)).$$

Then, along the trajectory of system (7.13) with $\vec{w}(k) = 0$, we have

$$\mathbb{E}\{\Delta V(\eta(k))\} = \mathbb{E}\{\eta^{\mathrm{T}}(k+1)P\eta(k+1) - \eta^{\mathrm{T}}(k)P\eta(k)\}$$

$$= \mathbb{E}\left\{\left[\vec{A}\eta(k) + \vec{g}(\vec{H}\eta(k)) + \hat{H}\sigma(\tilde{C}\vec{H}\eta(k)) + \sum_{i=1}^{n}(\beta_i(k) - \bar{\beta}_i)\vec{E}_i\right.\right.$$

$$\times \sigma(\tilde{C}\vec{H}\eta(k)) + \sum_{i=1}^{n}((1 - \beta_i(k))\gamma_i(k) - (1 - \bar{\beta}_i)\bar{\gamma}_i)\vec{E}_i\tilde{C}\vec{H}\eta(k)$$

$$\left.+ \sum_{i=1}^{n}((1 - \beta_i(k))(1 - \gamma_i(k)) - (1 - \bar{\beta}_i)(1 - \bar{\gamma}_i))\vec{E}_i\vec{F}\eta(k)\right]^{\mathrm{T}}$$

$$\times P\left[\vec{A}\eta(k) + \vec{g}(\vec{H}\eta(k)) + \hat{H}\sigma(\tilde{C}\vec{H}\eta(k)) + \sum_{i=1}^{n}(\beta_i(k) - \bar{\beta}_i)\right.$$

$$\times \vec{E}_i\sigma(\tilde{C}\vec{H}\eta(k)) + \sum_{i=1}^{n}((1 - \beta_i(k))\gamma_i(k) - (1 - \bar{\beta}_i)\bar{\gamma}_i)\vec{E}_i\tilde{C}\vec{H}\eta(k)$$

$$\left.+ \sum_{i=1}^{n}((1 - \beta_i(k))(1 - \gamma_i(k)) - (1 - \bar{\beta}_i)(1 - \bar{\gamma}_i))\vec{E}_i\vec{F}\eta(k)\right]$$

$$\left.- \eta^{\mathrm{T}}(k)P\eta(k)\right\}.$$

By noting (7.16), it can be obtained that

$$
\mathbb{E}\{\Delta V(\eta(k))\} = \mathbb{E}\left\{ [(\vec{A} + \hat{H}\tilde{L}\tilde{C}\vec{H})\eta(k) + \hat{H}\Psi(\tilde{C}\vec{H}\eta(k)) + \vec{g}(\vec{H}\eta(k))]^{\mathrm{T}} P \right.
$$

$$
\times [(\vec{A} + \hat{H}\tilde{L}\tilde{C}\vec{H})\eta(k) + \hat{H}\Psi(\tilde{C}\vec{H}\eta(k)) + \vec{g}(\vec{H}\eta(k))]
$$

$$
+ \sum_{i=1}^{n} \hat{\varphi}_i^2 \eta^{\mathrm{T}}(k)\vec{F}^{\mathrm{T}}\vec{E}_i^{\mathrm{T}} P \vec{E}_i \vec{F}\eta(k) + \sum_{i=1}^{n} \varphi_i^2 [\tilde{L}\tilde{C}\vec{H}\eta(k)
$$

$$
+ \Psi(\tilde{C}\vec{H}\eta(k))]^{\mathrm{T}}\vec{E}_i^{\mathrm{T}} P \vec{E}_i [\tilde{L}\tilde{C}\vec{H}\eta(k) + \Psi(\tilde{C}\vec{H}\eta(k))]
$$

$$
- 2\sum_{i=1}^{n} \phi_i^2 [\tilde{L}\tilde{C}\vec{H}\eta(k) + \Psi(\tilde{C}\vec{H}\eta(k))]^{\mathrm{T}}\vec{E}_i^{\mathrm{T}} P \vec{E}_i \tilde{C}\vec{H}\eta(k)
$$

$$
- 2\sum_{i=1}^{n} \bar{\phi}_i^2 [\tilde{L}\tilde{C}\vec{H}\eta(k) + \Psi(\tilde{C}\vec{H}\eta(k))]^{\mathrm{T}}\vec{E}_i^{\mathrm{T}} P \vec{E}_i \vec{F}\eta(k)
$$

$$
+ \sum_{i=1}^{n} \bar{\varphi}_i^2 \eta^{\mathrm{T}}(k)\vec{H}^{\mathrm{T}}\tilde{C}^{\mathrm{T}}\vec{E}_i^{\mathrm{T}} P \vec{E}_i \tilde{C}\vec{H}\eta(k) - \eta^{\mathrm{T}}(k)P\eta(k)
$$

$$
\left. - 2\sum_{i=1}^{n} \hat{\phi}_i^2 \eta^{\mathrm{T}}(k)\vec{H}^{\mathrm{T}}\tilde{C}^{\mathrm{T}}\vec{E}_i^{\mathrm{T}} P \vec{E}_i \vec{F}\eta(k) \right\}. \tag{7.21}
$$

From the elementary inequality $2a^{\mathrm{T}}b \leq a^{\mathrm{T}}a + b^{\mathrm{T}}b$, we have

$$
-2\sum_{i=1}^{n} \phi_i^2 [\tilde{L}\tilde{C}\vec{H}\eta(k) + \Psi(\tilde{C}\vec{H}\eta(k))]^{\mathrm{T}}\vec{E}_i^{\mathrm{T}} P \vec{E}_i \tilde{C}\vec{H}\eta(k)
$$

$$
\leq \sum_{i=1}^{n} \phi_i^2 [\eta^{\mathrm{T}}(k)\vec{H}^{\mathrm{T}}\tilde{C}^{\mathrm{T}}\tilde{L}^{\mathrm{T}}\vec{E}_i^{\mathrm{T}} P \vec{E}_i \tilde{L}\tilde{C}\vec{H}\eta(k)
$$

$$
+ 2\eta^{\mathrm{T}}(k)\vec{H}^{\mathrm{T}}\tilde{C}^{\mathrm{T}}\vec{E}_i^{\mathrm{T}} P \vec{E}_i \tilde{C}\vec{H}\eta(k)
$$

$$
+ \Psi^{\mathrm{T}}(\tilde{C}\vec{H}\eta(k))\vec{E}_i^{\mathrm{T}} P \vec{E}_i \Psi(\tilde{C}\vec{H}\eta(k))], \tag{7.22}
$$

$$
-2\sum_{i=1}^{n} \bar{\phi}_i^2 [\tilde{L}\tilde{C}\vec{H}\eta(k) + \Psi(\tilde{C}\vec{H}\eta(k))]^{\mathrm{T}}\vec{E}_i^{\mathrm{T}} P \vec{E}_i \vec{F}\eta(k)
$$

$$
\leq \sum_{i=1}^{n} \bar{\phi}_i^2 [\eta^{\mathrm{T}}(k)\vec{H}^{\mathrm{T}}\tilde{C}^{\mathrm{T}}\tilde{L}^{\mathrm{T}}\vec{E}_i^{\mathrm{T}} P \vec{E}_i \tilde{L}\tilde{C}\vec{H}\eta(k)
$$

$$
+ 2\eta^{\mathrm{T}}(k)\vec{F}^{\mathrm{T}}\vec{E}_i^{\mathrm{T}} P \vec{E}_i \vec{F}\eta(k)
$$

$$
+ \Psi^{\mathrm{T}}(\tilde{C}\vec{H}\eta(k))\vec{E}_i^{\mathrm{T}} P \vec{E}_i \Psi(\tilde{C}\vec{H}\eta(k))], \tag{7.23}
$$

$$
-2\sum_{i=1}^{n} \hat{\phi}_i^2 \eta^{\mathrm{T}}(k)\vec{H}^{\mathrm{T}}\tilde{C}^{\mathrm{T}}\vec{E}_i^{\mathrm{T}} P \vec{E}_i \vec{F}\eta(k)
$$

$$
\leq \sum_{i=1}^{n} \hat{\phi}_i^2 [\eta^{\mathrm{T}}(k)\vec{H}^{\mathrm{T}}\tilde{C}^{\mathrm{T}}\vec{E}_i^{\mathrm{T}} P \vec{E}_i \tilde{C}\vec{H}\eta(k)
$$

$$
+ \eta^{\mathrm{T}}(k)\vec{F}^{\mathrm{T}}\vec{E}_i^{\mathrm{T}} P \vec{E}_i \vec{F}\eta(k)], \tag{7.24}
$$

which result in

$$\mathbb{E}\{\Delta V(\eta(k))\} \le \mathbb{E}\{\xi^{\mathrm{T}}(k)\bar{\Gamma}\xi(k)\},$$

where

$$\xi(k) = \begin{bmatrix} \eta^{\mathrm{T}}(k) & \vec{g}^{\mathrm{T}}(\vec{H}\eta(k)) & \Psi^{\mathrm{T}}(\tilde{C}\vec{H}\eta(k)) \end{bmatrix}^{\mathrm{T}},$$

$$\bar{\Gamma} = \begin{bmatrix} \bar{\Upsilon}_{11} & * & * \\ P(\vec{A} + \hat{H}\tilde{L}\tilde{C}\vec{H}) & P & * \\ \Sigma_{31} & \hat{H}^{\mathrm{T}}P & \Sigma_{33} \end{bmatrix}. \tag{7.25}$$

Moreover, it follows from (7.15) and (7.17) that

$$\mathbb{E}\{\Delta V(\eta(k))\} \le \mathbb{E}\{\xi^{\mathrm{T}}(k)\bar{\Gamma}\xi(k) - \varepsilon_1\Psi^{\mathrm{T}}(\tilde{C}\vec{H}\eta(k))(\Psi(\tilde{C}\vec{H}\eta(k)) - \tilde{L}\tilde{C}\vec{H}\eta(k))$$

$$- \varepsilon_2(\vec{g}^{\mathrm{T}}(\vec{H}\eta(k))\vec{g}(\vec{H}\eta(k)) - \eta^{\mathrm{T}}(k)\vec{H}^{\mathrm{T}}\bar{E}_n\vec{H}\eta(k))\}$$

$$= \mathbb{E}\{\xi^{\mathrm{T}}(k)\Gamma\xi(k)\},$$

where

$$\Gamma = \begin{bmatrix} \bar{\Upsilon}_{11} + \varepsilon_2\vec{H}^{\mathrm{T}}\bar{E}_n\vec{H} & * & * \\ P(\vec{A} + \hat{H}\tilde{L}\tilde{C}\vec{H}) & P - \varepsilon_2 I & * \\ \Sigma_{31} + \varepsilon_1\hat{L}\tilde{C}\vec{H} & \hat{H}^{\mathrm{T}}P & \bar{\Upsilon}_{33} + \hat{H}^{\mathrm{T}}P\hat{H} - \varepsilon_1 I \end{bmatrix}. \tag{7.26}$$

We can obtain from (7.19), by considering the third leading principal submatrix, that $\Gamma < 0$ and, subsequently,

$$\mathbb{E}\{\Delta V(\eta(k))\} \le -\lambda_{\min}(-\Gamma)\|\xi(k)\|^2.$$

Finally, we can confirm from Lemma 1 of Wang and Ho [41] that the augmented filtering system (7.13) is exponentially mean-square stable.

To establish the H_∞ performance, we assume zero initial conditions and introduce the following index:

$$\mathbb{E}\{\Delta V(\eta(k))\} + \mathbb{E}\{\|\vec{z}(k)\|^2\} - \gamma^2\|\vec{w}(k)\|^2$$

$$= \mathbb{E}\{\xi^{\mathrm{T}}(k)\Gamma\xi(k) + \vec{w}^{\mathrm{T}}(k)\vec{B}^{\mathrm{T}}P\vec{B}\vec{w}(k) + 2\vec{w}^{\mathrm{T}}(k)\vec{B}^{\mathrm{T}}P[(\vec{A} + \hat{H}\tilde{L}\tilde{C}\vec{H})\eta(k)$$

$$+ \hat{H}\Psi(\tilde{C}\vec{H}\eta(k)) + \vec{g}(\vec{H}\eta(k))] + \eta^{\mathrm{T}}(k)\vec{M}^{\mathrm{T}}\vec{M}\eta(k) - \gamma^2\vec{w}^{\mathrm{T}}(k)\vec{w}(k)\}$$

$$= \mathbb{E}\{\hat{\xi}^{\mathrm{T}}(k)\bar{\Sigma}\hat{\xi}(k)\},$$

where

$$\hat{\xi}(k) = \begin{bmatrix} \eta^{\mathrm{T}}(k) & \vec{g}^{\mathrm{T}}(\vec{H}\eta(k)) & \Psi^{\mathrm{T}}(\tilde{C}\vec{H}\eta(k)) & \vec{w}^{\mathrm{T}}(k) \end{bmatrix}^{\mathrm{T}},$$

$$\bar{\Sigma} = \begin{bmatrix} \bar{\Upsilon}_{11} + \vec{M}^{\mathrm{T}}\vec{M} & * & * & * \\ P(\vec{A} + \hat{H}\tilde{L}\tilde{C}\vec{H}) & P & * & * \\ \Sigma_{31} & \hat{H}^{\mathrm{T}}P & \bar{\Upsilon}_{33} + \hat{H}^{\mathrm{T}}P\hat{H} & * \\ \vec{B}^{\mathrm{T}}P(\vec{A} + \hat{H}\tilde{L}\tilde{C}\vec{H}) & \vec{B}^{\mathrm{T}}P & \vec{B}^{\mathrm{T}}P\hat{H} & \vec{B}^{\mathrm{T}}P\vec{B} - \gamma^2 I \end{bmatrix}.$$

Again, it follows from the constraints (7.15) and (7.17) that

$$\mathbb{E}\{\Delta V(\eta(k))\} + \mathbb{E}\left\{\|\vec{z}(k)\|^2\right\} - \gamma^2\|\vec{w}(k)\|^2$$

$$\leq \mathbb{E}\{\hat{\xi}^{\mathrm{T}}(k)\bar{\Sigma}\hat{\xi}(k) - \varepsilon_1\Psi^{\mathrm{T}}(\tilde{C}\vec{H}\eta(k))(\Psi(\tilde{C}\vec{H}\eta(k)) - \tilde{L}\tilde{C}\vec{H}\eta(k))$$

$$- \varepsilon_2(\vec{g}^{\mathrm{T}}(\vec{H}\eta(k))\vec{g}(\vec{H}\eta(k)) - \eta^{\mathrm{T}}(k)\vec{H}^{\mathrm{T}}\bar{E}_n\vec{H}\eta(k))\}$$

$$= \mathbb{E}\{\hat{\xi}^{\mathrm{T}}(k)\Sigma\hat{\xi}(k)\}.$$

Furthermore, we can see from (7.19) in Theorem 7.2.3 that

$$\mathbb{E}\{\Delta V(\eta(k))\} + \mathbb{E}\{\|\vec{z}(k)\|^2\} - \gamma^2\|\vec{w}(k)\|^2 \leq 0$$

for all nonzero $\vec{w}(k)$.

By considering zero initial conditions, it follows from the above inequality that

$$\sum_{k=0}^{\infty} \mathbb{E}\{\|\vec{z}(k)\|^2\} \leq \gamma^2 \sum_{k=0}^{\infty} \|\vec{w}(k)\|^2,$$

which is equivalent to (7.18), and the proof is now complete.   □

Having conducted the filtering performance analysis in Theorem 7.2.2, we are now in a position to deal with the problem of designing distributed H_∞ filters. The solution to the distributed H_∞ filtering problem with both ROSSs and successive packet dropouts is obtained by the following theorem.

Theorem 7.2.3   *Let a positive scalar $\gamma > 0$ be given. For the nonlinear system (7.1) and sensors (7.3) with both ROSSs and successive packet dropouts, the filtering dynamics in (7.13) is exponentially mean-square stable and satisfies the H_∞ performance constraint (7.18) if there exist positive constant scalars ε_1 and ε_2, positive-definite matrices $S > 0$, $Q_i > 0$*

$(i = 1, 2, \ldots, n)$, and $R > 0$, and matrices $X \in \mathcal{T}_{n_x \times n_x}$ and $Y \in \mathcal{T}_{n_x \times n_y}$ satisfying

$$
\begin{bmatrix}
-S + \varepsilon_2 \bar{E}_n + \bar{M}^{\mathrm{T}} \bar{M} & * & * & * \\
\Pi_{21} & \Pi_{22} & * & * \\
\Pi_{31} & \Pi_{32} & -\bar{S} & * \\
\Pi_{41} & \Pi_{42} & 0 & -\ddot{S}
\end{bmatrix} < 0,
\tag{7.27}
$$

where

$$
\Pi_{21} = \left[(-\bar{M}^{\mathrm{T}} \bar{M})^{\mathrm{T}} \quad 0| \quad 0 \quad 0 \quad 0| \quad (\varepsilon_1 \hat{L} \tilde{C})^{\mathrm{T}} \quad 0 \quad 0 \right]^{\mathrm{T}}, \quad \Theta_I = \mathbf{1}_n \otimes I,
$$

$$
\Pi_{22} = \operatorname{diag}\{-Q + \bar{M}^{\mathrm{T}} \bar{M}, -R, -\varepsilon_2 I, -\varepsilon_2 I, -\varepsilon_2 I, -\varepsilon_1 I, -\gamma^2 I, -\gamma^2 I\},
$$

$$
\Pi_{31} = \begin{bmatrix}
S\bar{A} \\
Y[(I - \bar{\Lambda}_\beta)\bar{\Lambda}_\gamma + \bar{\Lambda}_\beta \tilde{L}]\tilde{C} \\
R[(I - \bar{\Lambda}_\beta)\bar{\Lambda}_\gamma + \bar{\Lambda}_\beta \tilde{L}]\tilde{C}
\end{bmatrix}, \quad
\Pi_{32} = \left[\bar{\Pi}_{311} \quad \bar{S} \quad \bar{\Pi}_{313} \right],
$$

$$
\bar{\Pi}_{311} = \begin{bmatrix}
0 & 0 \\
X & Y(I - \bar{\Lambda}_\beta)(I - \bar{\Lambda}_\gamma) \\
0 & R(I - \bar{\Lambda}_\beta)(I - \bar{\Lambda}_\gamma)
\end{bmatrix}, \quad
\bar{\Pi}_{313} = \begin{bmatrix}
0 & S\bar{B} & 0 \\
Y\bar{\Lambda}_\beta & 0 & Y\tilde{D} \\
R\bar{\Lambda}_\beta & 0 & R\tilde{D}
\end{bmatrix},
$$

$$
\Pi_{41} = \left[(\Lambda_\varphi \mathcal{W}\Theta_L)^{\mathrm{T}} \quad (\Lambda_{2\phi} \mathcal{W}\Theta_C)^{\mathrm{T}} \quad 0 \quad (\Lambda_\phi \mathcal{W}\Theta_L)^{\mathrm{T}} \quad 0 \right]^{\mathrm{T}}, \quad \Theta_C = \mathbf{1}_n \otimes \tilde{C},
$$

$$
\Pi_{42} = \left[0 \quad \bar{\Pi}_{412}| \quad 0 \quad 0 \quad 0| \quad \bar{\Pi}_{416} \quad 0 \quad 0 \right], \quad \bar{\Pi}_{412} = \left[0 \quad 0 \quad (\Lambda_{2\bar{\phi}} \mathcal{W}\Theta_I)^{\mathrm{T}} \quad 0 \quad 0 \right]^{\mathrm{T}},
$$

$$
\bar{\Pi}_{416} = \left[(\Lambda_\varphi \mathcal{W}\Theta_I)^{\mathrm{T}} \quad 0 \quad 0 \quad 0 \quad (\Lambda_\phi \mathcal{W}\Theta_I)^{\mathrm{T}} \right]^{\mathrm{T}}, \quad \ddot{S} = \operatorname{diag}\{S, Q, R\},
$$

$$
\Theta_{Hi} = \left[0 \quad (YE_i)^{\mathrm{T}} \quad (RE_i)^{\mathrm{T}} \right]^{\mathrm{T}}, \quad Q = \operatorname{diag}\{Q_1, Q_2, \ldots, Q_n\}, \quad \tilde{S} = I_{5n} \otimes \bar{S},
$$

$$
\Lambda_\varphi = \operatorname{diag}\{\varphi_1 I, \varphi_2 I, \ldots, \varphi_n I\}, \quad \mathcal{W} = \operatorname{diag}\{\Theta_{H1}, \Theta_{H2}, \ldots, \Theta_{Hn}\},
$$

$$
\Lambda_\phi = \operatorname{diag}\left\{ \sqrt{\phi_1^2 + \bar{\phi}_1^2} I, \sqrt{\phi_2^2 + \bar{\phi}_2^2} I, \ldots, \sqrt{\phi_n^2 + \bar{\phi}_n^2} I \right\}, \quad \Theta_L = \mathbf{1}_n \otimes \tilde{L}\tilde{C},
$$

$$
\Lambda_{2\phi} = \operatorname{diag}\left\{ \sqrt{2\phi_1^2 + \hat{\phi}_1^2 + \bar{\varphi}_1^2} I, \sqrt{2\phi_2^2 + \hat{\phi}_2^2 + \bar{\varphi}_2^2} I, \ldots, \sqrt{2\phi_n^2 + \hat{\phi}_n^2 + \bar{\varphi}_n^2} I \right\},
$$

$$
\Lambda_{2\bar{\phi}} = \operatorname{diag}\left\{ \sqrt{2\bar{\phi}_1^2 + \hat{\phi}_1^2 + \hat{\varphi}_1^2} I, \sqrt{2\bar{\phi}_2^2 + \hat{\phi}_2^2 + \hat{\varphi}_2^2} I, \ldots, \sqrt{2\bar{\phi}_n^2 + \hat{\phi}_n^2 + \hat{\varphi}_n^2} I \right\}, \tag{7.28}
$$

*and the other parameters are defined in (7.10). Moreover, if the above inequality is feasible,
two matrices $\bar{K}$ and $\bar{H}$ are given as follows:*

$$
\bar{K} = Q^{-1} X, \quad \bar{H} = Q^{-1} Y. \tag{7.29}
$$

*Therefore, the desired filter parameters K_{ij} and H_{ij} $(i = 1, 2, \ldots, n, j \in \mathcal{N}_i)$ can be obtained
from (7.11).*

Proof. By setting $P = \text{diag}\{S, Q, R\}$, applying Schur complement lemma [149], and noting (7.14), it can be seen that (7.19) is equivalent to

$$
\begin{bmatrix}
-S + \varepsilon_2 \bar{E}_n + \bar{M}^{\mathrm{T}}\bar{M} & * & * & * \\
\hat{\Pi}_{21} & \hat{\Pi}_{22} & * & * \\
\hat{\Pi}_{31} & \hat{\Pi}_{32} & -\bar{S} & * \\
\hat{\Pi}_{41} & \hat{\Pi}_{42} & 0 & -\tilde{S}
\end{bmatrix} < 0,
\tag{7.30}
$$

where

$$
\hat{\Pi}_{31} = \begin{bmatrix}
S\bar{A} \\
Q\bar{H}[(I - \bar{\Lambda}_\beta)\bar{\Lambda}_\gamma + \bar{\Lambda}_\beta\tilde{L}]\tilde{C} \\
R[(I - \bar{\Lambda}_\beta)\bar{\Lambda}_\gamma + \bar{\Lambda}_\beta\tilde{L}]\tilde{C}
\end{bmatrix}, \quad
\hat{\Pi}_{32} = \begin{bmatrix} \hat{\Pi}_{311} & \bar{S} & \hat{\Pi}_{313} \end{bmatrix},
$$

$$
\hat{\Pi}_{311} = \begin{bmatrix}
0 & 0 \\
Q\bar{K} & Q\bar{H}(I - \bar{\Lambda}_\beta)(I - \bar{\Lambda}_\gamma) \\
0 & R(I - \bar{\Lambda}_\beta)(I - \bar{\Lambda}_\gamma)
\end{bmatrix}, \quad
\hat{\Pi}_{313} = \begin{bmatrix}
0 & S\bar{B} & 0 \\
Q\bar{H}\bar{\Lambda}_\beta & 0 & Q\bar{H}\tilde{D} \\
R\bar{\Lambda}_\beta & 0 & R\tilde{D}
\end{bmatrix},
$$

$$
\hat{\Pi}_{41} = \begin{bmatrix} (\Lambda_\varphi\hat{\mathcal{W}}\Theta_L)^{\mathrm{T}} & (\Lambda_{2\phi}\hat{\mathcal{W}}\Theta_C)^{\mathrm{T}} & 0 & (\Lambda_\phi\hat{\mathcal{W}}\Theta_L)^{\mathrm{T}} & 0 \end{bmatrix}^{\mathrm{T}},
$$

$$
\hat{\Pi}_{42} = \begin{bmatrix} 0 & \hat{\Pi}_{412} & 0 & 0 & 0 & \hat{\Pi}_{416} & 0 & 0 \end{bmatrix}, \quad
\hat{\Pi}_{412} = \begin{bmatrix} 0 & 0 & (\Lambda_{2\bar{\phi}}\hat{\mathcal{W}}\Theta_I)^{\mathrm{T}} & 0 & 0 \end{bmatrix}^{\mathrm{T}},
$$

$$
\hat{\Pi}_{416} = \begin{bmatrix} (\Lambda_\varphi\hat{\mathcal{W}}\Theta_I)^{\mathrm{T}} & 0 & 0 & 0 & (\Lambda_\phi\hat{\mathcal{W}}\Theta_I)^{\mathrm{T}} \end{bmatrix}^{\mathrm{T}}, \quad
\hat{\mathcal{W}} = \text{diag}\{\hat{\Theta}_{H1}, \hat{\Theta}_{H2}, \ldots, \hat{\Theta}_{Hn}\},
$$

$$
\hat{\Theta}_{Hi} = \begin{bmatrix} 0 & (Q\bar{H}E_i)^{\mathrm{T}} & (RE_i)^{\mathrm{T}} \end{bmatrix}^{\mathrm{T}}.
\tag{7.31}
$$

Letting $Q = \text{diag}\{Q_1, Q_2, \ldots, Q_n\}$ and noting $Q\bar{K} = X$ and $Q\bar{H} = Y$, we can obtain (7.27) readily. In addition, from Lemma 7.2.1, it follows that $\bar{K} \in \mathcal{T}_{n_x \times n_x}$ and $\bar{H} \in \mathcal{T}_{n_x \times n_y}$, which completes the proof of this theorem. □

Remark 7.3 *It is well known that the main difficulties in designing distributed filters in sensor networks lie in the tight coupling among sensors in terms of both time and space. In this chapter, the filter parameters K_{ij} and H_{ij} ($i = 1, 2, \ldots, n$, $j \in \mathcal{N}_i$) are "assembled" to matrices $\bar{K}$ and $\bar{H}$ which should meet the constraints (7.12). Then, by Lemma 7.2.1, we can derive the conditions that $X \in \mathcal{T}_{n_x \times n_x}$ and $Y \in \mathcal{T}_{n_x \times n_y}$ are required to satisfy. Consequently, the distributed filters can be designed effectively.*

7.3 An Illustrative Example

In this section, we present a simulation example to illustrate the effectiveness of the proposed distributed filter design scheme for nonlinear systems with both ROSSs and successive packet dropouts over sensor networks.

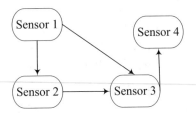

Figure 7.2 Topological structure of the sensor network

The sensor network shown in Figure 7.2 is represented by a directed graph $G = (V, \mathcal{E}, A)$ with the set of nodes $V = \{1, 2, 3, 4\}$, the set of edges

$$\mathcal{E} = \{(1, 1), (2, 1), (2, 2), (3, 1), (3, 2), (3, 3), (4, 3), (4, 4)\},$$

and the adjacency matrix

$$A = \begin{bmatrix} 1 & 0 & 0 & 0 \\ 1 & 1 & 0 & 0 \\ 1 & 1 & 1 & 0 \\ 0 & 0 & 1 & 1 \end{bmatrix}.$$

The nonlinear discrete system considered is modeled by (7.1) with the following parameters:

$$A = \begin{bmatrix} -0.6 & 0.2 \\ 0 & -0.8 \end{bmatrix}, \quad B = \begin{bmatrix} 0.5 & 1 \end{bmatrix}^{\mathrm{T}}, \quad M = \begin{bmatrix} 0.1 & 0.1 \end{bmatrix},$$

and the nonlinear function $f(x(k))$ is selected as

$$f(x(k)) = \begin{bmatrix} \dfrac{0.2x_1(k)}{2x_2^2(k) + 1} & 0.1 \sin(x_1(k))x_2(k) \end{bmatrix}^{\mathrm{T}}.$$

It is easy to see that the constraint (7.2) can be met with $E = \mathrm{diag}\{0.2, 0.15\}$. Consider the sensors with both ROSSs and successive packet dropouts described by (7.3) with the following parameters:

$$C_1 = \begin{bmatrix} 0.1 & 0 \end{bmatrix}, \quad C_2 = \begin{bmatrix} 0.2 & 0.1 \end{bmatrix}, \quad C_3 = \begin{bmatrix} 0.5 & 0.7 \end{bmatrix}, \quad C_4 = \begin{bmatrix} 0.1 & 0.2 \end{bmatrix},$$

$$D_1 = 1, \quad D_2 = 0.5, \quad D_3 = 0.7, \quad D_4 = 0.5.$$

In this example, the probabilities are taken as $\bar{\beta}_1 = 0.9, \bar{\beta}_2 = 0.8, \bar{\beta}_3 = 0.85, \bar{\beta}_4 = 0.7$ and $\bar{\gamma}_1 = 0.9, \bar{\gamma}_2 = 0.8, \bar{\gamma}_3 = 0.7, \bar{\gamma}_4 = 0.6$. Take the saturation level as $\vartheta_{\max} = 0.3$, and other parameters are chosen as $\underline{L}_1 = 0.3, \underline{L}_2 = 0.4, \underline{L}_3 = 0.2, \underline{L}_4 = 0.1, L_1 = 0.7, L_2 = 0.6,$

$L_3 = 0.8$, $L_4 = 0.9$. By solving (7.27) and (7.29) in Theorem 7.2.3, we can obtain the following parameters of the desired distributed filters:

$$K_{11} = \begin{bmatrix} 0.2997 & 0.2511 \\ 0.1260 & 0.1092 \end{bmatrix}, \quad K_{21} = \begin{bmatrix} 0.2238 & 0.1939 \\ 0.1512 & 0.1349 \end{bmatrix},$$

$$K_{22} = \begin{bmatrix} 0.2599 & 0.2554 \\ 0.1854 & 0.1812 \end{bmatrix}, \quad K_{31} = \begin{bmatrix} 0.2570 & 0.2254 \\ -0.0387 & 0.0405 \end{bmatrix},$$

$$K_{32} = \begin{bmatrix} 0.3055 & 0.2995 \\ 0.0622 & 0.0593 \end{bmatrix}, \quad K_{33} = \begin{bmatrix} -0.0533 & -0.0537 \\ 0.4484 & 0.4423 \end{bmatrix},$$

$$K_{43} = \begin{bmatrix} 0.3239 & 0.3177 \\ 0.0723 & 0.0694 \end{bmatrix}, \quad K_{44} = \begin{bmatrix} -0.0488 & -0.0491 \\ 0.5097 & 0.5051 \end{bmatrix},$$

$$H_{11} = \begin{bmatrix} 0.0844 & 0.1755 \end{bmatrix}^{\mathrm{T}}, \quad H_{21} = \begin{bmatrix} 0.0070 & 0.0604 \end{bmatrix}^{\mathrm{T}},$$

$$H_{22} = \begin{bmatrix} 0.1151 & 0.1578 \end{bmatrix}^{\mathrm{T}}, \quad H_{31} = \begin{bmatrix} -0.1054 & 0.1529 \end{bmatrix}^{\mathrm{T}},$$

$$H_{32} = \begin{bmatrix} 0.0761 & 0.2111 \end{bmatrix}^{\mathrm{T}}, \quad H_{33} = \begin{bmatrix} -0.0407 & 0.0249 \end{bmatrix}^{\mathrm{T}},$$

$$H_{43} = \begin{bmatrix} 0.0113 & -0.0017 \end{bmatrix}^{\mathrm{T}}, \quad H_{44} = \begin{bmatrix} 0.2516 & -0.0414 \end{bmatrix}^{\mathrm{T}},$$

and the optimal performance index given in (7.18) is $\gamma^* = 1.0214$. In the simulation, the exogenous disturbance inputs are selected as $w(k) = \exp(-0.2k)\sin(k)$ and $v(k) = [\sin(10k + 1)]/(3k + 1)$. The initial conditions are $x(0) = [0.4 \ 0.2]^{\mathrm{T}}$ and $\hat{x}_i(0) = [0 \ 0]^{\mathrm{T}}$ ($i = 1, 2, 3, 4$). Simulation results are shown in Figures 7.3–7.7. Figures 7.3–7.6 show the actual measurements

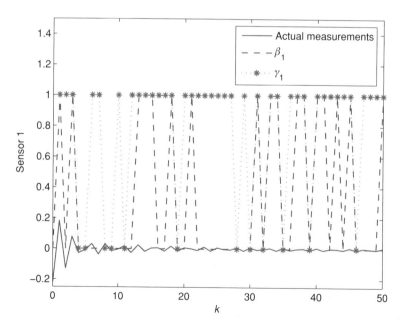

Figure 7.3   Measurements from Sensor 1

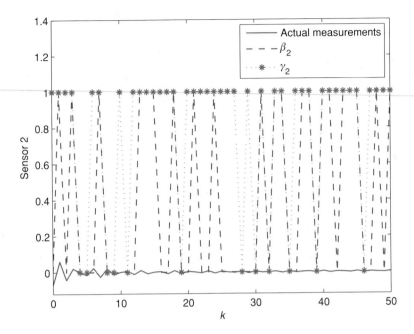

Figure 7.4 Measurements from Sensor 2

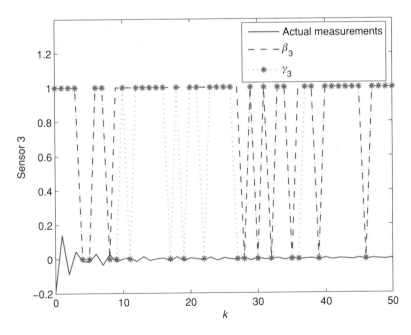

Figure 7.5 Measurements from Sensor 3

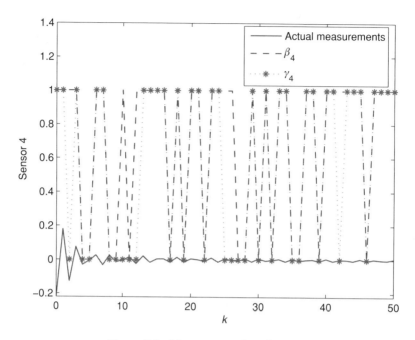

Figure 7.6   Measurements from Sensor 4

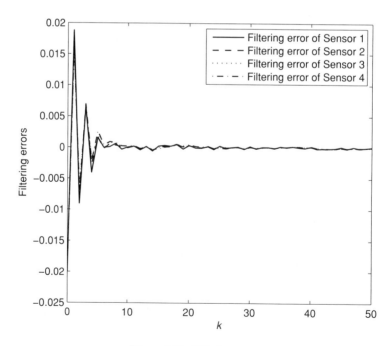

Figure 7.7   Filtering errors

and the binary signals β and γ for sensors 1 to 4. Figure 7.7 plots the filtering errors $z(k) - \hat{z}_i(k)$ ($i = 1, 2, 3, 4$). The simulation results have confirmed the effectiveness of the distributed filtering technique presented in this chapter.

7.4 Summary

In this chapter, we have dealt with the distributed H_∞ filtering problem for a class of nonlinear systems with randomly occurring incomplete information over sensor networks. The incomplete information considered includes both the ROSSs and successive packet dropouts. The issue of ROSSs has been addressed, and then the filtering dynamics has been analyzed by modeling both the ROSSs and successive packet dropouts in a unified framework. The distributed filters have been designed for the filtering dynamics to be exponentially mean-square stable and the filtering errors to satisfy the H_∞ performance constraint. Finally, an illustrative example has been provided that highlights the usefulness of the filtering approach developed.

8

Distributed Filtering with Quantization Errors: The Finite-Horizon Case

This chapter is concerned with the distributed finite-horizon filtering problem for a class of time-varying systems over lossy sensor networks. The time-varying system (target plant) is subject to RVNs caused by environmental circumstances. The lossy sensor network suffers from quantization errors and successive packet dropouts that are described in a unified framework. Two mutually independent sets of Bernoulli-distributed white sequences are introduced to govern the random occurrences of the RVNs and successive packet dropouts. Through available output measurements from not only the individual sensor but also its neighboring sensors according to the given topology, a sufficient condition is established for the desired distributed finite-horizon filter to ensure that the prescribed average filtering performance constraint is satisfied. The solution of the distributed filter gains is characterized by solving a set of RLMIs. A simulation example is provided to show the effectiveness of the proposed filtering scheme.

8.1 Problem Formulation

In this chapter, it is assumed that the sensor network has n sensor nodes which are distributed in space according to a fixed network topology represented by a directed graph $\mathcal{G} = (\mathcal{V}, \mathcal{E}, \mathcal{A})$ of order n with a set of nodes $\mathcal{V} = 1, 2, \ldots, n$, a set of edges $\mathcal{E} \in \mathcal{V} \times \mathcal{V}$, and a weighted adjacency matrix $\mathcal{A} = [a_{ij}]$ with nonnegative adjacency element a_{ij}. An edge of $\mathcal{G}$ is denoted by ordered pair (i, j). The adjacency elements associated with the edges of the graph are positive (i.e., $a_{ij} > 0 \iff (i, j) \in \mathcal{E}$), which means that sensor i can obtain information from sensor j. Also, we assume that $a_{ii} = 1$ for all $i \in \mathcal{V}$; therefore, (i, i) can be regarded as an additional edge. The set of neighbors of node $i \in \mathcal{V}$ plus the node itself are denoted by $\mathcal{N}_i = \{j \in \mathcal{V} : (i, j) \in \mathcal{E}\}$.

Filtering, Control and Fault Detection with Randomly Occurring Incomplete Information, First Edition.
Hongli Dong, Zidong Wang, and Huijun Gao.
© 2013 John Wiley & Sons, Ltd. Published 2013 by John Wiley & Sons, Ltd.

Let a finite time-horizon be denoted by $[0, N] := \{0, 1, 2, \ldots, N\}$. Consider a target plant described by the following discrete-time nonlinear time-varying stochastic system on $k \in [0, N]$:

$$\begin{cases} x(k + 1) = A(k)x(k) + \beta(k)f(x(k)) + (1 - \beta(k))g(x(k)) + G(k)w(k), \\ z(k) = M(k)x(k). \end{cases} \tag{8.1}$$

In this chapter, for every i $(1 \leq i \leq n)$, the model of sensor node i is given as follows:

$$y_i(k) = \gamma_i(k)[h(C_i(k)x(k)) + D_i(k)w(k)] + (1 - \gamma_i(k))y_i(k - 1), \quad i = 1, 2, \ldots, n, \tag{8.2}$$

where $x(k) \in \mathbb{R}^{n_x}$ represents the state vector which cannot be observed directly, $y_i(k) \in \mathbb{R}^{n_y}$ is the measurement output measured by sensor i from the plant, $z(k) \in \mathbb{R}^{n_z}$ is the output to be estimated, and $w(k) \in \mathbb{R}^{n_w}$ denotes the noise signal belonging to $l_2[0, \infty)$. $h(\cdot)$ is the round-off function which represents the quantization effect; $A(k)$, $G(k)$, $M(k)$, $C_i(k)$, and $D_i(k)$ are known, real, time-varying matrices with appropriate dimensions. $f(\cdot)$ and $g(\cdot)$ are continuously vector-valued functions.

Assumption 8.1   *The nonlinear functions f and g satisfy the following sector-bounded conditions:*

$$\begin{aligned} [f(x) - U_1 x]^T[f(x) - U_2 x] \leq 0, \quad \forall x \in \mathbb{R}^{n_x}, \\ [g(x) - U_3 x]^T[g(x) - U_4 x] \leq 0, \quad \forall x \in \mathbb{R}^{n_x}, \end{aligned} \tag{8.3}$$

where U_1, U_2, U_3, and $U_4 \in \mathbb{R}^{n_x \times n_x}$ are real matrices of appropriate dimensions, and $\mathcal{U}_1 = U_1 - U_2$ and $\mathcal{U}_2 = U_3 - U_4$ are symmetric positive-definite matrices.

The quantization error is assumed to be bounded as follows:

$$h(C_i(k)x(k)) - C_i(k)x(k) = v_i(k), \quad \|v_i(k)\| \leq \delta \text{ and } \delta > 0. \tag{8.4}$$

The two sequences of stochastic variables $\beta(k)$ and $\gamma_i(k)$ $(i = 1, 2, \ldots, n)$ in (8.1) and (8.2) are introduced to account for the probabilistic nature of the occurrence of the nonlinearities and packet dropouts. They are mutually independent Bernoulli-distributed white sequences taking values on 0 or 1 with

$$\begin{cases} \text{Prob}\{\beta(k) = 1\} = \bar{\beta} \\ \text{Prob}\{\beta(k) = 0\} = 1 - \bar{\beta} \end{cases} \text{ and } \begin{cases} \text{Prob}\{\gamma_i(k) = 1\} = \bar{\gamma}_i \\ \text{Prob}\{\gamma_i(k) = 0\} = 1 - \bar{\gamma}_i \end{cases},$$

respectively, where $\bar{\beta} \in [0, 1]$ and $\bar{\gamma}_i \in [0, 1]$ are known constants.

Remark 8.1   *The proposed measurement model in (8.2) provides a novel unified framework to account for the phenomenon of either successive packet dropouts or quantization at each* *time-point by resorting to the random variable $\gamma_i(k)$ ($i = 1, 2, \ldots, n$). At the kth time-point, if $\gamma_i(k) = 1$, the ith sensor node undergoes quantization; and if $\gamma_i(k) = 0$, the ith sensor node model (8.2) specializes to the one with a packet dropout (in this case, the latest measurement received in the buffers will be utilized if the current measurement is lost during packet transmissions). Considering the regressive nature of $y_i(k)$ in (8.2), the model (8.2) is comprehensive, taking into account both the probabilistic successive quantization errors and probabilistic successive packet dropouts in sensor networks.*

Remark 8.2 *Quantization error or round-off error exists very often in nearly all digital signal processing problems; whenever there is a need to represent a signal in digital form, this ordinarily involves rounding. The error signal is sometimes considered as an additional random signal called quantization noise because of its stochastic behavior. Usually, the quantization error is not significantly correlated with the signal and has an approximately uniform distribution. In (8.4), for simplicity, the quantization error is assumed to be a noise with bounded norm. As will be seen later, one of the purposes of the distributed filter design is to reduce the influence from the quantization errors to the filter performance by introducing the average H_∞-index for disturbance attenuation and rejection over the given finite horizon.*

In this chapter, the following filter structure is adopted on sensor node i:

$$
\begin{cases}
\hat{x}_i(k+1) = \sum_{j \in \mathcal{N}_i} a_{ij} K_{ij}(k)\hat{x}_j(k) + \sum_{j \in \mathcal{N}_i} a_{ij} H_{ij}(k)y_j(k), \\
\hat{z}_i(k) = M(k)\hat{x}_i(k),
\end{cases}
\tag{8.5}
$$

where $\hat{x}_i(k) \in \mathbb{R}^{n_x}$ is the state estimate on sensor node i and $\hat{z}_i(k) \in \mathbb{R}^{n_z}$ is the estimate of $z(k)$ from the filter on sensor node i. Here, $K_{ij}(k)$ and $H_{ij}(k)$ are the filter gain matrices on node i to be determined.

Letting $\tilde{z}_i(k) = z(k) - \hat{z}_i(k)$ ($i = 1, 2, \ldots, n$), we obtain the following system:

$$
\begin{cases}
x(k+1) = A(k)x(k) + \bar{\beta} f(x(k)) + (1 - \bar{\beta})g(x(k)) \\
\qquad + (\beta(k) - \bar{\beta})(f(x(k)) - g(x(k))) + G(k)w(k), \\
\hat{x}_i(k+1) = \sum_{j \in \mathcal{N}_i} a_{ij} K_{ij}(k)\hat{x}_j(k) + \sum_{j \in \mathcal{N}_i} \bar{\gamma}_j a_{ij} H_{ij}(k)(C_j(k)x(k) + D_j(k)w(k) \\
\qquad + v_j(k)) + \sum_{j \in \mathcal{N}_i}(\gamma_j(k) - \bar{\gamma}_j)a_{ij} H_{ij}(k)(C_j(k)x(k) + D_j(k)w(k) \\
\qquad + v_j(k) - y_j(k-1)) + \sum_{j \in \mathcal{N}_i}(1 - \bar{\gamma}_j)a_{ij} H_{ij}(k)y_j(k-1), \\
\tilde{z}_i(k) = M(k)[x(k) - \hat{x}_i(k)].
\end{cases}
\tag{8.6}
$$

Before proceeding further, the following definition is introduced.

Definition 8.1.1 *For a given disturbance attenuation level $\gamma > 0$ and some given positive-definite matrices $S_i > 0$ ($0 \leq i \leq n$), the filtering error $\tilde{z}_i(k)$ from (8.6) is said to satisfy the average H_∞ performance constraints if the following inequality holds:*

$$\frac{1}{n} \sum_{i=1}^{n} \mathbb{E}\{\|\tilde{z}_i\|_{[0,N-1]}^2\} < \gamma^2 \left\{ \|w\|_{[0,N-1]}^2 + \frac{1}{n} \sum_{i=1}^{n} \|v_i\|_{[0,N-1]}^2 \right.$$

$$\left. + \frac{1}{n} \sum_{i=1}^{n} (x(0) - \hat{x}_i(0))^{\mathrm{T}} S_i (x(0) - \hat{x}_i(0)) \right\} \tag{8.7}$$

Remark 8.3 *The average H_∞ performance (8.7) over the n filters for n sensors is a constraint adopted from classical H_∞ control theory [177]. It means that the average energy gains from the average energy of all disturbances on the target plant and sensor network (including initial state, process noise of the target plant, measurement noises of the sensor networks, and the quantization errors) to the average energy of all estimation errors over the given time horizon should be less than a given disturbance attenuation level γ. Such an average H_∞ performance index is more appropriate to quantify the overall performance of the distributed filters than the conventional central H_∞ performance constraint.*

Our aim in this chapter is to find the filter gain matrices $K_{ij}(k)$ and $H_{ij}(k)$ ($i = 1, 2, \ldots, n$, $j \in \mathcal{N}_i$) such that the filtering errors $\tilde{z}_i(k)$ ($i = 1, 2, \ldots, n$) from (8.6) satisfy the average H_∞ performance constraints (8.7).

For convenience of later analysis, we denote

$$\hat{x}(k) = \begin{bmatrix} \hat{x}_1^{\mathrm{T}}(k) & \hat{x}_2^{\mathrm{T}}(k) & \cdots & \hat{x}_n^{\mathrm{T}}(k) \end{bmatrix}^{\mathrm{T}}, \quad \bar{x}(k) = \mathbf{1}_n \otimes x(k),$$

$$\bar{\Lambda}_\gamma = \mathrm{diag}\{\bar{\gamma}_1, \bar{\gamma}_2, \ldots, \bar{\gamma}_n\}, \quad \mathcal{F}(\bar{x}(k)) = \mathbf{1}_n \otimes f(x(k)),$$

$$\hat{z}(k) = \begin{bmatrix} \hat{z}_1^{\mathrm{T}}(k) & \hat{z}_2^{\mathrm{T}}(k) & \cdots & \hat{z}_n^{\mathrm{T}}(k) \end{bmatrix}^{\mathrm{T}}, \quad \bar{M}(k) = I_n \otimes M(k),$$

$$\bar{z}(k) = \mathbf{1}_n \otimes z(k), \quad \mathcal{G}(\bar{x}(k)) = \mathbf{1}_n \otimes g(x(k)), \quad \bar{G}(k) = I_n \otimes G(k), \quad \bar{A}(k) = I_n \otimes A(k),$$

$$\tilde{y}(k) = \begin{bmatrix} y_1^{\mathrm{T}}(k) & y_2^{\mathrm{T}}(k) & \cdots & y_n^{\mathrm{T}}(k) \end{bmatrix}^{\mathrm{T}}, \quad \bar{v}(k) = \begin{bmatrix} v_1^{\mathrm{T}}(k) & v_2^{\mathrm{T}}(k) & \cdots & v_n^{\mathrm{T}}(k) \end{bmatrix}^{\mathrm{T}},$$

$$\tilde{C}(k) = \mathrm{diag}\{C_1(k), C_2(k), \ldots, C_n(k)\}, \quad \tilde{D}(k) = \mathrm{diag}\{D_1(k), D_2(k), \ldots, D_n(k)\},$$

$$\bar{w}(k) = \mathbf{1}_n \otimes w(k), \quad \tilde{v}(k) = \begin{bmatrix} \bar{w}^{\mathrm{T}}(k) & \bar{v}^{\mathrm{T}}(k) \end{bmatrix}^{\mathrm{T}},$$

where

$$\bar{K}(k) = [\bar{K}_{ij}(k)]_{n \times n} \quad \text{with} \quad \bar{K}_{ij}(k) = \begin{cases} a_{ij} K_{ij}(k), & i = 1, 2, \ldots, n; \quad j \in \mathcal{N}_i; \\ 0, & i = 1, 2, \ldots, n; \quad j \notin \mathcal{N}_i; \end{cases}$$

$$\bar{H}(k) = [\bar{H}_{ij}(k)]_{n \times n} \quad \text{with} \quad \bar{H}_{ij}(k) = \begin{cases} a_{ij} H_{ij}(k), & i = 1, 2, \ldots, n; \quad j \in \mathcal{N}_i; \\ 0, & i = 1, 2, \ldots, n; \quad j \notin \mathcal{N}_i. \end{cases} \tag{8.8}$$

Obviously, since $a_{ij} = 0$ when $j \notin \mathcal{N}_i$, $\bar{K}(k)$ and $\bar{H}(k)$ are two matrices that can be expressed as

$$\bar{K}(k) \in \mathcal{T}_{n_x \times n_x} \quad \text{and} \quad \bar{H}(k) \in \mathcal{T}_{n_x \times n_y}, \tag{8.9}$$

where $\mathcal{T}_{p \times q} = \{\bar{T} = [T_{ij}] \in \mathbb{R}^{np \times nq} \mid T_{ij} \in \mathbb{R}^{p \times q}, \ T_{ij} = 0 \ \text{if} \ j \notin \mathcal{N}_i\}$.

Letting $\eta(k) = [\bar{x}^{\mathrm{T}}(k) \ \ \hat{x}^{\mathrm{T}}(k) \ \ \tilde{y}^{\mathrm{T}}(k-1)]^{\mathrm{T}}$ and $\tilde{z}(k) = \bar{z}(k) - \hat{z}(k)$, the following augmented system is obtained that governs the filtering dynamics for the sensor network:

$$\begin{cases} \eta(k+1) = \mathcal{A}(k)\eta(k) + (\mathcal{L}_1 + (\beta(k) - \bar{\beta})\mathcal{L}_2)\mathcal{F}(\vec{H}\eta(k)) + \mathcal{G}(k)\tilde{v}(k) \\ \qquad\quad + \sum_{i=1}^{n} (\gamma_i(k) - \bar{\gamma}_i)(\mathcal{C}_i(k)\eta(k) + \mathcal{D}_i(k)\tilde{v}(k)), \\ \tilde{z}(k) = \mathcal{M}(k)\eta(k), \end{cases} \tag{8.10}$$

where

$$\mathcal{A}(k) = \begin{bmatrix} \bar{A}(k) & 0 & 0 \\ \bar{H}(k)\bar{\Lambda}_\gamma \tilde{C}(k) & \bar{K}(k) & \bar{H}(k)(I - \bar{\Lambda}_\gamma) \\ \bar{\Lambda}_\gamma \tilde{C}(k) & 0 & I - \bar{\Lambda}_\gamma \end{bmatrix}, \quad \mathcal{L}_1 = \begin{bmatrix} \bar{\beta}I & (1-\bar{\beta})I \\ 0 & 0 \\ 0 & 0 \end{bmatrix},$$

$$\mathcal{G}(k) = \begin{bmatrix} \bar{G}(k) & 0 \\ \bar{H}(k)\bar{\Lambda}_\gamma \tilde{D}(k) & \bar{H}(k)\bar{\Lambda}_\gamma \\ \bar{\Lambda}_\gamma \tilde{D}(k) & \bar{\Lambda}_\gamma \end{bmatrix}, \quad \mathcal{C}_i(k) = \begin{bmatrix} 0 & 0 & 0 \\ \bar{H}(k)E_i\tilde{C}(k) & 0 & -\bar{H}(k)E_i \\ E_i\tilde{C}(k) & 0 & -E_i \end{bmatrix},$$

$$\mathcal{D}_i(k) = \begin{bmatrix} 0 & 0 \\ \bar{H}(k)E_i\tilde{D}(k) & \bar{H}(k)E_i \\ E_i\tilde{D}(k) & E_i \end{bmatrix}, \quad \mathcal{L}_2 = \begin{bmatrix} I & -I \\ 0 & 0 \\ 0 & 0 \end{bmatrix}, \quad \vec{H} = [I \ \ 0 \ \ 0], \quad E_i = e_i e_i^{\mathrm{T}},$$

$$\mathcal{M}(k) = [\bar{M}(k) \ \ -\bar{M}(k) \ \ 0], \quad \mathcal{F}(\vec{H}\eta(k)) = [\mathcal{F}^{\mathrm{T}}(\vec{H}\eta(k)) \ \ \mathcal{G}^{\mathrm{T}}(\vec{H}\eta(k))]^{\mathrm{T}}. \tag{8.11}$$

Also, from (8.3), we have

$$\begin{aligned} [\mathcal{F}(\vec{H}\eta(k)) - \bar{U}_1\vec{H}\eta(k)]^{\mathrm{T}}[\mathcal{F}(\vec{H}\eta(k)) - \bar{U}_2\vec{H}\eta(k)] &\leq 0 \\ [\mathcal{G}(\vec{H}\eta(k)) - \bar{U}_3\vec{H}\eta(k)]^{\mathrm{T}}[\mathcal{G}(\vec{H}\eta(k)) - \bar{U}_4\vec{H}\eta(k)] &\leq 0 \end{aligned} \tag{8.12}$$

where

$$\bar{U}_1 := I_n \otimes U_1, \quad \bar{U}_2 := I_n \otimes U_2, \quad \bar{U}_3 := I_n \otimes U_3, \quad \bar{U}_4 := I_n \otimes U_4.$$

Hence, we have

$$[\mathcal{F}(\vec{H}\eta(k)) - \vec{U}_1\vec{H}\eta(k)]^{\mathrm{T}}[\mathcal{F}(\vec{H}\eta(k)) - \vec{U}_2\vec{H}\eta(k)] \leq 0, \tag{8.13}$$

where

$$\bar{\mathcal{U}}_1 := \begin{bmatrix} \bar{U}_1^T & \bar{U}_3^T \end{bmatrix}^T, \quad \bar{\mathcal{U}}_2 := \begin{bmatrix} \bar{U}_2^T & \bar{U}_4^T \end{bmatrix}^T.$$

The average H_∞ performance constraints (8.7) can be rewritten as follows:

$$J := \mathbb{E}\|\tilde{z}\|_{[0,N-1]}^2 - \gamma^2 \{\|\tilde{v}\|_{[0,N-1]}^2 + \bar{e}^T(0)R\bar{e}(0)\} < 0, \tag{8.14}$$

where $\bar{e}(0) = \bar{x}(0) - \hat{x}(0)$ and $R = \text{diag}\{S_1, S_2, \ldots, S_n\}$.

8.2   Main Results

In this section, we investigate both the filter analysis and design problems for the distributed finite-horizon filtering of system (8.1) with n sensors whose topology is determined by the given graph $\mathcal{G} = (\mathcal{V}, \mathcal{E}, \mathcal{A})$.

We are now in a position to provide the analysis results in the following theorem.

Theorem 8.2.1 *Consider the filtering dynamics in (8.10) and suppose that the filter parameters $K_{ij}(k)$ and $H_{ij}(k)$ in (8.5) are given. For a positive scalar $\gamma > 0$ and a sequence of positive-definite matrices $S_i > 0$ $(i = 1, 2, \ldots, n)$, the average H_∞ performance requirement defined in (8.14) is achieved for all nonzero $\tilde{v}(k)$ if, with the initial condition $\eta^T(0)P(0)\eta(0) \leqslant \gamma^2 \bar{e}^T(0)R\bar{e}(0)$, there exists a sequence of positive-definite matrices $\{P(k)\}_{0\leqslant k\leqslant N+1}$ satisfying the following recursive matrix inequalities:*

$$\Sigma_k := \begin{bmatrix} \Sigma_{11k} & * & * \\ \Sigma_{21k} & \Sigma_{22k} & * \\ \Sigma_{31k} & \mathcal{G}^T(k)P(k+1)\mathcal{L}_1 & \Sigma_{33k} \end{bmatrix} < 0, \tag{8.15}$$

for all $0 \leq k \leq N$, where

$$\Sigma_{11k} = \mathcal{A}^T(k)P(k+1)\mathcal{A}(k) + \sum_{i=1}^{n} \sigma_i^2 \mathcal{C}_i^T(k)P(k+1)\mathcal{C}_i(k) - \varepsilon_1 \tilde{\mathcal{U}}_1 + \mathcal{M}^T(k)\mathcal{M}(k) - P(k),$$

$$\Sigma_{21k} = \mathcal{L}_1^T P(k+1)\mathcal{A}(k) - \varepsilon_1 \tilde{\mathcal{U}}_2,$$

$$\Sigma_{22k} = \mathcal{L}_1^T P(k+1)\mathcal{L}_1 + \varsigma^2 \mathcal{L}_2^T P(k+1)\mathcal{L}_2 - \varepsilon_1 I,$$

$$\Sigma_{31k} = \mathcal{G}^T(k)P(k+1)\mathcal{A}(k) + \sum_{i=1}^{n} \sigma_i^2 \mathcal{D}_i^T(k)P(k+1)\mathcal{C}_i(k),$$

$$\Sigma_{33k} = \mathcal{G}^T(k)P(k+1)\mathcal{G}(k) + \sum_{i=1}^{n} \sigma_i^2 \mathcal{D}_i^T(k)P(k+1)\mathcal{D}_i(k) - \gamma^2 I,$$

$$\sigma_i = \sqrt{\bar{\gamma}_i(1-\bar{\gamma}_i)} \quad (i = 1, 2, \ldots, n), \quad \varsigma = \sqrt{\bar{\beta}(1-\bar{\beta})},$$

$$\tilde{\mathcal{U}}_1 = \frac{\vec{H}^T(\bar{\mathcal{U}}_1^T\bar{\mathcal{U}}_2 + \bar{\mathcal{U}}_2^T\bar{\mathcal{U}}_1)\vec{H}}{2}, \quad \tilde{\mathcal{U}}_2 = -\frac{(\bar{\mathcal{U}}_1 + \bar{\mathcal{U}}_2)\vec{H}}{2}.$$

Proof. Define

$$J(k) = \eta^{\mathrm{T}}(k+1)P(k+1)\eta(k+1) - \eta^{\mathrm{T}}(k)P(k)\eta(k). \tag{8.16}$$

It follows from (8.10) that

$$
\begin{aligned}
\mathbb{E}\{J(k)\} = \mathbb{E}\Bigg\{ & \Bigg[\mathcal{A}(k)\eta(k) + (\mathcal{L}_1 + (\beta(k) - \bar{\beta})\mathcal{L}_2)\mathcal{F}(\vec{H}\eta(k)) + \mathcal{G}(k)\tilde{v}(k) + \mathcal{D}_i(k)\tilde{v}(k) \\
& + \sum_{i=1}^{n}(\gamma_i(k) - \bar{\gamma}_i)\mathcal{C}_i(k)\eta(k) \Bigg]^{\mathrm{T}} P(k+1) \Bigg[\mathcal{A}(k)\eta(k) + (\mathcal{L}_1 + (\beta(k) - \bar{\beta})\mathcal{L}_2) \\
& \times \mathcal{F}(\vec{H}\eta(k)) + \mathcal{G}(k)\tilde{v}(k) + \sum_{i=1}^{n}(\gamma_i(k) - \bar{\gamma}_i)(\mathcal{C}_i(k)\eta(k) + \mathcal{D}_i(k)\tilde{v}(k)) \Bigg] \\
& - \eta^{\mathrm{T}}(k)P(k)\eta(k) \Bigg\} \\
= \mathbb{E}\Bigg\{ & [\mathcal{A}(k)\eta(k) + \mathcal{L}_1\mathcal{F}(\vec{H}\eta(k)) + \mathcal{G}(k)\tilde{v}(k)]^{\mathrm{T}} P(k+1)[\mathcal{A}(k)\eta(k) \\
& + \mathcal{L}_1\mathcal{F}(\vec{H}\eta(k)) + \mathcal{G}(k)\tilde{v}(k)] + \varsigma^2\mathcal{F}^{\mathrm{T}}(\vec{H}\eta(k))\mathcal{L}_2^{\mathrm{T}} P(k+1)\mathcal{L}_2\mathcal{F}(\vec{H}\eta(k)) \\
& + \sum_{i=1}^{n}\sigma_i^2[\mathcal{C}_i(k)\eta(k) + \mathcal{D}_i(k)\tilde{v}(k)]^{\mathrm{T}} P(k+1)[\mathcal{C}_i(k)\eta(k) + \mathcal{D}_i(k)\tilde{v}(k)] \\
& - \eta^{\mathrm{T}}(k)P(k)\eta(k) \Bigg\}.
\end{aligned}
$$

Adding the zero term $\tilde{z}^{\mathrm{T}}(k)\tilde{z}(k) - \gamma^2\tilde{v}^{\mathrm{T}}(k)\tilde{v}(k) - \tilde{z}^{\mathrm{T}}(k)\tilde{z}(k) + \gamma^2\tilde{v}^{\mathrm{T}}(k)\tilde{v}(k)$ to $\mathbb{E}\{J(k)\}$ results in

$$
\begin{aligned}
\mathbb{E}\{J(k)\} &= \mathbb{E}\left\{ \begin{bmatrix} \eta^{\mathrm{T}}(k) & \mathcal{F}^{\mathrm{T}}(\vec{H}\eta(k)) & \tilde{v}^{\mathrm{T}}(k) \end{bmatrix} \bar{\Sigma}_k \begin{bmatrix} \eta(k) \\ \mathcal{F}(\vec{H}\eta(k)) \\ \tilde{v}(k) \end{bmatrix} - \tilde{z}^{\mathrm{T}}(k)\tilde{z}(k) + \gamma^2\tilde{v}^{\mathrm{T}}(k)\tilde{v}(k) \right\} \\
&= \mathbb{E}\left\{ \xi^{\mathrm{T}}(k)\bar{\Sigma}_k\xi(k) - \tilde{z}^{\mathrm{T}}(k)\tilde{z}(k) + \gamma^2\tilde{v}^{\mathrm{T}}(k)\tilde{v}(k) \right\},
\end{aligned} \tag{8.17}
$$

where

$$
\bar{\Sigma}_k = \begin{bmatrix} \Sigma_{11k} + \varepsilon_1\tilde{\mathcal{U}}_1 & * & * \\ \Sigma_{21k} + \varepsilon_1\tilde{\mathcal{U}}_2 & \Sigma_{22k} + \varepsilon_1 I & * \\ \Sigma_{31k} & \mathcal{G}^{\mathrm{T}}(k)P(k+1)\mathcal{L}_1 & \Sigma_{33k} \end{bmatrix}, \tag{8.18}
$$

$$\xi(k) = [\eta^{\mathrm{T}}(k) \quad \mathcal{F}^{\mathrm{T}}(\vec{H}\eta(k)) \quad \tilde{v}^{\mathrm{T}}(k)]^{\mathrm{T}}.$$

Moreover, it follows from the constraint (8.13) that

$$
\begin{aligned}
\mathbb{E}\{J(k)\} &\leq \mathbb{E}\{\xi^{\mathrm{T}}(k)\bar{\Sigma}_k\xi(k) - \varepsilon_1[\mathcal{F}(\vec{H}\eta(k)) - \bar{\mathcal{U}}_1\vec{H}\eta(k)]^{\mathrm{T}}[\mathcal{F}(\vec{H}\eta(k)) - \bar{\mathcal{U}}_2\vec{H}\eta(k)] \\
&\quad - \tilde{z}^{\mathrm{T}}(k)\tilde{z}(k) + \gamma^2\tilde{v}^{\mathrm{T}}(k)\tilde{v}(k)\} \\
&= \mathbb{E}\{\xi^{\mathrm{T}}(k)\Sigma_k\xi(k) - \tilde{z}^{\mathrm{T}}(k)\tilde{z}(k) + \gamma^2\tilde{v}^{\mathrm{T}}(k)\tilde{v}(k)\}.
\end{aligned}
\tag{8.19}
$$

Summing up (8.19) on both sides from 0 to $N-1$ with respect to k, we obtain

$$
\begin{aligned}
\sum_{k=0}^{N-1}\mathbb{E}\{J(k)\} &= \mathbb{E}\left\{\eta^{\mathrm{T}}(N)P(N)\eta(N)\right\} - \eta^{\mathrm{T}}(0)P(0)\eta(0) \\
&\leq \mathbb{E}\left\{\sum_{k=0}^{N-1}\xi^{\mathrm{T}}(k)\Sigma_k\xi(k)\right\} - \mathbb{E}\left\{\sum_{k=0}^{N-1}(\tilde{z}^{\mathrm{T}}(k)\tilde{z}(k) - \gamma^2\tilde{v}^{\mathrm{T}}(k)\tilde{v}(k))\right\}.
\end{aligned}
$$

Hence, the average $\mathcal{H}_\infty$ performance index defined in (8.14) is given by

$$
J \leq \mathbb{E}\left\{\sum_{k=0}^{N-1}\xi^{\mathrm{T}}(k)\Sigma_k\xi(k)\right\} - \mathbb{E}\left\{\eta^{\mathrm{T}}(N)P(N)\eta(N)\right\} + \eta^{\mathrm{T}}(0)P(0)\eta(0) - \gamma^2\bar{e}^{\mathrm{T}}(0)R\bar{e}(0).
\tag{8.20}
$$

Noting that $P(N) > 0$ and the initial condition $\eta^{\mathrm{T}}(0)P(0)\eta(0) \leqslant \gamma^2\bar{e}^{\mathrm{T}}(0)R\bar{e}(0)$, we have $J < 0$ when (8.15) holds. The proof is now complete. $\qquad\square$

The following lemma is needed for obtaining our main results.

Lemma 8.2.2 [143] Let $P = \mathrm{diag}\{P_1, P_2, \ldots, P_n\}$ with $P_i \in \mathbb{R}^{p\times p}$ $(1 \leq i \leq n)$ being invertible matrices. If $X = PW$ for $W \in \mathbb{R}^{np\times nq}$, then we have $W \in \mathcal{T}_{p\times q} \Longleftrightarrow X \in \mathcal{T}_{p\times q}$.

Having conducted the filtering performance analysis in Theorem 8.2.1, we are now in a position to deal with the problem of designing distributed finite-horizon filters. The solution to the distributed finite-horizon filtering problem over lossy sensor networks is obtained by the following theorem.

Theorem 8.2.3 *Given a positive scalar $\gamma > 0$ and positive-definite matrices $S_i = S_i^{\mathrm{T}} > 0$ $(1 \leq i \leq n)$. For the target plant (8.1) with randomly varying nonlinearities and sensor network (8.2) with both quantization and successive packet dropouts, the finite-horizon filter design problem addressed is solved if there exist positive-definite matrices $\{P_i(k)\}_{0\leq k\leq N+1}$ $(1 \leq i \leq n)$, a family of matrices $\{Q(k)\}_{0\leq k\leq N} \in \mathcal{T}_{n_x\times n_y}$, and a positive constant scalar ε_1 satisfying the initial condition*

$$
\eta^{\mathrm{T}}(0)P(0)\eta(0) \leqslant \gamma^2\bar{e}^{\mathrm{T}}(0)R\bar{e}(0)
\tag{8.21}
$$

and the RLMIs

$$
\begin{bmatrix}
\Pi_{11}(k) & * & * \\
\Pi_{21}(k) & \Pi_{22}(k) & * \\
\Pi_{31}(k) & \Pi_{32}(k) & \Pi_{33}(k)
\end{bmatrix} < 0
\tag{8.22}
$$

for all $0 \le k \le N$, where

$$
\Pi_{11}(k) = \mathcal{M}^{\mathrm{T}}(k)\mathcal{M}(k) - P(k) - \varepsilon_1 \tilde{\mathcal{U}}_1, \quad \Pi_{21}(k) = [-\varepsilon_1 \tilde{\mathcal{U}}_2^{\mathrm{T}} \quad 0]^{\mathrm{T}}, \quad \vec{E} = [0 \quad I \quad 0]^{\mathrm{T}},
$$

$$
\Pi_{22}(k) = \mathrm{diag}\{-\varepsilon_1 I, -\gamma^2 I\}, \quad \mathcal{G}_0(Pk) = P(k+1)\mathcal{G}_0(k) + Q(k)\vec{D}(k),
$$

$$
\Pi_{31}(k) =
\begin{bmatrix}
P(k+1)\mathcal{A}_0(k) + Q(k)\vec{C}(k) \\
\bar{C}(k) \\
0
\end{bmatrix},
\quad
\mathcal{A}_0(k) =
\begin{bmatrix}
\bar{A}(k) & 0 & 0 \\
0 & 0 & 0 \\
\bar{\Lambda}_\gamma \tilde{C}(k) & 0 & I - \bar{\Lambda}_\gamma
\end{bmatrix},
$$

$$
\Pi_{32}(k) =
\begin{bmatrix}
P(k+1)\mathcal{L}_1 & \mathcal{G}_0(Pk) \\
0 & \bar{D}(k) \\
\varsigma P(k+1)\mathcal{L}_2 & 0
\end{bmatrix},
\quad
\mathcal{C}_{oi}(k) =
\begin{bmatrix}
0 & 0 & 0 \\
0 & 0 & 0 \\
E_i \tilde{C}(k) & 0 & -E_i
\end{bmatrix},
$$

$$
\Pi_{33}(k) = \mathrm{diag}\{-P(k+1), -\bar{P}(k+1), -P(k+1)\}, \quad \bar{P}(k+1) = I_n \otimes P(k+1),
$$

$$
\bar{C}(k) = [\sigma_1 \mathcal{C}_{P1}^{\mathrm{T}} \quad \cdots \quad \sigma_n \mathcal{C}_{Pn}^{\mathrm{T}}]^{\mathrm{T}}, \quad \mathcal{C}_{Pi} = P(k+1)\mathcal{C}_{oi}(k) + Q(k)\hat{C}_i(k),
$$

$$
\bar{D}(k) = [\sigma_1 \mathcal{D}_{P1}^{\mathrm{T}} \quad \cdots \quad \sigma_n \mathcal{D}_{Pn}^{\mathrm{T}}]^{\mathrm{T}}, \quad \mathcal{D}_{Pi} = P(k+1)\mathcal{D}_{oi}(k) + Q(k)\hat{D}_i(k),
$$

$$
\vec{C}(k) =
\begin{bmatrix}
0 & I & 0 \\
\bar{\Lambda}_\gamma \tilde{C}(k) & 0 & I - \bar{\Lambda}_\gamma
\end{bmatrix},
\quad
\vec{D}(k) =
\begin{bmatrix}
0 & 0 \\
\bar{\Lambda}_\gamma \tilde{D}(k) & \bar{\Lambda}_\gamma
\end{bmatrix},
$$

$$
\hat{D}_i(k) =
\begin{bmatrix}
0 & 0 \\
E_i \tilde{D}(k) & E_i,
\end{bmatrix},
\quad
\mathcal{G}_0(k) =
\begin{bmatrix}
\bar{G}(k) & 0 \\
0 & 0 \\
\bar{\Lambda}_\gamma \tilde{D}(k) & \bar{\Lambda}_\gamma
\end{bmatrix},
$$

$$
\hat{C}_i(k) =
\begin{bmatrix}
0 & 0 & 0 \\
E_i \tilde{C}(k) & 0 & -E_i
\end{bmatrix},
\quad
\mathcal{D}_{oi}(k) =
\begin{bmatrix}
0 & 0 \\
0 & 0 \\
E_i \tilde{D}(k) & E_i
\end{bmatrix},
$$

$$
P(k) = \mathrm{diag}\{P_1(k), P_2(k), \ldots, P_n(k)\}, \quad R = \mathrm{diag}\{S_1, S_2, \ldots, S_n\}.
\tag{8.23}
$$

Moreover, if the inequalities (8.21) and (8.22) are feasible, the two matrices $\bar{K}(k)$ and $\bar{H}(k)$ are given as follows:

$$
[\bar{K}(k) \quad \bar{H}(k)] = (\vec{E}^{\mathrm{T}} P(k+1)\vec{E})^{-1} \vec{E}^{\mathrm{T}} Q(k) \quad (0 \le k \le N).
\tag{8.24}
$$

Accordingly, the desired filter parameters K_{ij} and H_{ij} ($i = 1, 2, \ldots, n, j \in \mathcal{N}_i$) can be obtained from (8.8).

Proof. In order to avoid partitioning the positive define matrices $\{P(k)\}_{0 \le k \le N+1}$ and reduce unnecessary conservatism, we rewrite the parameters in Theorem 8.2.1 in the following form:

$$\mathcal{A}(k) = \mathcal{A}_0(k) + \vec{E}\mathcal{L}(k)\vec{C}(k), \quad \mathcal{G}(k) = \mathcal{G}_0(k) + \vec{E}\mathcal{L}(k)\vec{D}(k),$$
$$\mathcal{C}_i(k) = \mathcal{C}_{0i}(k) + \vec{E}\mathcal{L}(k)\hat{C}_i(k), \quad \mathcal{D}_i(k) = \mathcal{D}_{0i}(k) + \vec{E}\mathcal{L}(k)\hat{D}_i(k), \tag{8.25}$$

where

$$\mathcal{L}(k) = [\bar{K}(k) \quad \bar{H}(k)]. \tag{8.26}$$

Let $P(k) = \text{diag}\{P_1(k), P_2(k), \ldots, P_n(k)\}$. Noticing (8.25) and using the Schur complement lemma, (8.22) can be obtained by (8.15) after some straightforward algebraic manipulations. In addition, from Lemma 8.2.2, it is easy to find the way to satisfy the conditions $\bar{K} \in \mathcal{T}_{n_x \times n_x}$ and $\bar{H} \in \mathcal{T}_{n_x \times n_y}$. The proof of this theorem is now complete. □

By means of Theorem 8.2.3, we can summarize the distributed filter design (DFD) algorithm as follows.

Algorithm DFD

Step 1. Given the average $\mathcal{H}_\infty$ performance index γ, the positive-definite matrix R, the initial conditions $x(0)$, $\hat{x}_i(0)$, and $y_i(-1)$, select the initial value for matrix $\{P(0)\}$ which satisfies the initial condition (8.21) and set $k = 0$.

Step 2. Obtain the positive-definite matrix $\{P(k + 1)\}$ and matrix $Q(k)$ for the sampling instant k by solving the RLMIs (8.22).

Step 3. Derive the distributed filter parameter matrices $\bar{K}(k)$ and $\bar{H}(k)$ by solving (8.24). Accordingly, the desired filter parameters K_{ij} and H_{ij} ($i = 1, 2, \ldots, n, j \in \mathcal{N}_i$) can be obtained from (8.8).

Step 4. If $k < N$, then go to Step 2, else go to Step 5.

Step 5. Stop.

Remark 8.4 *The main results obtained in Theorem 8.2.3 are quite general and encompass the following typically encountered features in wireless sensor networks: (1) coupling between the nodes according to a certain topology; (2) quantization errors and successive packet dropouts occurring in the channels from the plant to the sensors; (3) RVNs appearing in the target plant; and (4) average H_∞ performance guaranteeing the disturbance rejection capability of the filtering dynamics. Consequently, the distributed finite-horizon filters are designed by employing the RLMI approach that provides a better filtering accuracy since more information about the system under consideration is considered.*

8.3 An Illustrative Example

In this section, we present a simulation example to illustrate the effectiveness of the proposed distributed finite-horizon filter design scheme for randomly varying nonlinear systems with both quantization and successive packet dropouts for sensor networks.

The sensor network is represented by a directed graph $G = (V, \mathcal{E}, A)$ with a set of nodes $V = \{1, 2, 3, 4, 5\}$, a set of edges

$$\mathcal{E} = \{(1, 1), (1, 2), (1, 5), (2, 2), (2, 3), (3, 1), (3, 3), (4, 2), (4, 4), (5, 2), (5, 5)\},$$

and the following adjacency matrix:

$$A = \begin{bmatrix} 1 & 1 & 0 & 0 & 1 \\ 0 & 1 & 1 & 0 & 0 \\ 1 & 0 & 1 & 0 & 0 \\ 0 & 1 & 0 & 1 & 0 \\ 0 & 1 & 0 & 0 & 1 \end{bmatrix}.$$

The time-varying nonlinear discrete system considered is modeled by (8.1) with the parameters

$$A(k) = \begin{bmatrix} -0.6\sin(4k) & 0.2 \\ 0 & -0.8 \end{bmatrix}, \quad G(k) = [0.5 \quad 0.8]^{\mathrm{T}}, \quad M(k) = [0.1 \quad 0.1]$$

and the nonlinear functions

$$f(x(k)) = \begin{bmatrix} -0.5x_1(k) + 0.4x_2(k) \\ 0.1x_1(k) + \dfrac{\sin x_1(k)}{\sqrt{x_1^2(k) + x_2^2(k) + 10}} \end{bmatrix}, \quad g(x(k)) = \begin{bmatrix} \dfrac{x_1(k)}{4x_1^2(k) + 10} + 0.2x_2(k) \\ 0.7x_1(k) + 0.2x_2(k) \end{bmatrix}.$$

It is not difficult to verify that the above nonlinear functions f and g satisfy (8.3) with

$$U_1 = \begin{bmatrix} -0.2 & 0.4 \\ 0.05 & 0 \end{bmatrix}, \quad U_2 = \begin{bmatrix} -0.8 & 0.4 \\ 0.15 & 0 \end{bmatrix}, \quad U_3 = \begin{bmatrix} 0 & 0.3 \\ 0.2 & 0.2 \end{bmatrix}, \quad U_4 = \begin{bmatrix} 0 & 0.1 \\ 1.2 & 0.2 \end{bmatrix}.$$

The concerned sensors with both quantization and successive missing measurements are modeled by (8.2) with the following parameters:

$$C_1(k) = [0.2 \quad 0.1\sin(2k)], \quad D_1(k) = 0.1, \quad C_2(k) = [0.3 \quad 0.2], \quad D_2(k) = 1,$$

$$C_3(k) = [0.1 \quad 0.4\sin(2k)], \quad D_3(k) = 0.5, \quad C_4(k) = [0.3\sin(4k) \quad 0], \quad D_4(k) = 0.4,$$

$$C_5(k) = [0.2\sin(3k) \quad 0.1\sin(2k)], \quad D_5(k) = 1.$$

In this example, the probabilities are taken as $\bar{\beta} = 0.8$, $\bar{\gamma}_1 = 0.9$, $\bar{\gamma}_2 = 0.8$, $\bar{\gamma}_3 = 0.7$, $\bar{\gamma}_4 = 0.6$, and $\bar{\gamma}_5 = 0.86$. Moreover, the disturbance attenuation level is chosen as $\gamma = 1$. The positive-definite matrices are given as $S_i = \mathrm{diag}\{2, 2\}$ ($i = 1, 2, \ldots, 5$). Choosing the initial values $x(0) = [0.26 \quad -0.2]^{\mathrm{T}}$, $\hat{x}_i(0) = 0$ ($i = 1, 2, \ldots, 5$) and $y_i(-1) = 0$ ($i = 1, 2, \ldots, 5$), we can find the initial positive-definite matrix $P(0) = I$ to satisfy the initial condition (8.21). According to the given *DFD* algorithm, the RLMIs (8.22) in Theorem 8.2.3 can be solved recursively using Matlab (with the YALMIP 3.0). Accordingly, we can obtain the parameters

Table 8.1 Distributed filter parameters

k	0	1	$\cdots$
$K_{11}(k)$	$\begin{bmatrix} -0.0147 & -0.0126 \\ -0.0124 & -0.0124 \end{bmatrix}$	$\begin{bmatrix} 0.0030 & 0.0021 \\ 0.0023 & 0.0020 \end{bmatrix}$	$\cdots$
$K_{12}(k)$	$\begin{bmatrix} 0.3327 & 0.3412 \\ -0.0329 & -0.0338 \end{bmatrix}$	$\begin{bmatrix} 0.7784 & 0.7756 \\ 0.8707 & 0.8675 \end{bmatrix}$	$\cdots$
$K_{15}(k)$	$\begin{bmatrix} -0.0000 & 0.0000 \\ 0.0000 & -0.0000 \end{bmatrix}$	$\begin{bmatrix} -0.0000 & 0.0000 \\ 0.0000 & -0.0000 \end{bmatrix}$	$\cdots$
$K_{22}(k)$	$\begin{bmatrix} -0.2960 & -0.3035 \\ -0.5717 & -0.5858 \end{bmatrix}$	$\begin{bmatrix} 0.8973 & 0.8940 \\ 0.7370 & 0.7342 \end{bmatrix}$	$\cdots$
$K_{23}(k)$	$\begin{bmatrix} 0.2327 & -0.2567 \\ 0.1804 & -0.3098 \end{bmatrix}$	$\begin{bmatrix} -0.4074 & -0.3670 \\ -0.4292 & -0.3904 \end{bmatrix}$	$\cdots$
$K_{31}(k)$	$\begin{bmatrix} -0.0258 & -0.0557 \\ -0.3376 & -0.3128 \end{bmatrix}$	$\begin{bmatrix} -0.0019 & -0.0024 \\ 0.0001 & -0.0001 \end{bmatrix}$	$\cdots$
$K_{33}(k)$	$\begin{bmatrix} 0.2072 & 0.0913 \\ -0.0994 & 0.0118 \end{bmatrix}$	$\begin{bmatrix} -0.1612 & -0.0513 \\ -0.0564 & -0.0430 \end{bmatrix}$	$\cdots$
$K_{42}(k)$	$\begin{bmatrix} 0.2943 & 0.3004 \\ 0.2826 & 0.2896 \end{bmatrix}$	$\begin{bmatrix} 0.9508 & 0.9493 \\ 0.6878 & 0.6850 \end{bmatrix}$	$\cdots$
$K_{44}(k)$	$\begin{bmatrix} -0.2374 & -0.5779 \\ -0.1982 & -0.5253 \end{bmatrix}$	$\begin{bmatrix} -0.0201 & -0.0196 \\ -0.0195 & -0.0190 \end{bmatrix}$	$\cdots$
$K_{52}(k)$	$\begin{bmatrix} 0.2681 & 0.2748 \\ 0.2990 & 0.3054 \end{bmatrix}$	$\begin{bmatrix} 0.6057 & 0.5916 \\ 0.6040 & 0.5902 \end{bmatrix}$	$\cdots$
$K_{55}(k)$	$0_{2\times 2}$	$0_{2\times 2}$	$\cdots$
$H_{11}(k)$	$[0.0079 \quad -0.0021]^{\mathrm{T}}$	$[0.0131 \quad 0.0425]^{\mathrm{T}}$	$\cdots$
$H_{12}(k)$	$[0.0162 \quad -0.0016]^{\mathrm{T}}$	$[0.0247 \quad 0.0277]^{\mathrm{T}}$	$\cdots$
$H_{15}(k)$	$0_{2\times 1}$	$0_{2\times 1}$	$\cdots$
$H_{22}(k)$	$[-0.0014 \quad -0.0029]^{\mathrm{T}}$	$[0.0285 \quad 0.0237]^{\mathrm{T}}$	$\cdots$
$H_{23}(k)$	$[-0.0058 \quad -0.0092]^{\mathrm{T}}$	$[0.0643 \quad 0.0494]^{\mathrm{T}}$	$\cdots$
$H_{31}(k)$	$[0.9840 \quad 0.2144]^{\mathrm{T}}$	$[0.0077 \quad 0.0025]^{\mathrm{T}}$	$\cdots$
$H_{33}(k)$	$[-0.0059 \quad 0.0015]^{\mathrm{T}}$	$[0.0537 \quad 0.0081]^{\mathrm{T}}$	$\cdots$
$H_{42}(k)$	$[0.1686 \quad 0.1643]^{\mathrm{T}}$	$[0.2420 \quad 0.2431]^{\mathrm{T}}$	$\cdots$
$H_{44}(k)$	$[0.0703 \quad 0.0677]^{\mathrm{T}}$	$[0.0460 \quad 0.0376]^{\mathrm{T}}$	$\cdots$
$H_{52}(k)$	$[0.1561 \quad 0.1558]^{\mathrm{T}}$	$[0.1635 \quad 0.1645]^{\mathrm{T}}$	$\cdots$
$H_{55}(k)$	$0_{2\times 1}$	$0_{2\times 1}$	$\cdots$

of the desired distributed filters, which are listed in Table 8.1, and the optimized performance index $\gamma^* = 0.7468$.

In the simulation, the exogenous disturbance input is selected as $w(k) = \exp(-k)$ and the quantization errors given by $v_i(k) = \frac{\sin(10k+1)}{3k+1}$ ($i = 1, 2, \ldots, 5$). Simulation results are shown in Figures 8.1 and 8.2. Figure 8.1 shows the output $z(k)$ and its estimates from the filters i. Figure 8.2 plots the filtering errors $z(k) - \hat{z}_i(k)$ ($i = 1, 2, \ldots, 5$). The simulation results have

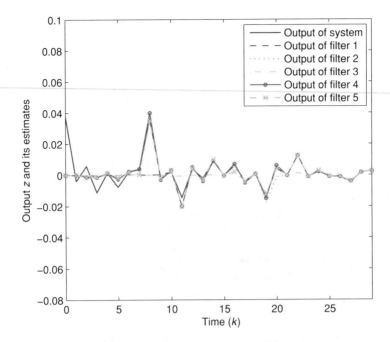

Figure 8.1   Output $z(k)$ and its estimates

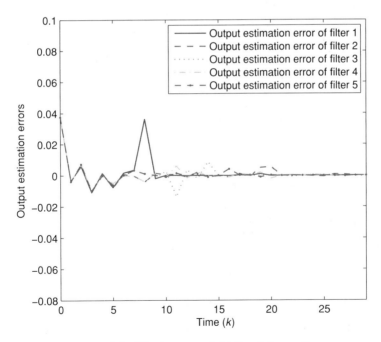

Figure 8.2   Filtering errors $\tilde{z}_i(k)$ $(i = 1, 2, \ldots, 5)$

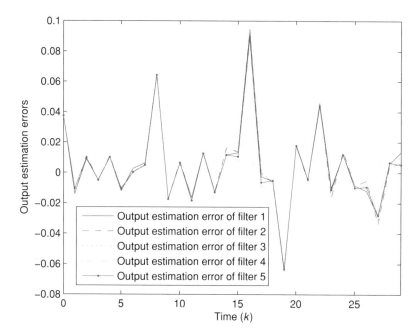

Figure 8.3   Filtering errors with successive packet dropouts

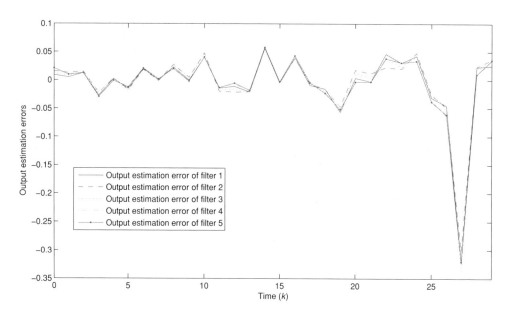

Figure 8.4   Filtering errors with quantization errors and successive packet dropouts

confirmed the effectiveness of the distributed finite-horizon filtering technique presented in this chapter.

To further illustrate the effects of the quantization errors and successive packet dropouts on the performance of the distributed filters, we then consider the case that $\bar{\gamma}_i = 0$ ($i = 1, 2, \ldots, 5$); that is, the ith sensor suffers from the packet dropouts only. The other system data remain the same as before. In this case, the filtering errors $z(k) - \hat{z}_i(k)$ ($i = 1, 2, \ldots, 5$) are shown in Figure 8.3, and the optimized performance index is calculated as $\gamma^* = 0.9233$. Furthermore, when $\bar{\gamma}_1 = 0.9$, $\bar{\gamma}_2 = 0.8$, $\bar{\gamma}_3 = 0.7$, $\bar{\gamma}_4 = 0.6$, and $\bar{\gamma}_5 = 0.86$ (corresponding to the case that the sensors undergo both quantization errors and successive packet dropouts), the filtering errors $z(k) - \hat{z}_i(k)$ ($i = 1, 2, \ldots, 5$) are shown in Figure 8.4 and the optimized performance index can be obtained as $\gamma^* = 1.5641$. All the illustrations demonstrate the effectiveness of the filtering approach developed. Moreover, by comparing Figure 8.2 with Figure 8.4 and noting the optimized performance indices in these three cases, it can be concluded that, when the sensor network only suffers from successive packet dropouts or quantization errors, the proposed distributed filtering scheme has given a better disturbance rejection attenuation level than that for the case when both successive packet dropouts and quantization errors occur.

8.4 Summary

In this chapter, we have dealt with the distributed finite-horizon filtering problem for a class of discrete time-varying systems with RVNs over lossy sensor networks involving quantization errors and successive packet dropouts. The issue of RVNs has been addressed, and then the filtering dynamics has been analyzed by modeling quantization and successive packet dropouts in a unified framework. A new distributed finite-horizon filtering technique by means of a set of RLMIs has been proposed to satisfy the prescribed average filtering performance constraint. Finally, an illustrative example has been provided that highlights the usefulness of the filtering approach developed.

9

Distributed Filtering for Markov Jump Nonlinear Time-Delay Systems

This chapter is concerned with the distributed H_∞ filtering problem for a class of discrete-time Markovian jump nonlinear time-delay systems with deficient statistics of modes transitions. The system measurements are collected through a lossy sensor network subject to randomly occurring quantization errors (ROQEs) and randomly occurring packet dropouts (ROPDs). The system model (dynamical plant) includes the mode-dependent Lipschitz-like nonlinearities. The description of deficient statistics of modes transitions is comprehensive and accounts for known, unknown, and uncertain TPs. Two sets of Bernoulli-distributed white sequences are introduced to govern the phenomena of ROQEs and ROPDs in the lossy sensor network. We aim to design the distributed H_∞ filters through available system measurements from not only the individual sensor but also its neighboring sensors according to a given topology. The stability analysis is first carried out to obtain sufficient conditions for ensuring stochastic stability as well as the prescribed average H_∞ performance constraint for the dynamics of the estimation errors, and then a filter design scheme is outlined by explicitly characterizing the filter gains in terms of some matrix inequalities. Simulation results demonstrate the effectiveness of the proposed filtering scheme.

9.1 Problem Formulation

9.1.1 Deficient Statistics of Markovian Modes Transitions

Before proceeding to the system description, we first discuss the deficient statistics of the mode transition in order to facilitate further technical development. Let $r(k)$ $(k \geq 0)$ be a Markov chain taking values in a finite state space $S = \{1, 2, \ldots, s\}$ with TP matrix $\Psi = [\lambda_{lj}]$ given by

$$\text{Prob}\,\{r(k+1) = j | r(k) = l\} = \lambda_{lj}, \ \forall l, j \in S,$$

Filtering, Control and Fault Detection with Randomly Occurring Incomplete Information, First Edition.
Hongli Dong, Zidong Wang, and Huijun Gao.
© 2013 John Wiley & Sons, Ltd. Published 2013 by John Wiley & Sons, Ltd.

where $\lambda_{lj} \geq 0$ $(l, j \in S)$ is the TP from l to j and $\sum_{j=1}^{s} \lambda_{lj} = 1$, $\forall l \in S$. Accordingly, the Markovian TP matrix Ψ is defined by

$$\Psi = \begin{bmatrix} \lambda_{11} & \lambda_{12} & \cdots & \lambda_{1s} \\ \lambda_{21} & \lambda_{22} & \cdots & \lambda_{2s} \\ & & \ddots & \\ \lambda_{s1} & \lambda_{s2} & \cdots & \lambda_{ss} \end{bmatrix}. \tag{9.1}$$

Generally speaking, for some practical systems, the statistics of modes transitions is deficient (imperfectly known); that is, some elements in the TP matrix Ψ are unknown or uncertain within given intervals in terms of certain statistics characters. As such, it is assumed in this chapter that the TP matrix $\Psi = [\lambda_{lj}]_{s \times s}$ belongs to a given polytope, namely $\Psi \in P_\Psi$, where P_Ψ is a given convex-bounded polyhedral domain described by M vertices $\Psi^{(r)}$ $(r = 1, 2, \ldots, M)$ as follows:

$$P_\Psi := \left\{ \Psi \mid \Psi = \sum_{r=1}^{M} a_r \Psi^{(r)}, \quad \sum_{r=1}^{M} a_r = 1, \quad a_r \geq 0 \right\}. \tag{9.2}$$

Here, $\Psi^{(r)} = [\lambda_{lj}^{(r)}]_{s \times s}$ $(l, j \in S, r = 1, 2, \ldots M)$ are given TP matrices that still contain unknown elements. It is easy to see that the convex combination of these TP matrices is also a possible TP matrix.

For notation clarity, for any $l \in S$, we denote $S := S_{\mathcal{K}}^l \cup S_{\mathcal{UC}}^l \cup S_{\mathcal{UK}}^l$ where

$$S_{\mathcal{K}}^l := \{j : \lambda_{lj} \text{ is known}\}, \quad S_{\mathcal{UC}}^l := \{j : \lambda_{lj} \text{ is uncertain}\}, \quad S_{\mathcal{UK}}^l := \{j : \lambda_{lj} \text{ is unknown}\}. \tag{9.3}$$

Also, we define $\lambda_{\mathcal{K}}^l := \sum_{j \in S_{\mathcal{K}}^l} \lambda_{lj}$ and $\lambda_{\mathcal{UC}}^l := \sum_{j \in S_{\mathcal{UC}}^l} \lambda_{lj}^{(r)}$, $\forall r = 1, 2, \ldots, M$.

Remark 9.1 *In practice, it is usually difficult or expensive to obtain all or part of the TPs even for a simple system. As such, the ideal requirement on the precise knowledge of TPs would inevitably limit the application of established results, and this has led to the recent investigations on MJSs with deficient statistics of modes transitions. For example, in (9.1), some TPs have been assumed to be imperfectly known; that is, some TPs are uncertain but within given intervals (i.e., uncertain TPs), and some do not even have such available intervals/bounds (i.e., unknown TPs). The descriptions of uncertain and unknown TPs allow different levels of the limitation to better reflect the actual situation. Furthermore, the concept of defective statistics of modes transitions was proposed by Zhao et al. [110] for a class of neural networks. In this chapter, the concept of deficient statistics of modes transitions will be applied for the distributed filtering problems, which is of more practical significance in a sensor networked environment.*

9.1.2   The Network Topology

The sensor network has n sensor nodes which are distributed in space according to a fixed network topology represented by a directed graph $\mathcal{G} = (\mathcal{V}, \mathcal{E}, \mathcal{A})$ of order n with a set of nodes

$\mathcal{V} = 1, 2, \ldots, n$, a set of edges $\mathcal{E} \in \mathcal{V} \times \mathcal{V}$, and the weighted adjacency matrix $\mathcal{A} = [a_{ij}]$ with nonnegative adjacency element a_{ij}. An edge of $\mathcal{G}$ is denoted by ordered pair (i, j). The adjacency elements associated with the edges of the graph are positive (i.e., $a_{ij} > 0 \Longleftrightarrow (i, j) \in \mathcal{E}$) which means that sensor i can obtain information from sensor j. Also, we assume that $a_{ii} = 1$ for all $i \in \mathcal{V}$, and therefore (i, i) can be regarded as an additional edge. The set of neighbors of node $i \in \mathcal{V}$ plus the node itself are denoted by $\mathcal{N}_i = \{j \in \mathcal{V} : (i, j) \in \mathcal{E}\}$.

9.1.3　Descriptions of the Target Plant and the Sensor Network

Consider the following class of discrete Markovian jump nonlinear time-delay systems (dynamical plants):

$$\begin{cases} x(k+1) = f(r(k), x(k)) + g(r(k), x(k - \tau(k))) + G(r(k))w(k), \\ z(k) = M(r(k))x(k), \\ x(j) = \varphi(j), \qquad j = -d_M, -d_M + 1, \ldots, 0, \end{cases} \tag{9.4}$$

where $x(k) \in \mathbb{R}^{n_x}$ represents the state vector, $z(k) \in \mathbb{R}^{n_z}$ is the output to be estimated, $w(k) \in \mathbb{R}^{n_w}$ is the disturbance input which belongs to $l_2[0, \infty)$, and $\varphi(j)$ is the initial state of the system. For fixed system mode, $G(r(k))$ and $M(r(k))$ are constant matrices with appropriate dimensions. $\tau(k)$ is a *known* time-varying communication delay with the following constraint:

$$d_m \leq \tau(k) \leq d_M$$

where d_m and d_M are constant positive scalars representing the lower and upper bounds on the communication delays, respectively.

The nonlinear functions $f(\cdot, \cdot)$ and $g(\cdot, \cdot)$ are assumed to satisfy $f(r(k), 0) = 0$, $g(r(k), 0) = 0$, and the following conditions:

$$\begin{aligned} \| f(r(k), x(k) + \sigma(k)) - f(r(k), x(k)) - A(r(k))\sigma(k)\| &\leq b_1(r(k))\|\sigma(k)\|, \\ \|g(r(k), x(k - \tau(k)) + \sigma(k)) - g(r(k), x(k - \tau(k))) - A_d(r(k))\sigma(k)\| &\leq b_2(r(k))\|\sigma(k)\|, \end{aligned} \tag{9.5}$$

where, for fixed system mode, $A(r(k))$ and $A_d(r(k))$ are known matrices, and $b_1(r(k))$ and $b_2(r(k))$ are known positive scalars.

Remark 9.2　*In (9.5), the nonlinear functions $f(r(k), x(k))$ and $g(r(k), x(k - \tau(k)))$, which are called the Lipschitz-like nonlinearities, quantify the distance between the nonlinear system in (9.4) and the corresponding linear model with $A(r(k))$ and $A_d(r(k))$ as its system parameter matrices. Different from Refs [41, 178], in this chapter, such nonlinearities are dependent on the Markovian jumping mode and, therefore, more general.*

In this chapter, the n sensors are modeled by

$$\begin{aligned} y_i(k) &= \alpha_i(k)h(C_i(r(k))x(k)) + (1 - \alpha_i(k))\beta_i(k)C_i(r(k))x(k) + D_i(r(k))w(k), \\ & \qquad i = 1, 2, \ldots, n, \end{aligned} \tag{9.6}$$

where $y_i(k) \in \mathbb{R}^{n_y}$ is the measurement output collected by sensor i from the plant, $h(\cdot)$ is the round-off function representing the quantization effect, $\alpha_i(k)$ and $\beta_i(k)$ ($i = 1, 2, \ldots, n$) are Bernoulli-distributed white sequences taking values on 0 and 1 with

$$
\begin{aligned}
\text{Prob}\{\alpha_i(k) = 1\} &= \bar{\alpha}_i, \quad \text{Prob}\{\alpha_i(k) = 0\} = 1 - \bar{\alpha}_i, \\
\text{Prob}\{\beta_i(k) = 1\} &= \bar{\beta}_i, \quad \text{Prob}\{\beta_i(k) = 0\} = 1 - \bar{\beta}_i,
\end{aligned}
\tag{9.7}
$$

where $\bar{\alpha}_i \in [0, 1]$ and $\bar{\beta}_i \in [0, 1]$ are known constants. Throughout the chapter, we assume that the $r(k)$, $\alpha_i(k)$, and $\beta_i(k)$ are mutually independent in all i ($1 \leq i \leq n$). Moreover, all the matrices mentioned above are known matrices with respect to fixed system mode $r(k)$.

The quantization error is bounded as follows:

$$
h(C_i(r(k))x(k)) - C_i(r(k))x(k) = v_i(k), \quad \|v_i(k)\| \leq \delta \quad \text{and} \quad \delta > 0.
\tag{9.8}
$$

Remark 9.3 *The proposed measurement model in (9.6) provides a novel yet unified framework by resorting to the random variables $\alpha_i(k)$ and $\beta_i(k)$ to account for the randomly occurring phenomena of both packet dropouts and quantizations. It is well known that quantization error or round-off error exists very often in nearly all digital signal processing problems; whenever there is a need to represent a signal in digital form, this ordinarily involves rounding. The error signal is sometimes considered as an additional random signal called quantization noise because of its stochastic behavior. Usually, the quantization error is not significantly correlated with the signal, and has an approximately uniform distribution. In (9.8), for simplicity, the quantization error is assumed to be a noise with bounded norm. As will be seen later, one of the purposes of the distributed filter design is to reduce the influence from the quantization errors to the filter performance by introducing the average H_∞-index for disturbance attenuation and rejection over the given finite horizon.*

The distributed filter on the ith sensor node is of the following form:

$$
\begin{cases}
\hat{x}_i(k + 1) = f(r(k), \hat{x}_i(k)) + g(r(k), \hat{x}_i(k - \tau(k))) + \displaystyle\sum_{j \in \mathcal{N}_i} a_{ij} K_{ij}(r(k)) \\
\qquad \times [y_j(k) - \bar{\alpha}_j C_j(r(k))\hat{x}_j(k) - (1 - \bar{\alpha}_j)\bar{\beta}_j C_j(r(k))\hat{x}_j(k)], \\
\hat{z}_i(k) = M(r(k))\hat{x}_i(k),
\end{cases}
\tag{9.9}
$$

where $\hat{x}_i(k) \in \mathbb{R}^{n_x}$ is the state estimate from sensor node i and $\hat{z}_i(k) \in \mathbb{R}^{n_z}$ is the estimate of $z(k)$ from the filter on sensor node i. Here, matrices $K_{ij}(r(k))$ are the filter parameters on node i to be determined.

For notational simplification, we use the following notation:

$$
\begin{aligned}
m_1(r(k), k) &= f(r(k), x(k)) - f(r(k), \hat{x}_i(k)) - A(r(k))(x(k) - \hat{x}_i(k)), \\
m_2(r(k), k - \tau(k)) &= g(r(k), x(k - \tau(k))) - g(r(k), \hat{x}_i(k - \tau(k))) \\
&\quad - A_d(r(k))(x(k - \tau(k)) - \hat{x}_i(k - \tau(k))).
\end{aligned}
\tag{9.10}
$$

Letting $e_i(k) = x(k) - \hat{x}_i(k)$ and $\tilde{z}_i(k) = z(k) - \hat{z}_i(k)$ $(i = 1, 2, \ldots, n)$, we obtain the following system that governs the filtering error dynamics for the sensor network:

$$e_i(k+1) = A(r(k))e_i(k) + A_d(r(k))e_i(k - \tau(k)) + m_1(r(k), k) + m_2(r(k), k - \tau(k))$$

$$- \sum_{j \in \mathcal{N}_i} a_{ij} K_{ij}(r(k)) \bar{\alpha}_j v_j(k) + \left[G(r(k)) - \sum_{j \in \mathcal{N}_i} a_{ij} K_{ij}(r(k)) D_j(r(k)) \right] w(k)$$

$$- \sum_{j \in \mathcal{N}_i} a_{ij} K_{ij}(r(k))(\alpha_j(k) - \bar{\alpha}_j) v_j(k) - \sum_{j \in \mathcal{N}_i} a_{ij} K_{ij}(r(k)) \bar{\alpha}_j C_j(r(k)) e_j(k)$$

$$- \sum_{j \in \mathcal{N}_i} a_{ij} K_{ij}(r(k))[(1 - \alpha_j(k))\beta_j(k) - (1 - \bar{\alpha}_j)\bar{\beta}_j] C_j(r(k)) x(k)$$

$$- \sum_{j \in \mathcal{N}_i} a_{ij} K_{ij}(r(k))(\alpha_j(k) - \bar{\alpha}_j) C_j(r(k)) x(k)$$

$$- \sum_{j \in \mathcal{N}_i} a_{ij} K_{ij}(r(k))(1 - \bar{\alpha}_j)\bar{\beta}_j C_j(r(k)) e_j(k),$$

$$\tilde{z}_i(k) = M(r(k)) e_i(k). \tag{9.11}$$

For convenience of later analysis, we denote

$$\bar{A}(r(k)) = I_n \otimes A(r(k)), \quad \bar{A}_d(r(k)) = I_n \otimes A_d(r(k)),$$

$$e(k) = \begin{bmatrix} e_1^T(k) & e_2^T(k) & \cdots & e_n^T(k) \end{bmatrix}^T, \quad \bar{x}(k) = \mathbf{1}_n \otimes x(k), \quad E_i = \kappa_i \kappa_i^T,$$

$$\bar{z}(k) = \begin{bmatrix} \tilde{z}_1^T(k) & \tilde{z}_2^T(k) & \cdots & \tilde{z}_n^T(k) \end{bmatrix}^T, \quad M_1(r(k), k) = \mathbf{1}_n \otimes m_1(r(k), k),$$

$$M_2(r(k), k - \tau(k)) = \mathbf{1}_n \otimes m_2(r(k), k - \tau(k)), \quad \bar{\Lambda}_\alpha = \text{diag}\{\bar{\alpha}_1 I, \bar{\alpha}_2 I, \ldots, \bar{\alpha}_n I\}, \tag{9.12}$$

$$\bar{D}(r(k)) = \begin{bmatrix} D_1^T(r(k)) & D_2^T(r(k)) & \cdots & D_n^T(r(k)) \end{bmatrix}^T, \quad \bar{M}(r(k)) = I_n \otimes M(r(k)),$$

$$\bar{C}(r(k)) = \text{diag}\{C_1(r(k)), C_2(r(k)), \ldots, C_n(r(k))\}, \quad \bar{G}(r(k)) = \mathbf{1}_n \otimes G(r(k)),$$

$$\bar{\Lambda}_\beta = \text{diag}\{\bar{\beta}_1 I, \bar{\beta}_2 I, \ldots, \bar{\beta}_n I\}, \quad v(k) = \begin{bmatrix} v_1^T(k) & v_2^T(k) & \cdots & v_n^T(k) \end{bmatrix}^T,$$

where

$$\bar{K}(r(k)) = \begin{bmatrix} \bar{K}_{ij}(r(k)) \end{bmatrix}_{n \times n},$$

$$\bar{K}_{ij}(r(k)) = \begin{cases} a_{ij} K_{ij}(r(k)), & i = 1, 2, \ldots, n; \quad j \in \mathcal{N}_i, \\ 0, & i = 1, 2, \ldots, n; \quad j \notin \mathcal{N}_i. \end{cases} \tag{9.13}$$

Obviously, since $a_{ij} = 0$ when $j \notin \mathcal{N}_i$, $\bar{K}(r(k))$ is a matrix that can be expressed as

$$\bar{K}(r(k)) \in \mathcal{T}_{n_x \times n_y}, \tag{9.14}$$

where $\mathcal{T}_{p \times q} = \{\bar{U} = [U_{ij}] \in \mathbb{R}^{np \times nq} \mid U_{ij} \in \mathbb{R}^{p \times q}, \ U_{ij} = 0 \quad \text{if } j \notin \mathcal{N}_i\}$. Then, we have the following error dynamics compact form:

$$
\begin{aligned}
e(k+1) = {}& (\bar{A}(r(k)) - \bar{K}(r(k))\bar{\Lambda}_\alpha \bar{C}(r(k)) - \bar{K}(r(k))(I - \bar{\Lambda}_\alpha)\bar{\Lambda}_\beta \bar{C}(r(k)))e(k) \\
& + \bar{A}_d(r(k))e(k - \tau(k)) + M_1(r(k), k) + M_2(r(k), k - \tau(k)) \\
& + [\bar{G}(r(k)) - \bar{K}(r(k))\bar{D}(r(k))]w(k) - \bar{K}(r(k))\bar{\Lambda}_\alpha v(k) \\
& - \sum_{i=1}^{n}(\alpha_i(k) - \bar{\alpha}_i)\bar{K}(r(k))E_i v(k) - \sum_{i=1}^{n}(\alpha_i(k) - \bar{\alpha}_i)\bar{K}(r(k)) \\
& \times E_i \bar{C}(r(k))\bar{x}(k) - \sum_{i=1}^{n}[(1 - \alpha_i(k))\beta_i(k) - (1 - \bar{\alpha}_i)\bar{\beta}_i] \\
& \times \bar{K}(r(k))E_i \bar{C}(r(k))\bar{x}(k) \\
\bar{z}(k) = {}& \bar{M}(r(k))e(k).
\end{aligned}
\tag{9.15}
$$

Note that the set S comprises various operation modes of the system in (9.15) and the Markov chain $\{r(k), k \in [0, \infty)\}$ takes values in the finite set $S = \{1, 2, \ldots, s\}$. For presentation purposes, for each possible $r(k) = l$ ($l \in S$), a matrix $N(r(k))$ will be denoted by N_l.

By letting $\eta(k) = [\bar{x}^{\mathrm{T}}(k) \quad e^{\mathrm{T}}(k)]^{\mathrm{T}}$, the error dynamics governed by (9.15) can be rewritten as

$$
\begin{aligned}
\eta(k+1) = {}& \mathcal{Y}_l - \sum_{i=1}^{n}(\alpha_i(k) - \bar{\alpha}_i)(\mathcal{K}_{li}v(k) - \mathcal{C}_{li}\eta(k)) \\
& - \sum_{i=1}^{n}((1 - \alpha_i(k))\beta_i(k) - (1 - \bar{\alpha}_i)\bar{\beta}_i)\mathcal{C}_{li}\eta(k), \\
\bar{z}(k) = {}& \mathcal{M}_l \eta(k),
\end{aligned}
\tag{9.16}
$$

where

$$
\begin{aligned}
\mathcal{Y}_l &= -\bar{\mathcal{C}}_l \eta(k) + \mathcal{A}_{dl}\eta(k - \tau(k)) + \mathcal{H}\mathcal{F}_l(\eta(k)) + \mathcal{G}_l w(k) + \bar{\mathcal{K}}_l v(k), \\
\mathcal{C}_{li} &= \begin{bmatrix} 0 & 0 \\ \bar{K}_l E_i \bar{C}_l & 0 \end{bmatrix}, \quad \mathcal{H} = \begin{bmatrix} I & I & 0 & 0 \\ 0 & 0 & I & I \end{bmatrix}, \quad \mathcal{G}_l = \begin{bmatrix} \bar{G}_l \\ \bar{G}_l - \bar{K}_l \bar{D}_l \end{bmatrix}, \\
\bar{\mathcal{K}}_l &= \begin{bmatrix} 0 & -(\bar{K}_l \bar{\Lambda}_\alpha)^{\mathrm{T}} \end{bmatrix}^{\mathrm{T}}, \quad \mathcal{A}_{dl} = I_2 \otimes \bar{A}_{dl}, \quad \mathcal{K}_{li} = \begin{bmatrix} 0 & (\bar{K}_l E_i)^{\mathrm{T}} \end{bmatrix}^{\mathrm{T}}, \\
\bar{\mathcal{C}}_l &= \mathrm{diag}\{-\bar{A}_l, -\bar{A}_l + \bar{K}_l \bar{\Lambda}_\alpha \bar{C}_l + \bar{K}_l(I - \bar{\Lambda}_\alpha)\bar{\Lambda}_\beta \bar{C}_l\}, \\
\mathcal{F}_l(\eta(k)) &= \begin{bmatrix} \bar{f}_l^{\mathrm{T}}(x(k)) & \bar{g}_l^{\mathrm{T}}(x(k - \tau(k))) & M_{1l}^{\mathrm{T}}(k) & M_{2l}^{\mathrm{T}}(k - \tau(k)) \end{bmatrix}^{\mathrm{T}}, \\
\mathcal{M}_l &= \begin{bmatrix} 0 & \bar{M}_l \end{bmatrix}, \quad \bar{f}_l(x(k)) = \mathbf{1}_n \otimes (f_l(x(k)) - A_l x(k)), \\
\bar{g}_l(x(k - \tau(k))) &= \mathbf{1}_n \otimes (g_l(x(k - \tau(k))) - A_{dl}x(k - \tau(k))).
\end{aligned}
\tag{9.17}
$$

Moreover, it follows from (9.5) that

$$\|\mathcal{F}_l(\eta(k))\|^2 \le b_{1l}^2 \|\pi_1 \eta(k)\|^2 + b_{2l}^2 \|\pi_2 \eta(k - \tau(k))\|^2, \tag{9.18}$$

where

$$\pi_1 = \begin{bmatrix} I & 0 & 0 & 0 \\ 0 & 0 & I & 0 \end{bmatrix}^{\mathrm{T}}, \quad \pi_2 = \begin{bmatrix} 0 & I & 0 & 0 \\ 0 & 0 & 0 & I \end{bmatrix}^{\mathrm{T}}. \tag{9.19}$$

Before proceeding further, we introduce the following definition.

Definition 9.1.1　*The discrete-time stochastic system (9.16) is said to be stochastically stable if, in case of $w(k) = 0$ and $v(k) = 0$, for any initial conditions $(\varphi(i), r(0))$, the following holds:*

$$\mathbb{E}\left\{ \sum_{k=0}^{\infty} \|\eta(k)\|^2 \ \middle| \ \varphi(i), r(0) \right\} < \infty.$$

Our aim in this chapter is to design a filter of the form (9.9) on each node i of the sensor network for the dynamical plant (9.4). In other words, we are going to find the filter parameters K_{ijl} $(l \in S)$ such that the following two requirements are satisfied simultaneously:

- *Stochastic stability.* The zero-solution of the error dynamics system (9.16) is stochastically stable.
- *Average H_∞ performance.* Under the zero initial condition, for a given disturbance attenuation level $\gamma > 0$ and all nonzero $w(k)$ and $v(k)$, the filtering error $\bar{z}(k)$ from (9.16) satisfies the following condition (which is called average H_∞ performance constraint in this chapter):

$$J = \frac{1}{n}\mathbb{E}\left\{ \sum_{k=0}^{\infty} \bar{z}^{\mathrm{T}}(k)\bar{z}(k) \right\} - \gamma^2 \left(\sum_{k=0}^{\infty} w^{\mathrm{T}}(k)w(k) + \frac{1}{n}\sum_{k=0}^{\infty} v^{\mathrm{T}}(k)v(k) \right) < 0. \tag{9.20}$$

9.2　Main Results

In this section, we investigate both the filter analysis and design problems for the distributed H_∞ filtering of system (9.4) with n sensors whose topology is determined by the given graph $\mathcal{G} = (\mathcal{V}, \mathcal{E}, \mathcal{A})$.

First, we propose the following average H_∞ performance analysis results with completely known TPs.

Theorem 9.2.1　*For given filter parameters K_{ijl} $(l \in S)$ and a prescribed disturbance attenuation level $\gamma > 0$, the filtering dynamics in (9.16) is stochastically stable and also satisfies* *the average H_∞ performance constraint (9.20) if there exist a set of positive-definite matrix $P_l > 0$ ($l \in S$), a positive-definite matrix $Q > 0$, and a positive scalar ε satisfying*

$$\Gamma_l = \begin{bmatrix} \Gamma_{11l} & * & * & * & * \\ -\mathcal{A}_{dl}^{\mathrm{T}} \bar{P}_l \bar{C}_l & \Gamma_{22l} & * & * & * \\ -\mathcal{H}^{\mathrm{T}} \bar{P}_l \bar{C}_l & \mathcal{H}^{\mathrm{T}} \bar{P}_l \mathcal{A}_{dl} & \mathcal{H}^{\mathrm{T}} \bar{P}_l \mathcal{H} - \varepsilon I & * & * \\ -\mathcal{G}_l^{\mathrm{T}} \bar{P}_l \bar{C}_l & \mathcal{G}_l^{\mathrm{T}} \bar{P}_l \mathcal{A}_{dl} & \mathcal{G}_l^{\mathrm{T}} \bar{P}_l \mathcal{H} & \mathcal{G}_l^{\mathrm{T}} \bar{P}_l \mathcal{G}_l - \gamma^2 I & * \\ \Gamma_{51l} & \bar{\mathcal{K}}_l^{\mathrm{T}} \bar{P}_l \mathcal{A}_{dl} & \bar{\mathcal{K}}_l^{\mathrm{T}} \bar{P}_l \mathcal{H} & \bar{\mathcal{K}}_l^{\mathrm{T}} \bar{P}_l \mathcal{G}_l & \Gamma_{55l} \end{bmatrix} < 0, \qquad (9.21)$$

where

$$\Gamma_{11l} = \hat{\Gamma}_{11l} + \frac{1}{n} \mathcal{M}_l^{\mathrm{T}} \mathcal{M}_l + \varepsilon b_{1l} \pi_1^{\mathrm{T}} \pi_1, \quad \Gamma_{22l} = \mathcal{A}_{dl}^{\mathrm{T}} \bar{P}_l \mathcal{A}_{dl} + \varepsilon b_{2l} \pi_2^{\mathrm{T}} \pi_2 - Q,$$

$$\Gamma_{51l} = -\bar{\mathcal{K}}_l^{\mathrm{T}} \bar{P}_l \bar{C}_l - \sum_{i=1}^{n} (\phi_i^2 + \hat{\phi}_i^2) \mathcal{K}_{li}^{\mathrm{T}} \bar{P}_l \mathcal{C}_{li}, \ \phi_i^2 = \bar{\alpha}_i (1 - \bar{\alpha}_i),$$

$$\Gamma_{55l} = \bar{\mathcal{K}}_l^{\mathrm{T}} \bar{P}_l \bar{\mathcal{K}}_l + \sum_{i=1}^{n} \phi_i^2 \mathcal{K}_{li}^{\mathrm{T}} \bar{P}_l \mathcal{K}_{li} - \frac{1}{n} \gamma^2 I, \ \bar{P}_l = \sum_{j \in S} \lambda_{lj} P_j,$$

$$\hat{\Gamma}_{11l} = \bar{\mathcal{C}}_l^{\mathrm{T}} \bar{P}_l \bar{\mathcal{C}}_l + \sum_{i=1}^{n} (\phi_i^2 + \bar{\phi}_i^2 + 2\hat{\phi}_i^2) \mathcal{C}_{li}^{\mathrm{T}} \bar{P}_l \mathcal{C}_{li} + (d_M - d_m + 1) Q - P_l,$$

$$\bar{\phi}_i^2 = (1 - \bar{\alpha}_i) \bar{\beta}_i - (1 - \bar{\alpha}_i)^2 \bar{\beta}_i^2, \ \hat{\phi}_i^2 = \bar{\alpha}_i (1 - \bar{\alpha}_i) \bar{\beta}_i.$$

Proof. Let $\Theta(k) := \{\eta(k), \eta(k-1), \ldots, \eta(k-d_M)\}$ and consider the following Lyapunov functional candidate for system (9.16):

$$V(\Theta(k), r(k)) = V_1(\Theta(k), r(k)) + V_2(\Theta(k), r(k)) + V_3(\Theta(k), r(k))$$

$$= \eta^{\mathrm{T}}(k) P(r(k)) \eta(k) + \sum_{i=k-\tau(k)}^{k-1} \eta^{\mathrm{T}}(i) Q \eta(i) + \sum_{m=-d_M+1}^{-d_m} \sum_{i=k+m}^{k-1} \eta^{\mathrm{T}}(i) Q \eta(i),$$

$$(9.22)$$

where $P(r(k)) > 0$ and $Q > 0$ are matrices to be determined. Calculate the difference of $V_i(\Theta(k), r(k))$ ($i = 1, 2, 3$) along the solution of system (9.16) and take the mathematical expectation conditions $\Theta(k)$ and $r(k)$. Then, for all $l \in S$, we have

$$\mathbb{E}\{\Delta V_1(\Theta(k), r(k)) \mid \Theta(k), r(k) = l\} = \mathbb{E}\{V_1(\Theta(k+1), r(k+1)) | \Theta(k), l\} - V_1(\Theta(k), l)$$

$$= \mathbb{E}\left\{ \eta^{\mathrm{T}}(k+1) \sum_{j \in S} \lambda_{lj} P_j \eta(k+1) - \eta^{\mathrm{T}}(k) P_l \eta(k) \ \middle| \ \Theta(k), l \right\}$$

$$
= \mathbb{E}\left\{\left(\left[\mathcal{Y}_l - \sum_{i=1}^{n}(\alpha_i(k) - \bar{\alpha}_i)(\mathcal{K}_{li}v(k) - \mathcal{C}_{li}\eta(k)) - \sum_{i=1}^{n}((1 - \alpha_i(k))\beta_i(k)\right.\right.\right.
$$

$$
\left. - (1 - \bar{\alpha}_i)\bar{\beta}_i)\mathcal{C}_{li}\eta(k)\right]^{\mathrm{T}} \bar{P}_l \left[\mathcal{Y}_l - \sum_{i=1}^{n}(\alpha_i(k) - \bar{\alpha}_i)(\mathcal{K}_{li}v(k) - \mathcal{C}_{li}\eta(k))\right.
$$

$$
\left.\left.\left. - \sum_{i=1}^{n}((1 - \alpha_i(k))\beta_i(k) - (1 - \bar{\alpha}_i)\bar{\beta}_i)\mathcal{C}_{li}\eta(k)\right] - \eta^{\mathrm{T}}(k)P_l\eta(k)\right) \;\middle|\; \Theta(k), l\right\}
$$

$$
= \mathbb{E}\left\{\left(\mathcal{Y}_l^{\mathrm{T}} \bar{P}_l \mathcal{Y}_l - \eta^{\mathrm{T}}(k)P_l\eta(k) + \sum_{i=1}^{n}\phi_i^2(\mathcal{K}_{li}v(k) - \mathcal{C}_{li}\eta(k))^{\mathrm{T}}\bar{P}_l(\mathcal{K}_{li}v(k)\right.\right.
$$

$$
\left. - \mathcal{C}_{li}\eta(k)) - 2\sum_{i=1}^{n}\hat{\phi}_i^2(\mathcal{K}_{li}v(k) - \mathcal{C}_{li}\eta(k))^{\mathrm{T}}\bar{P}_l\mathcal{C}_{li}\eta(k)\right.
$$

$$
\left.\left. + \sum_{i=1}^{n}\bar{\phi}_i^2(\mathcal{C}_{li}\eta(k))^{\mathrm{T}}\bar{P}_l(\mathcal{C}_{li}\eta(k))\right)\;\middle|\;\Theta(k), l\right\}. \tag{9.23}
$$

Similarly, by noting $d_m \leq \tau(k) \leq d_M$, one has

$$
\mathbb{E}\{\Delta V_2(\Theta(k), r(k))|\Theta(k), r(k) = l\}
$$

$$
= \mathbb{E}\left\{\left(\eta^{\mathrm{T}}(k)Q\eta(k) - \eta^{\mathrm{T}}(k - \tau(k))Q\eta(k - \tau(k)) + \sum_{i=k+1-\tau(k+1)}^{k-1}\eta^{\mathrm{T}}(i)Q\eta(i)\right.\right.
$$

$$
\left.\left. - \sum_{i=k+1-\tau(k)}^{k-1}\eta^{\mathrm{T}}(i)Q\eta(i)\right)\;\middle|\;\Theta(k), l\right\}
$$

$$
\leq \mathbb{E}\left\{\left(\eta^{\mathrm{T}}(k)Q\eta(k) - \eta^{\mathrm{T}}(k - \tau(k))Q\eta(k - \tau(k))\right.\right.
$$

$$
\left.\left. + \sum_{i=k+1-d_M}^{k-d_m}\eta^{\mathrm{T}}(i)Q\eta(i)\right)\;\middle|\;\Theta(k), l\right\},
$$

$$
\mathbb{E}\{\Delta V_3(\Theta(k), r(k))|\Theta(k), r(k) = l\}
$$

$$
= \mathbb{E}\left\{\sum_{m=-d_M+1}^{-d_m}\left(\eta^{\mathrm{T}}(k)Q\eta(k) - \eta^{\mathrm{T}}(k + m)Q\eta(k + m)\right)\;\middle|\;\Theta(k), l\right\}
$$

$$
= \mathbb{E}\left\{\left((d_M - d_m)\eta^{\mathrm{T}}(k)Q\eta(k) - \sum_{i=k+1-d_M}^{k-d_m}\eta^{\mathrm{T}}(i)Q\eta(i)\right)\;\middle|\;\Theta(k), l\right\}. \tag{9.24}
$$

Therefore, combining (9.22)–(9.24), one immediately obtains

$$
\mathbb{E}\{\Delta V(\Theta(k), r(k))|\Theta(k), r(k) = l\}
$$

$$
= \mathbb{E}\left\{\left((-\bar{\mathcal{C}}_l\eta(k) + \mathcal{A}_{dl}\eta(k - \tau(k)) + \mathcal{H}\mathcal{F}_l(\eta(k)) + \mathcal{G}_l w(k) + \bar{\mathcal{K}}_l v(k))^{\mathrm{T}}\bar{P}_l\right.\right.
$$

$$
\times (-\bar{\mathcal{C}}_l\eta(k) + \mathcal{A}_{dl}\eta(k - \tau(k)) + \mathcal{H}\mathcal{F}_l(\eta(k)) + \mathcal{G}_l w(k) + \bar{\mathcal{K}}_l v(k))
$$

$$
- \eta^{\mathrm{T}}(k)P_l\eta(k) + \sum_{i=1}^{n}\phi_i^2(\mathcal{K}_{li}v(k) - \mathcal{C}_{li}\eta(k))^{\mathrm{T}}\bar{P}_l(\mathcal{K}_{li}v(k) - \mathcal{C}_{li}\eta(k))
$$

$$
+ \sum_{i=1}^{n}\bar{\phi}_i^2(\mathcal{C}_{li}\eta(k))^{\mathrm{T}}\bar{P}_l(\mathcal{C}_{li}\eta(k)) - 2\sum_{i=1}^{n}\hat{\phi}_i^2(\mathcal{K}_{li}v(k) - \mathcal{C}_{li}\eta(k))^{\mathrm{T}}\bar{P}_l\mathcal{C}_{li}\eta(k)
$$

$$
\left.\left. + (d_M - d_m + 1)\eta^{\mathrm{T}}(k)Q\eta(k) - \eta^{\mathrm{T}}(k - \tau(k))Q\eta(k - \tau(k))\right)\right|\Theta(k), l\right\}. \quad (9.25)
$$

In the following, we first prove the stochastic stability of the system (9.16) with $w(k) = 0$ and $v(k) = 0$. It follows from (9.25) that

$$
\mathbb{E}\{\Delta V(\Theta(k), r(k))|\Theta(k), r(k) = l\} \le \mathbb{E}\{\hat{\xi}^{\mathrm{T}}(k)\hat{\Gamma}_l\hat{\xi}(k)\}, \quad (9.26)
$$

where $\hat{\xi}(k) = [\eta^{\mathrm{T}}(k) \quad \eta^{\mathrm{T}}(k - \tau(k)) \quad \mathcal{F}_l^{\mathrm{T}}(\eta(k))]^{\mathrm{T}}$,

$$
\hat{\Gamma}_l = \begin{bmatrix} \hat{\Gamma}_{11l} & * & * \\ -\mathcal{A}_{dl}^{\mathrm{T}}\bar{P}_l\bar{\mathcal{C}}_l & \mathcal{A}_{dl}^{\mathrm{T}}\bar{P}_l\mathcal{A}_{dl} - Q & * \\ -\mathcal{H}^{\mathrm{T}}\bar{P}_l\bar{\mathcal{C}}_l & \mathcal{H}^{\mathrm{T}}\bar{P}_l\mathcal{A}_{dl} & \mathcal{H}^{\mathrm{T}}\bar{P}_l\mathcal{H} \end{bmatrix}.
$$

Moreover, it follows from (9.18) that

$$
\mathbb{E}\{\Delta V(\Theta(k), r(k))|\Theta(k), r(k) = l\}
$$

$$
\le \mathbb{E}\{\hat{\xi}^{\mathrm{T}}(k)\hat{\Gamma}_l\hat{\xi}(k) - \varepsilon(\mathcal{F}_l^{\mathrm{T}}(\eta(k))\mathcal{F}_l(\eta(k)) - b_{1l}^2\eta^{\mathrm{T}}(k)\pi_1^{\mathrm{T}}\pi_1\eta(k)
$$

$$
- b_{2l}^2\eta^{\mathrm{T}}(k - \tau(k))\pi_2^{\mathrm{T}}\pi_2\eta(k - \tau(k)))\}
$$

$$
= \mathbb{E}\{\hat{\xi}^{\mathrm{T}}(k)\bar{\Gamma}_l\hat{\xi}(k)\}, \quad (9.27)
$$

where

$$
\bar{\Gamma}_l = \begin{bmatrix} \hat{\Gamma}_{11l} + \varepsilon b_{1l}^2\pi_1^{\mathrm{T}}\pi_1 & * & * \\ -\mathcal{A}_{dl}^{\mathrm{T}}\bar{P}_l\bar{\mathcal{C}}_l & \Gamma_{22l} & * \\ -\mathcal{H}^{\mathrm{T}}\bar{P}_l\bar{\mathcal{C}}_l & \mathcal{H}^{\mathrm{T}}\bar{P}_l\mathcal{A}_{dl} & \mathcal{H}^{\mathrm{T}}\bar{P}_l\mathcal{H} - \varepsilon I \end{bmatrix}.
$$

It can be obtained from (9.21) that $\bar{\Gamma}_l < 0$ and, subsequently,

$$
\mathbb{E}\{V(\Theta(k+1), r(k+1))|\Theta(k), l\} - V(\Theta(k), l)
$$
$$
< -\lambda_{\min}(-\bar{\Gamma}_l)\hat{\xi}^{\mathrm{T}}(k)\hat{\xi}(k) \leq -\lambda_{\min}(-\bar{\Gamma}_l)\eta^{\mathrm{T}}(k)\eta(k)
$$
$$
\leq -\rho\eta^{\mathrm{T}}(k)\eta(k), \tag{9.28}
$$

where $\rho := \inf\{\lambda_{\min}(-\bar{\Gamma}_l), l \in S\}$. From (9.28), we obtain that, for any $T \geq 1$,

$$
\mathbb{E}\{V(\Theta(T+1), r(T+1))t\} - \mathbb{E}\{V(\Theta(0), r(0))\} < -\rho \sum_{k=0}^{T} \mathbb{E}\{\eta^{\mathrm{T}}(k)\eta(k)\},
$$

which yields the following for $T \geq 1$:

$$
\sum_{k=0}^{T} \mathbb{E}\{\eta^{\mathrm{T}}(k)\eta(k)\} < \frac{1}{\rho}(\mathbb{E}\{V(\Theta(0), r(0))\} - \mathbb{E}\{V(\Theta(T+1), r(T+1))\})
$$
$$
\leq \frac{1}{\rho}\mathbb{E}\{V(\Theta(0), r(0))\}. \tag{9.29}
$$

The above implies that $\sum_{k=0}^{\infty} \mathbb{E}\{\eta^{\mathrm{T}}(k)\eta(k)\} < \frac{1}{\rho}\mathbb{E}\{V(\Theta(0), r(0))\} < \infty$. According to Definition 9.1.1, the error dynamics system (9.16) is stochastically stable.

Let us now deal with the average $\mathcal{H}_\infty$ performance of system (9.16). Assume zero initial conditions and introduce the following index:

$$
\bar{J} := \mathbb{E}\{\Delta V(\Theta(k), r(k))\} + \frac{1}{n}\mathbb{E}\{\|\bar{z}(k)\|^2\} - \gamma^2(\|w(k)\|^2 + \frac{1}{n}\|v(k)\|^2)
$$
$$
= \mathbb{E}\left\{\hat{\xi}^{\mathrm{T}}(k)\hat{\Gamma}_l\hat{\xi}(k) + 2(\mathcal{G}_l w(k) + \bar{\mathcal{K}}_l v(k))^{\mathrm{T}}\bar{P}_l(-\bar{C}_l \eta(k) + \mathcal{A}_{dl}\eta(k - \tau(k))\right.
$$
$$
+ \mathcal{H}\mathcal{F}_l(\eta(k))) + (\mathcal{G}_l w(k) + \bar{\mathcal{K}}_l v(k))^{\mathrm{T}}\bar{P}_l(\mathcal{G}_l w(k) + \bar{\mathcal{K}}_l v(k))
$$
$$
+ \sum_{i=1}^{n} \phi_i^2(v^{\mathrm{T}}(k)\mathcal{K}_{li}^{\mathrm{T}}\bar{P}_l(\mathcal{K}_{li}v(k) - 2\mathcal{C}_{li}\eta(k))) - 2\sum_{i=1}^{n} \hat{\phi}_i^2(\mathcal{K}_{li}v(k))^{\mathrm{T}}\bar{P}_l\mathcal{C}_{li}\eta(k)
$$
$$
+ \frac{1}{n}\eta^{\mathrm{T}}(k)\mathcal{M}_l^{\mathrm{T}}\mathcal{M}_l\eta(k) - \gamma^2 w^{\mathrm{T}}(k)w(k) - \frac{\gamma^2}{n}v^{\mathrm{T}}(k)v(k)\right\}
$$
$$
= \mathbb{E}\{\xi^{\mathrm{T}}(k)\check{\Gamma}_l\xi(k)\}, \tag{9.30}
$$

where $\xi(k) = [\eta^{\mathrm{T}}(k) \quad \eta^{\mathrm{T}}(k - \tau(k)) \quad \mathcal{F}_l^{\mathrm{T}}(\eta(k)) \quad w^{\mathrm{T}}(k) \quad v^{\mathrm{T}}(k)]^{\mathrm{T}}$ and

$$
\check{\Gamma}_l = \begin{bmatrix}
\hat{\Gamma}_{11l} + \frac{1}{n}\mathcal{M}_l^{\mathrm{T}}\mathcal{M}_l & * & * & * & * \\
-\mathcal{A}_{dl}^{\mathrm{T}}\bar{P}_l\bar{C}_l & \mathcal{A}_{dl}^{\mathrm{T}}\bar{P}_l\mathcal{A}_{dl} - Q & * & * & * \\
-\mathcal{H}^{\mathrm{T}}\bar{P}_l\bar{C}_l & \mathcal{H}^{\mathrm{T}}\bar{P}_l\mathcal{A}_{dl} & \mathcal{H}^{\mathrm{T}}\bar{P}_l\mathcal{H} & * & * \\
-\mathcal{G}_l^{\mathrm{T}}\bar{P}_l\bar{C}_l & \mathcal{G}_l^{\mathrm{T}}\bar{P}_l\mathcal{A}_{dl} & \mathcal{G}_l^{\mathrm{T}}\bar{P}_l\mathcal{H} & \mathcal{G}_l^{\mathrm{T}}\bar{P}_l\mathcal{G}_l - \gamma^2 I & * \\
\Gamma_{51l} & \bar{\mathcal{K}}_l^{\mathrm{T}}\bar{P}_l\mathcal{A}_{dl} & \bar{\mathcal{K}}_l^{\mathrm{T}}\bar{P}_l\mathcal{H} & \bar{\mathcal{K}}_l^{\mathrm{T}}\bar{P}_l\mathcal{G}_l & \Gamma_{55l}
\end{bmatrix} < 0.
$$

Again, it follows from the constraint (9.18) that

$$
\begin{aligned}
\bar{J} &\leq \mathbb{E}\{\xi^{\mathrm{T}}(k)\check{\Gamma}_l\xi(k) - \varepsilon(\mathcal{F}_l^{\mathrm{T}}(\eta(k))\mathcal{F}_l(\eta(k)) \\
&\quad - b_{1l}^2\eta^{\mathrm{T}}(k)\pi_1^{\mathrm{T}}\pi_1\eta(k) - b_{2l}^2\eta^{\mathrm{T}}(k - \tau(k))\pi_2^{\mathrm{T}}\pi_2\eta(k - \tau(k)))\} \\
&= \mathbb{E}\{\xi^{\mathrm{T}}(k)\Gamma_l\xi(k)\}.
\end{aligned} \tag{9.31}
$$

It can be obtained from (9.21) that $\Gamma_l < 0$ and, subsequently,

$$
\bar{J} = \mathbb{E}\{\Delta V(\Theta(k), r(k))\} + \frac{1}{n}\mathbb{E}\left\{\|\bar{z}(k)\|^2\right\} - \gamma^2(\|w(k)\|^2 + \frac{1}{n}\|v(k)\|^2) < 0. \tag{9.32}
$$

By considering zero initial conditions, it follows from the above inequality that

$$
\frac{1}{n}\sum_{k=0}^{\infty}\mathbb{E}\{\|\bar{z}(k)\|^2\} - \gamma^2\left(\sum_{k=0}^{\infty}\|w(k)\|^2 + \frac{1}{n}\sum_{k=0}^{\infty}\|v(k)\|^2\right) < 0, \tag{9.33}
$$

which is equivalent to (9.20), and the proof is now complete. □

Based on the analysis results with the desired distributed filters, we are now ready to solve the filter design problem for system (9.16) in the following theorem with the deficient TPs.

Theorem 9.2.2 *Consider system (9.4) with the deficient TP matrix described in (9.3). Let $\gamma > 0$ be a given disturbance attenuation level. The filtering dynamics in (9.16) is stochastically stable and also satisfies the average H_∞ performance constraint (9.20) if there exist symmetric positive-definite matrices P_{il} ($i = 1, 2, \ldots, n, l \in S$), Q, matrices N_{lj} ($l, j \in S$), and a positive scalar ε satisfying*

$$
\Sigma_l = \begin{bmatrix} \Sigma_{1l} & * \\ \Sigma_{2l} & \Sigma_{3l} \end{bmatrix} < 0, \tag{9.34}
$$

where

$$\begin{cases} R_j = \dfrac{1}{\lambda_{\mathcal{K}}^l} P_{\mathcal{K}}^l = \dfrac{1}{\lambda_{\mathcal{K}}^l} \sum_{j \in S_{\mathcal{K}}^l} \lambda_{lj} P_j, & \forall j \in S_{\mathcal{K}}^l \\[3mm] R_j = \dfrac{1}{\lambda_{\mathcal{UC}}^l} P_{\mathcal{UC}}^l = \dfrac{1}{\lambda_{\mathcal{UC}}^l} \sum_{j \in S_{\mathcal{UC}}^l} \lambda_{lj}^{(r)} P_j, & \forall j \in S_{\mathcal{UC}}^l \\[3mm] R_j = P_j, & \forall j \in S_{\mathcal{UK}}^l \end{cases} \tag{9.35}$$

and

$$\Sigma_{1l} = \mathrm{diag}\left\{ (d_M - d_m + 1)Q + \frac{1}{n}\mathcal{M}_l^{\mathrm{T}}\mathcal{M}_l + \varepsilon b_{1l}^2 \pi_1^{\mathrm{T}}\pi_1 - P_l, -Q + \varepsilon b_{2l}^2 \pi_2^{\mathrm{T}}\pi_2, \right.$$

$$\left. - \varepsilon I, -\gamma^2 I, -\frac{\gamma^2}{n}I \right\}, \quad \Sigma_{2l} = [\Sigma_{21l} \quad \Sigma_{22l} \quad \Sigma_{23l}], \quad \Sigma_{22l} = \left[(\Sigma_{221l})^{\mathrm{T}} \quad 0 \right]^{\mathrm{T}},$$

$$\Sigma_{221l} = \left[-R_j \mathcal{A}_{dl} \quad - R_j \mathcal{H} \quad - R_j \check{\mathcal{G}}_l + N_{lj}\check{D}_l \right], \quad \vec{C}_{Nlj} = -R_j \hat{A}_l + N_{lj}\vec{C}_l,$$

$$\Sigma_{21l} = \left[\vec{C}_{Nlj}^{\mathrm{T}} \quad (\Lambda_{\bar\phi} \tilde{N}_{lj}\tilde{C}_l)^{\mathrm{T}} \quad (\Lambda_{\bar\phi} \tilde{N}_{lj}\tilde{C}_l)^{\mathrm{T}} \quad (\Lambda_{\hat\phi} \tilde{N}_{lj}\tilde{C}_l)^{\mathrm{T}} \quad (\Lambda_{\hat\phi} \tilde{N}_{lj}\tilde{C}_l)^{\mathrm{T}} \right]^{\mathrm{T}},$$

$$\Sigma_{23l} = \left[(N_{lj}\check{\Lambda}_{\alpha})^{\mathrm{T}} \quad - (\Lambda_{\bar\phi} \tilde{N}_{lj}\tilde{E}\Theta_l)^{\mathrm{T}} \quad 0 \quad - (\Lambda_{\hat\phi} \tilde{N}_{lj}\tilde{E}\Theta_l)^{\mathrm{T}} \quad 0 \right]^{\mathrm{T}},$$

$$\Sigma_{3l} = \mathrm{diag}\{-R_j, -\hat{R}_j, -\hat{R}_j, -\hat{R}_j, -\hat{R}_j\}, \quad \hat{A}_l = I_2 \otimes \bar{A}_l,$$

$$\vec{C}_l = \mathrm{diag}\{0, \bar{\Lambda}_{\alpha}\bar{C}_l + (I - \bar{\Lambda}_{\alpha})\bar{\Lambda}_{\beta}\bar{C}_l\}, \quad \tilde{N}_{lj} = I_n \otimes N_{lj},$$

$$\tilde{C}_l = [\check{C}_{l1}^{\mathrm{T}} \quad \check{C}_{l2}^{\mathrm{T}} \quad \cdots \quad \check{C}_{ln}^{\mathrm{T}}]^{\mathrm{T}}, \quad \check{D}_l = [0 \quad \bar{D}_l^{\mathrm{T}}]^{\mathrm{T}}, \quad \check{\Lambda}_{\alpha} = [0 \quad \bar{\Lambda}_{\alpha}]^{\mathrm{T}},$$

$$\tilde{E} = \mathrm{diag}\{\check{E}_1, \check{E}_2, \ldots, \check{E}_n\}, \quad \Theta_l = \mathbf{1}_n \otimes I, \quad \hat{R}_j = I_n \otimes R_j,$$

$$\Lambda_{\phi} = \mathrm{diag}\{\phi_1 I, \phi_2 I, \ldots, \phi_n I\}, \quad \Lambda_{\bar\phi} = \mathrm{diag}\{\bar\phi_1 I, \bar\phi_2 I, \ldots, \bar\phi_n I\},$$

$$\Lambda_{\hat\phi} = \mathrm{diag}\{\hat\phi_1 I, \hat\phi_2 I, \ldots, \hat\phi_n I\}, \quad \check{C}_{li} = \begin{bmatrix} 0 & 0 \\ E_i \bar{C}_l & 0 \end{bmatrix}, \quad \check{E}_i = \begin{bmatrix} 0 \\ E_i \end{bmatrix},$$

$$P_l = \mathrm{diag}\{P_{1l}, P_{2l}, \ldots, P_{nl}\} \quad (l \in S), \quad \check{\mathcal{G}}_l = \mathbf{1}_2 \otimes \bar{G}_l.$$

Moreover, if the above inequalities are feasible, the distributed filter matrices $\bar{K}_l$ are given as follows:

$$\check{K}_l = R_j^{-1} N_{lj} \quad (l, j \in S), \tag{9.36}$$

where

$$\check{K}_l = \mathrm{diag}\{I, \bar{K}_l\}. \tag{9.37}$$

This way, the desired filter parameters K_{ij} $(i = 1, 2, \ldots, n, j \in \mathcal{N}_i)$ can be obtained from (9.13).

Proof. In order to avoid partitioning the positive define matrices P_l ($l \in S$) and Q, we rewrite the parameters in Theorem 9.2.1 in the following form:

$$\bar{C}_l = -\hat{A}_l + \check{K}_l \vec{C}_l, \quad \bar{K}_l = -\check{K}_l \check{\Lambda}_\alpha, \quad \mathcal{G}_l = \check{\mathcal{G}}_l - \check{K}_l \check{D}_l,$$
$$\mathcal{K}_{li} = \check{K}_l \check{E}_i, \quad \mathcal{C}_{li} = \check{K}_l \check{C}_{li}.$$

$$(9.38)$$

Noticing (9.38) and the Schur complement lemma, it follows from (9.21) that

$$\hat{\Sigma}_l = \begin{bmatrix} \Sigma_{1l} & * \\ \hat{\Sigma}_{2l} & \hat{\Sigma}_{3l} \end{bmatrix} < 0,$$

$$(9.39)$$

where

$$\hat{\Sigma}_{2l} = \begin{bmatrix} \hat{\Sigma}_{21l} & \hat{\Sigma}_{22l} & \hat{\Sigma}_{23l} \end{bmatrix}, \quad \hat{\Sigma}_{22l} = \begin{bmatrix} (\hat{\Sigma}_{221l})^{\mathrm{T}} & 0 \end{bmatrix}^{\mathrm{T}},$$

$$\hat{\Sigma}_{221l} = \begin{bmatrix} -\bar{P}_l \mathcal{A}_{dl} & -\bar{P}_l \mathcal{H} & -\bar{P}_l (\check{\mathcal{G}}_l - \check{K}_l \check{D}_l) \end{bmatrix}, \quad \vec{\mathcal{C}}_{Pl} = \bar{P}_l (-\hat{A}_l + \check{K}_l \vec{C}_l),$$

$$\hat{\Sigma}_{21l} = \begin{bmatrix} \vec{\mathcal{C}}_{Pl}^{\mathrm{T}} & (\Lambda_\phi \hat{P}_l \check{K}_l \check{C}_l)^{\mathrm{T}} & (\Lambda_{\tilde{\phi}} \hat{P}_l \check{K}_l \check{C}_l)^{\mathrm{T}} & (\Lambda_{\hat{\phi}} \hat{P}_l \check{K}_l \check{C}_l)^{\mathrm{T}} & (\Lambda_{\hat{\phi}} \hat{P}_l \check{K}_l \check{C}_l)^{\mathrm{T}} \end{bmatrix}^{\mathrm{T}},$$

$$\hat{\Sigma}_{23l} = \begin{bmatrix} (\bar{P}_l \check{K}_l \check{\Lambda}_\alpha)^{\mathrm{T}} & -(\Lambda_\phi \hat{P}_l \check{K}_l \tilde{E} \Theta_l)^{\mathrm{T}} & 0 & -(\Lambda_{\hat{\phi}} \hat{P}_l \check{K}_l \tilde{E} \Theta_l)^{\mathrm{T}} & 0 \end{bmatrix}^{\mathrm{T}},$$

$$\hat{\Sigma}_{3l} = \mathrm{diag}\{-\bar{P}_l, -\hat{P}_l, -\hat{P}_l, -\hat{P}_l, -\hat{P}_l\}, \quad \hat{P}_l = I_n \otimes \bar{P}_l, \quad \check{K}_l = I_n \otimes \check{K}_l.$$

Now, we decompose the deficient TP matrix as follows:

$$\bar{P}_l = \sum_{j=1}^{s} \lambda_{lj} P_j = P_{\mathcal{K}}^l + \sum_{j \in S_{UC}^l} \left(\sum_{r=1}^{M} a_r \lambda_{lj}^{(r)} \right) P_j + \sum_{j \in S_{UK}^l} \lambda_{lj} P_j,$$

where $\sum_{r=1}^{M} a_r \lambda_{lj}^{(r)}$ ($\forall j \in S_{UC}^l$) represents an uncertain element in the polyhedral uncertainty description. As $\sum_{r=1}^{M} a_r = 1$ and a_r can take values arbitrarily in $[0, 1]$, $\hat{\Sigma}_l$ in (9.39) can be rewritten as

$$\hat{\Sigma}_l = \sum_{r=1}^{M} a_r \begin{bmatrix} \Sigma_{1l} & * \\ \bar{\Sigma}_{2l} & \bar{\Sigma}_{3l} \end{bmatrix} = \sum_{r=1}^{M} a_r \bar{\Sigma}_l,$$

$$(9.40)$$

where

$$\bar{\Sigma}_{2l} = \begin{bmatrix} \bar{\Sigma}_{21l} & \bar{\Sigma}_{22l} & \bar{\Sigma}_{23l} \end{bmatrix}, \quad \bar{\Sigma}_{22l} = \begin{bmatrix} (\bar{\Sigma}_{221l})^{\mathrm{T}} & 0 \end{bmatrix}^{\mathrm{T}},$$

$$\bar{\Sigma}_{221l} = \begin{bmatrix} -P_b^l \mathcal{A}_{dl} & -P_b^l \mathcal{H} & -P_b^l (\check{\mathcal{G}}_l - \check{K}_l \check{D}_l) \end{bmatrix}, \quad \hat{A}_{Pl} = P_b^l (-\hat{A}_l + \check{K}_l \vec{C}_l),$$

$$\bar{\Sigma}_{21l} = \begin{bmatrix} (\hat{A}_{Pl})^{\mathrm{T}} & (\Lambda_\phi \hat{P}_b^l \check{K}_l \check{C}_l)^{\mathrm{T}} & (\Lambda_{\tilde{\phi}} \hat{P}_b^l \check{K}_l \check{C}_l)^{\mathrm{T}} & (\Lambda_{\hat{\phi}} \hat{P}_b^l \check{K}_l \check{C}_l)^{\mathrm{T}} & (\Lambda_{\hat{\phi}} \hat{P}_b^l \check{K}_l \check{C}_l)^{\mathrm{T}} \end{bmatrix}^{\mathrm{T}},$$

$$\bar{\Sigma}_{23l} = \begin{bmatrix} (P_b^l \check{K}_l \check{\Lambda}_\alpha)^{\mathrm{T}} & -(\Lambda_\phi \hat{P}_b^l \check{K}_l \tilde{E} \Theta_l)^{\mathrm{T}} & 0 & -(\Lambda_{\hat{\phi}} \hat{P}_b^l \check{K}_l \tilde{E} \Theta_l)^{\mathrm{T}} & 0 \end{bmatrix}^{\mathrm{T}},$$

$$\bar{\Sigma}_{3l} = \mathrm{diag}\{-P_b^l, -\hat{P}_b^l, -\hat{P}_b^l, -\hat{P}_b^l, -\hat{P}_b^l\}, \quad \hat{P}_b^l = I_n \otimes P_b^l,$$

$$P_b^l = P_{\mathcal{K}}^l + P_{UC}^l + P_{UK}^l = \sum_{j \in S_{\mathcal{K}}^l} \lambda_{lj} P_j + \sum_{j \in S_{UC}^l} \lambda_{lj}^{(r)} P_j + \sum_{j \in S_{UK}^l} \lambda_{lj} P_j.$$

Therefore, we have $\bar{\Sigma}_l < 0$.

Note that $\bar{\Sigma}_l$ in (9.40) can be rewritten as

$$\bar{\Sigma}_l = \sum_{j \subset S_K^l} \lambda_{lj} \begin{bmatrix} \Sigma_{1l} & * \\ \check{\Sigma}_{2l} & \check{\Sigma}_{3l} \end{bmatrix} + \sum_{j \in S_{uc}^l} \lambda_{lj}^{(r)} \begin{bmatrix} \Sigma_{1l} & * \\ \tilde{\Sigma}_{2l} & \tilde{\Sigma}_{3l} \end{bmatrix} + \sum_{j \in S_{uK}^l} \lambda_{lj} \begin{bmatrix} \Sigma_{1l} & * \\ \vec{\Sigma}_{2l} & \vec{\Sigma}_{3l} \end{bmatrix} < 0, \quad (9.41)$$

where

$$\check{\Sigma}_{2l} = \begin{bmatrix} \check{\Sigma}_{21l} & \check{\Sigma}_{22l} & \check{\Sigma}_{23l} \end{bmatrix}, \quad \check{\Sigma}_{22l} = \begin{bmatrix} (\check{\Sigma}_{221l})^T & 0 \end{bmatrix}^T, \quad \hat{A}_{PC2l} = P_{c2}^l(-\hat{A}_l + \check{K}_l\check{C}_l),$$

$$\check{\Sigma}_{221l} = \begin{bmatrix} -P_{c1}^l A_{dl} & -P_{c1}^l \mathcal{H} & -P_{c1}^l(\check{G}_l - \check{K}_l\check{D}_l) \end{bmatrix}, \quad \hat{A}_{PC1l} = P_{c1}^l(-\hat{A}_l + \check{K}_l\check{C}_l),$$

$$\check{\Sigma}_{21l} = \begin{bmatrix} \hat{A}_{PC1l}^T & (\Lambda_\phi \hat{P}_{c1}^l \check{K}_l\check{C}_l)^T & (\Lambda_{\check{\phi}} \hat{P}_{c1}^l \check{K}_l\check{C}_l)^T & (\Lambda_{\hat{\phi}} \hat{P}_{c1}^l \check{K}_l\check{C}_l)^T & (\Lambda_{\hat{\phi}} \hat{P}_{c1}^l \check{K}_l\check{C}_l)^T \end{bmatrix}^T,$$

$$\check{\Sigma}_{23l} = \begin{bmatrix} (P_{c1}^l \check{K}_l \check{\Lambda}_\alpha)^T & -(\Lambda_\phi \hat{P}_{c1}^l \check{K}_l\tilde{E}\Theta_l)^T & 0 & -(\Lambda_{\hat{\phi}} \hat{P}_{c1}^l \check{K}_l\tilde{E}\Theta_l)^T & 0 \end{bmatrix}^T,$$

$$\check{\Sigma}_{3l} = \text{diag}\{-P_{c1}^l, -\hat{P}_{c1}^l, -\hat{P}_{c1}^l, -\hat{P}_{c1}^l, -\hat{P}_{c1}^l\}, \quad \hat{P}_{c1}^l = I_n \otimes P_{c1}^l,$$

$$P_{c1}^l = \left(\sum_{j \in S_K^l} \lambda_{lj} \right)^{-1} \sum_{j \in S_K^l} \lambda_{lj} P_j, \quad \tilde{\Sigma}_{2l} = \begin{bmatrix} \tilde{\Sigma}_{21l} & \tilde{\Sigma}_{22l} & \tilde{\Sigma}_{23l} \end{bmatrix},$$

$$\tilde{\Sigma}_{22l} = \begin{bmatrix} (\tilde{\Sigma}_{221l})^T & 0 \end{bmatrix}^T, \quad \tilde{\Sigma}_{221l} = \begin{bmatrix} -P_{c2}^l A_{dl} & -P_{c2}^l \mathcal{H} & -P_{c2}^l(\check{G}_l - \check{K}_l\check{D}_l) \end{bmatrix},$$

$$\tilde{\Sigma}_{21l} = \begin{bmatrix} \hat{A}_{PC2l}^T & (\Lambda_\phi \hat{P}_{c2}^l \check{K}_l\check{C}_l)^T & (\Lambda_{\check{\phi}} \hat{P}_{c2}^l \check{K}_l\check{C}_l)^T & (\Lambda_{\hat{\phi}} \hat{P}_{c2}^l \check{K}_l\check{C}_l)^T & (\Lambda_{\hat{\phi}} \hat{P}_{c2}^l \check{K}_l\check{C}_l)^T \end{bmatrix}^T,$$

$$\tilde{\Sigma}_{23l} = \begin{bmatrix} (P_{c2}^l \check{K}_l \check{\Lambda}_\alpha)^T & -(\Lambda_\phi \hat{P}_{c2}^l \check{K}_l\tilde{E}\Theta_l)^T & 0 & -(\Lambda_{\hat{\phi}} \hat{P}_{c2}^l \check{K}_l\tilde{E}\Theta_l)^T & 0 \end{bmatrix}^T,$$

$$\tilde{\Sigma}_{3l} = \text{diag}\{-P_{c2}^l, -\hat{P}_{c2}^l, -\hat{P}_{c2}^l, -\hat{P}_{c2}^l, -\hat{P}_{c2}^l\}, \quad \hat{P}_{c2}^l = I_n \otimes P_{c2}^l,$$

$$P_{c2}^l = \left(\sum_{j \in S_{uc}^l} \lambda_{lj}^{(r)} \right)^{-1} \sum_{j \in S_{uc}^l} \lambda_{lj}^{(r)} P_j,$$

$$\vec{\Sigma}_{2l} = \begin{bmatrix} \vec{\Sigma}_{21l} & \vec{\Sigma}_{22l} & \vec{\Sigma}_{23l} \end{bmatrix}, \quad \vec{\Sigma}_{22l} = \begin{bmatrix} (\tilde{\Sigma}_{221l})^T & 0 \end{bmatrix}^T,$$

$$\vec{\Sigma}_{221l} = \begin{bmatrix} -P_j A_{dl} & -P_j \mathcal{H} & -P_j(\check{G}_l - \check{K}_l\check{D}_l) \end{bmatrix}, \quad \hat{A}_{PKl} = P_j(-\hat{A}_l + \check{K}_l\check{C}_l),$$

$$\vec{\Sigma}_{21l} = \begin{bmatrix} \hat{A}_{PKl}^T & (\Lambda_\phi \hat{P}_j \check{K}_l\check{C}_l)^T & (\Lambda_{\check{\phi}} \hat{P}_j \check{K}_l\check{C}_l)^T & (\Lambda_{\hat{\phi}} \hat{P}_j \check{K}_l\check{C}_l)^T & (\Lambda_{\hat{\phi}} \hat{P}_j \check{K}_l\check{C}_l)^T \end{bmatrix}^T,$$

$$\vec{\Sigma}_{23l} = \begin{bmatrix} (P_j \check{K}_l \check{\Lambda}_\alpha)^T & -(\Lambda_\phi \hat{P}_j \check{K}_l\tilde{E}\Theta_l)^T & 0 & -(\Lambda_{\hat{\phi}} \hat{P}_j \check{K}_l\tilde{E}\Theta_l)^T & 0 \end{bmatrix}^T,$$

$$\vec{\Sigma}_{3l} = \text{diag}\{-P_j, -\hat{P}_j, -\hat{P}_j, -\hat{P}_j, -\hat{P}_j\}, \quad \hat{P}_j = I_n \otimes P_j.$$

By setting $N_{lj} = R_j \check{K}_l$ $(l, j \in S)$, where R_j is defined in (9.35) of Theorem 9.2.2, we can obtain that inequalities (9.34) guaranteeing that (9.21) holds. In addition, letting $P_l = \text{diag}\{P_{1l}, P_{2l}, \ldots, P_{nl}\}$ $(l \in S)$, from Lemma 8.2.2, it follows that $\bar{K}_l \in \mathcal{T}_{n_x \times n_y}$, which completes the proof of this theorem. $\qquad\square$

Remark 9.4   *It should be pointed out that the main results presented in Theorem 9.2.2 are quite comprehensive and reflect deficient statistics of modes transitions, Lipschitz-like nonlinearities, time-delays, ROQEs, and ROPDs. Of course, we could easily extend our main results to the various simplified cases; for example, the cases when there are no deficient statistics of modes transitions, and/or no external stochastic disturbances, and/or no time delays, and/or no ROQEs/ROPDs, and so on. The corresponding corollaries are omitted here in order to keep the chapter concise.*

9.3   An Illustrative Example

In this section, we present a simulation example to illustrate the effectiveness of the proposed distributed filter design scheme for Markovian jump nonlinear time-delay systems with both ROQEs and ROPDs in sensor networks.

The sensor network is represented by a directed graph $\mathcal{G} = (\mathcal{V}, \mathcal{E}, \mathcal{A})$ with a set of nodes $\mathcal{V} = \{1, 2, 3, 4, 5\}$, a set of edges

$$\mathcal{E} = \{(1, 1), (1, 2), (1, 5), (2, 2), (2, 3), (3, 1), (3, 3), (4, 2), (4, 4), (5, 2), (5, 5)\},$$

and the following adjacency matrix:

$$\mathcal{A} = \begin{bmatrix} 1 & 1 & 0 & 0 & 1 \\ 0 & 1 & 1 & 0 & 0 \\ 1 & 0 & 1 & 0 & 0 \\ 0 & 1 & 0 & 1 & 0 \\ 0 & 1 & 0 & 0 & 1 \end{bmatrix}.$$

Assume that the system involves four modes and the TP matrix comprises three vertices $\Psi^{(r)}$ ($r = 1, 2, 3$). The second lines of $\Psi^{(r)}$, $\Psi_2^{(r)}$, are given by

$$\Psi_2^{(1)} = [?\quad 0.1\quad ?\quad 0.3], \quad \Psi_2^{(2)} = [?\quad 0.4\quad ?\quad 0.2], \quad \Psi_2^{(3)} = [?\quad 0.2\quad ?\quad 0.5],$$

and other rows in the three vertices are defined with the same elements, which are listed as follows:

$$\Psi_1^{(r)} = [?\quad 0.1\quad ?\quad ?], \quad \Psi_3^{(r)} = [?\quad 0.3\quad 0.6\quad ?], \quad \Psi_4^{(r)} = [0.1\quad 0.2\quad 0.2\quad 0.5].$$

For simplicity, the TP matrix in the polytope uncertainty description can be rewritten in the following form:

$$\Psi = \begin{bmatrix} ? & 0.1 & ? & ? \\ ? & [0.1, 0.4] & ? & [0.2, 0.5] \\ ? & 0.3 & 0.6 & ? \\ 0.1 & 0.2 & 0.2 & 0.5 \end{bmatrix},$$

and the other system data are as follows:

Mode 1:

$$f_1(x(k)) = \begin{bmatrix} 0.48x_1(k) + 0.2x_2(k) + 0.12\sin(x_2(k)) \\ 0.03x_1(k) + 0.50x_2(k) \end{bmatrix}, \quad G_1 = [0.1 \quad -0.3]^T,$$

$$g_1(x(k-\tau(k))) = \begin{bmatrix} 0.1x_1(k-\tau(k)) \\ 0.2x_2(k-\tau(k)) + 0.1\cos(x_2(k-\tau(k))) \end{bmatrix},$$

$$C_{11} = C_{21} = C_{31} = [0.2 \quad 0.4], \quad C_{41} = C_{51} = [0.4 \quad 0.5],$$

$$D_{11} = 0.2, \quad D_{21} = D_{31} = D_{41} = D_{51} = 0.1, M_1 = [0.2 \quad 0.4].$$

Mode 2:

$$f_2(x(k)) = \begin{bmatrix} 0.28x_1(k) + 0.4x_2(k) + 0.12\sin(x_2(k)) \\ 0.05x_1(k) + 0.40x_2(k) \end{bmatrix}, \quad G_2 = [0.2 \quad -0.3]^T,$$

$$g_2(x(k-\tau(k))) = \begin{bmatrix} 0.1x_1(k-\tau(k)) \\ 0.2x_2(k-\tau(k)) + 0.1\cos(x_2(k-\tau(k))) \end{bmatrix},$$

$$C_{12} = C_{22} = C_{32} = [0.1 \quad 0.3], \quad C_{42} = C_{52} = [0.2 \quad 0.5],$$

$$D_{12} = 0.3, \quad D_{22} = D_{32} = D_{42} = D_{52} = 0.1, \quad M_2 = [0.2 \quad 0.4].$$

Mode 3:

$$f_3(x(k)) = \begin{bmatrix} 0.18x_1(k) + 0.5x_2(k) + 0.2\sin(x_2(k)) \\ 0.3x_1(k) + 0.50x_2(k) \end{bmatrix}, \quad G_3 = [0.4 \quad -0.1]^T,$$

$$g_3(x(k-\tau(k))) = \begin{bmatrix} 0.2x_1(k-\tau(k)) \\ 0.3x_2(k-\tau(k)) + 0.2\cos(x_2(k-\tau(k))) \end{bmatrix},$$

$$C_{13} = C_{23} = [0.2 \quad 0.2], \quad C_{33} = C_{43} = C_{53} = [0.2 \quad 0.3],$$

$$D_{13} = D_{23} = D_{33} = D_{43} = D_{53} = 0.1, \quad M_3 = [0.1 \quad 0.5].$$

Mode 4:

$$f_4(x(k)) = \begin{bmatrix} 0.5x_1(k) + 0.1x_2(k) + 0.4\sin(x_2(k)) \\ 0.1x_1(k) + 0.3x_2(k) \end{bmatrix}, \quad G_4 = [0.4 \quad 0.2]^T,$$

$$g_4(x(k-\tau(k))) = \begin{bmatrix} 0.4x_1(k-\tau(k)) \\ 0.1x_2(k-\tau(k)) + 0.3\cos(x_2(k-\tau(k))) \end{bmatrix},$$

$$C_{14} = C_{24} = [0.5 \quad 0.2], \quad C_{34} = C_{44} = C_{54} = [0.4 \quad 0.2],$$

$$D_{14} = D_{24} = D_{34} = D_{44} = D_{54} = 0.2, \quad M_4 = [0.1 \quad 0.4].$$

In view of (9.5), the other system parameters can be obtained as follows:

$$A_1 = \begin{bmatrix} 0.48 & 0.2 \\ 0.03 & 0.5 \end{bmatrix}, \quad A_2 = \begin{bmatrix} 0.28 & 0.4 \\ 0.05 & 0.4 \end{bmatrix}, \quad A_3 = \begin{bmatrix} 0.18 & 0.5 \\ 0.3 & 0.5 \end{bmatrix}, \quad A_4 = \begin{bmatrix} 0.5 & 0.1 \\ 0.1 & 0.3 \end{bmatrix},$$

$A_{d1} = \text{diag}\{0.1, 0.2\}, \quad A_{d2} = \text{diag}\{0.1, 0.2\}, \quad A_{d3} = \text{diag}\{0.2, 0.3\},$

$A_{d4} = \text{diag}\{0.4, 0.1\}, b_{11} = 0.2, \quad b_{21} = 0.12, \quad b_{12} = 0.2, \quad b_{22} = 0.12,$

$b_{13} = 0.22, \quad b_{23} = 0.22, \quad b_{14} = 0.42, \quad b_{24} = 0.4.$

Assume that the time-varying communication delays satisfy $2 \leq \tau(k) \leq 6$ and the probabilities are taken as $\bar{\alpha}_1 = 0.9, \bar{\alpha}_2 = 0.8, \bar{\alpha}_3 = 0.7, \bar{\alpha}_4 = 0.6, \bar{\alpha}_5 = 0.9$, and $\bar{\beta}_1 = 0.9, \bar{\beta}_2 = 0.8, \bar{\beta}_3 = 0.85, \bar{\beta}_4 = 0.7, \bar{\beta}_5 = 0.9$, respectively.

Our purpose here is to design distributed filters of the form in (9.9) on each node i of the sensor network for system (9.4) such that the resulting filtering error system is stochastically stable and has a guaranteed average H_∞ performance. By solving (9.34) in Theorem 9.2.2, we can obtain the parameters of the desired distributed filters, which are listed in Table 9.1. The performance index given in (9.20) is $\gamma = 1.2037$.

Table 9.1   Distributed filter parameters

$r(k)$	1	2	3	4
$K_{11}(r(k))$	$\begin{bmatrix} 0.8947 \\ 0.8246 \end{bmatrix}$	$\begin{bmatrix} 1.2947 \\ 0.9774 \end{bmatrix}$	$\begin{bmatrix} 1.2047 \\ 2.0953 \end{bmatrix}$	$\begin{bmatrix} 0.4423 \\ 0.6794 \end{bmatrix}$
$K_{12}(r(k))$	$\begin{bmatrix} -0.0007 \\ -0.0014 \end{bmatrix}$	$\begin{bmatrix} -0.0016 \\ -0.0039 \end{bmatrix}$	$\begin{bmatrix} -0.0000 \\ -0.0000 \end{bmatrix}$	$\begin{bmatrix} 0.0020 \\ 0.0043 \end{bmatrix}$
$K_{15}(k)$	$\begin{bmatrix} -0.0001 \\ -0.0006 \end{bmatrix}$	$\begin{bmatrix} -0.0003 \\ -0.0009 \end{bmatrix}$	$\begin{bmatrix} 0.0001 \\ -0.0001 \end{bmatrix}$	$\begin{bmatrix} 0.0028 \\ 0.0047 \end{bmatrix}$
$K_{22}(r(k))$	$\begin{bmatrix} 0.8190 \\ 0.8737 \end{bmatrix}$	$\begin{bmatrix} 1.1842 \\ 0.8999 \end{bmatrix}$	$\begin{bmatrix} 1.2196 \\ 2.2876 \end{bmatrix}$	$\begin{bmatrix} 1.2181 \\ 0.5497 \end{bmatrix}$
$K_{23}(r(k))$	$\begin{bmatrix} -0.0174 \\ 0.0025 \end{bmatrix}$	$\begin{bmatrix} -0.0339 \\ -0.0099 \end{bmatrix}$	$\begin{bmatrix} 0.0205 \\ -0.0224 \end{bmatrix}$	$\begin{bmatrix} 0.0502 \\ 0.0989 \end{bmatrix}$
$K_{31}(r(k))$	$\begin{bmatrix} 0.0001 \\ -0.0021 \end{bmatrix}$	$\begin{bmatrix} 0.0008 \\ -0.0049 \end{bmatrix}$	$\begin{bmatrix} 0.0002 \\ -0.0001 \end{bmatrix}$	$\begin{bmatrix} 0.0023 \\ 0.0019 \end{bmatrix}$
$K_{33}(r(k))$	$\begin{bmatrix} 0.9412 \\ 0.8391 \end{bmatrix}$	$\begin{bmatrix} 1.3936 \\ 1.0309 \end{bmatrix}$	$\begin{bmatrix} 0.8714 \\ 1.6042 \end{bmatrix}$	$\begin{bmatrix} 0.8222 \\ 0.6742 \end{bmatrix}$
$K_{42}(r(k))$	$\begin{bmatrix} 0.0001 \\ -0.0012 \end{bmatrix}$	$\begin{bmatrix} 0.0008 \\ -0.0024 \end{bmatrix}$	$\begin{bmatrix} 0.0000 \\ -0.0000 \end{bmatrix}$	$\begin{bmatrix} 0.0017 \\ -0.0003 \end{bmatrix}$
$K_{44}(r(k))$	$\begin{bmatrix} 0.6735 \\ 0.6749 \end{bmatrix}$	$\begin{bmatrix} 0.8724 \\ 0.6516 \end{bmatrix}$	$\begin{bmatrix} 0.9286 \\ 1.7425 \end{bmatrix}$	$\begin{bmatrix} 1.6277 \\ 0.7151 \end{bmatrix}$
$K_{52}(r(k))$	$\begin{bmatrix} 0.0001 \\ -0.0012 \end{bmatrix}$	$\begin{bmatrix} 0.0008 \\ -0.0024 \end{bmatrix}$	$\begin{bmatrix} 0.0000 \\ 0.0001 \end{bmatrix}$	$\begin{bmatrix} 0.0033 \\ 0.0010 \end{bmatrix}$
$K_{55}(r(k))$	$\begin{bmatrix} 0.6741 \\ 0.5821 \end{bmatrix}$	$\begin{bmatrix} 0.8517 \\ 0.6423 \end{bmatrix}$	$\begin{bmatrix} 0.8573 \\ 1.5038 \end{bmatrix}$	$\begin{bmatrix} 0.4628 \\ 0.8088 \end{bmatrix}$

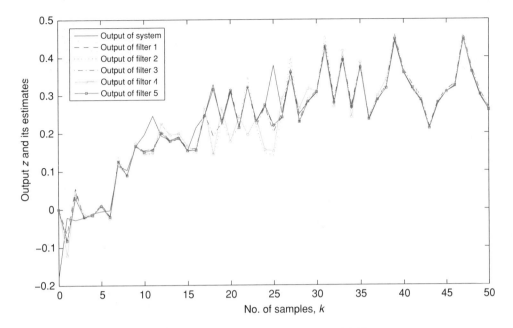

Figure 9.1   Output $z(k)$ and its estimates $\hat{z}_i(k)$

In the simulation, the initial values of the states are $x(0) = [0.2 \quad -0.6]^{\mathrm{T}}$ and $\hat{x}_i(0) = [0 \quad 0]^{\mathrm{T}}$ ($i = 1, 2, \ldots, 5$), the exogenous disturbance input is selected as $w(k) = \exp(-k)$ and the quantization errors are $v_i(k) = \frac{\sin(10k+1)}{3k+1}$ ($i = 1, 2, \ldots, 5$). To demonstrate the mode switches, we consider the real TP matrix as follows:

$$\Psi = \begin{bmatrix} 0.3 & 0.2 & 0.2 & 0.3 \\ 0.4 & 0.3 & 0.1 & 0.2 \\ 0.1 & 0.3 & 0.5 & 0.1 \\ 0.1 & 0.3 & 0.3 & 0.3 \end{bmatrix}.$$

Simulation results are shown in Figures 9.1–9.4. Figure 9.1 shows the output $z(k)$ and its estimates from the filters i ($i = 1, 2, \ldots, 5$). Figure 9.2 plots the filtering errors $z(k) - \hat{z}_i(k)$ ($i = 1, 2, \ldots, 5$). The actual state response $x_1(k)$ and its estimates from the filters 1, 2, 3, 4, and 5 are depicted in Figure 9.3, and the actual state response $x_2(k)$ and its estimates from the filters 1, 2, 3, 4, and 5 are plotted in Figure 9.4. The simulation results confirm the distributed filters designed perform very well, as expected.

9.4   Summary

In this chapter, we have dealt with the distributed H_∞ filtering problem for a class of discrete-time Markovian jump nonlinear time-delay systems with deficient statistics of modes transitions, ROQEs, packet dropouts, and stochastic disturbances in sensor networks. The nonlinearities are described in a Lipschitz-like manner, and the deficient statistics of modes transitions,

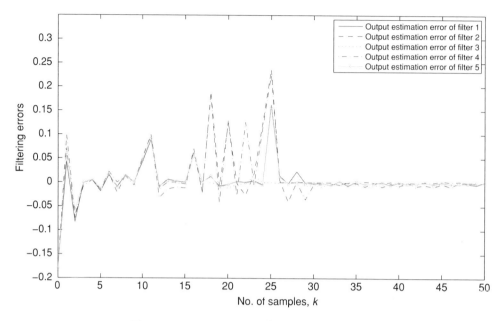

Figure 9.2   Filtering errors $\vec{z}_i(k)$ ($k = 1, 2, 3, 4, 5$)

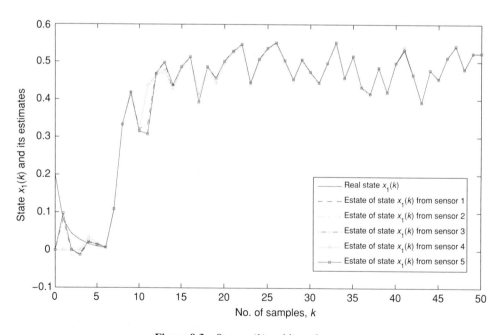

Figure 9.3   State $x_1(k)$ and its estimates

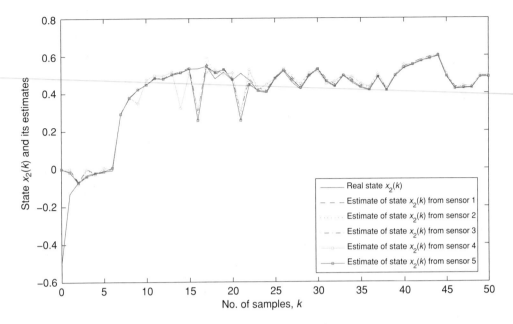

Figure 9.4   State $x_2(k)$ and its estimates

as a combination of TPs contributing to the practicability of MJSs, have been investigated in a unified framework which comprises exactly known TPs, uncertain TPs, and partially unknown TPs. Sufficient conditions have been derived for the filtering dynamics system under consideration to satisfy the stochastically stable and the average H_∞ performance constraint. Furthermore, the explicit expression of the desired filter gains have been derived. Finally, an illustrative example has highlighted the effectiveness of the filtering technology presented in this chapter.

10

A New Finite-Horizon H_∞ Filtering Approach to Mobile Robot Localization

In this chapter, a new stochastic H_∞ filtering approach is proposed to deal with the localization problem of the mobile robots modeled by a class of discrete nonlinear time-varying systems subject to missing measurements and quantization effects. The missing measurements are modeled via a diagonal matrix consisting of a series of mutually independent random variables satisfying certain probabilistic distributions on the interval [0, 1]. The measured output is quantized by a logarithmic quantizer. Attention is focused on the design of a stochastic H_∞ filter such that the H_∞ estimation performance is guaranteed over a given finite horizon in the simultaneous presence of plant nonlinearities (in the robot kinematic model and the distance measurements), probabilistic missing measurements, quantization effects, linearization error, and external non-Gaussian disturbances. A *necessary and sufficient* condition is first established for the existence of the desired time-varying filters in virtue of the solvability of certain coupled recursive Riccati difference equations (RDEs). Owing to its recursive nature, the RDE approach developed is shown to be suitable for online application without the need of increasing the problem size. Both theoretical analysis and simulation results are provided to demonstrate the effectiveness of the proposed localization approach.

10.1 Mobile Robot Kinematics and Absolute Measurement

10.1.1 Kinematic Model

Consider a mobile robot depicted in the coordinate system XOY as shown in Figure 10.1, where the position of the robot is described by a vector $z = (x, y)$, and the orientation of the

Filtering, Control and Fault Detection with Randomly Occurring Incomplete Information, First Edition.
Hongli Dong, Zidong Wang, and Huijun Gao.
© 2013 John Wiley & Sons, Ltd. Published 2013 by John Wiley & Sons, Ltd.

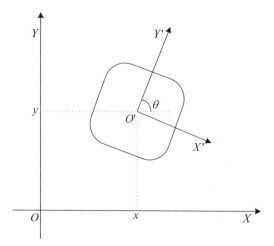

Figure 10.1   Scheme of the mobile robot

robot is denoted by an angle θ between the coordinate axis X and the robot forward axis Y'. The kinematic model of the robot is described by the following equations [179, 180]:

$$\begin{cases} \dot{x}(t) = v(t)\cos\theta(t) \\ \dot{y}(t) = v(t)\sin\theta(t) \\ \dot{\theta}(t) = \widehat{\omega}(t) \end{cases}, \tag{10.1}$$

where $v(t)$ and $\widehat{\omega}(t)$ are, respectively, the displacement and angular velocities of the robot, both of which can be obtained in present technology by employing odometric measures with satisfied accurateness. It is assumed that the displacement and angular velocities of the robot received by the odometric measures are constant over the sampling period. Then, the continuous-time system (10.1) can be discretized to the following system:

$$\begin{cases} x_{k+1} = x_k + \Delta T v_k \cos\theta_k \\ y_{k+1} = y_k + \Delta T v_k \sin\theta_k \\ \theta_{k+1} = \theta_k + \Delta T \widehat{\omega}_k \end{cases}, \tag{10.2}$$

where ΔT is the sample time.

Setting $z_k = [\, x_k^{\mathrm{T}} \quad y_k^{\mathrm{T}} \quad \theta_k^{\mathrm{T}} \,]^{\mathrm{T}}$ and

$$u_k = \begin{bmatrix} \Delta T v_k \\ \Delta T \widehat{\omega}_k \end{bmatrix} := \begin{bmatrix} u_{1,k} \\ u_{2,k} \end{bmatrix},$$

system (10.2) can be rewritten as

$$z_{k+1} = f(z_k, u_k), \tag{10.3}$$

where

$$f(z_k, u_k) = z_k + \begin{bmatrix} u_{1,k} \cos \theta_k \\ u_{1,k} \sin \theta_k \\ u_{2,k} \end{bmatrix}.$$ (10.4)

By expanding the nonlinear function $f(z_k, u_k)$ in a Taylor series about the filtered estimate $\hat{z}_k$, (10.3) can be further reorganized as

$$z_{k+1} = A_k z_k + w_k,$$ (10.5)

where

$$A_k = \begin{bmatrix} \dfrac{\partial f_x}{\partial x_k} & \dfrac{\partial f_x}{\partial y_k} & \dfrac{\partial f_x}{\partial \theta_k} \\[2mm] \dfrac{\partial f_y}{\partial x_k} & \dfrac{\partial f_y}{\partial y_k} & \dfrac{\partial f_y}{\partial \theta_k} \\[2mm] \dfrac{\partial f_\theta}{\partial x_k} & \dfrac{\partial f_\theta}{\partial y_k} & \dfrac{\partial f_\theta}{\partial \theta_k} \end{bmatrix}\Bigg|_{z_k = \hat{z}_k} = \begin{bmatrix} 1 & 0 & -u_{1,k} \sin \theta_k \\ 0 & 1 & u_{1,k} \cos \theta_k \\ 0 & 0 & 1 \end{bmatrix}\Bigg|_{z_k = \hat{z}_k}$$ (10.6)

and $w_k = f(\hat{z}_k, u_k) - A_k \hat{z}_k + \sigma_z$. Here, σ_z represents the higher order terms of the Taylor series expansions of the nonlinear function $f(z_k, u_k)$.

Remark 10.1 *It is well recognized that the evolution of the system (10.5) depends primarily on the system matrix A_k, and the nonlinear term w_k (also known as linearization error) plays a relatively less important role. As such, a conventional way is to treat the nonlinear term w_k as one of the sources for disturbances. On the other hand, it is inevitable that the system states are contaminated by external noises. For mathematical convenience, from now on, we slightly abuse the notation by using w_k to include both the linearization errors and the external environmental noises. Moreover, w_k is assumed to belong to $l_2[0, \infty)$.*

10.1.2 Measurement Model with Quantization and Missing Observations

In this subsection, we shall establish a new measurement model for the mobile robot in order to account for both the missing measurements and quantization effects.

As shown in Figure 10.2, the point M is chosen as a marker and then the distance from the robot's planar Cartesian coordinates (x_k, y_k) to the marker M's coordinates (x_M, y_M) at a time instant k can be expressed as follows:

$$d_k = \sqrt{(x_M - x_k)^2 + (y_M - y_k)^2}.$$ (10.7)

The azimuth φ_k at time k can be related to the current system state variables x_k, y_k, and θ_k as follows:

$$\varphi_k = \theta_k - \arctan \frac{y_M - y_k}{x_M - x_k}.$$ (10.8)

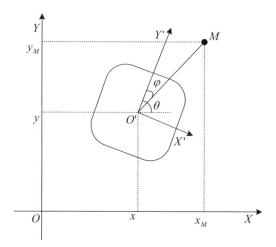

Figure 10.2   Absolute measurements

In robotics applications, both the distance d_k and the azimuth φ_k are treated as measurements. Consequently, from (10.7) and (10.8), the measurement equation is obtained as follows:

$$m_k = g(z_k), \tag{10.9}$$

where

$$g(z_k) = \begin{bmatrix} \sqrt{(x_M - x_k)^2 + (y_M - y_k)^2} \\ \theta_k - \arctan\dfrac{y_M - y_k}{x_M - x_k} \end{bmatrix}. \tag{10.10}$$

Again, using Taylor series expansions, the measurement equation (10.9) can be rewritten as

$$m_k = C_k z_k + \xi_k, \tag{10.11}$$

where

$$C_k = \begin{bmatrix} \dfrac{\partial g_d}{\partial x_k} & \dfrac{\partial g_d}{\partial y_k} & \dfrac{\partial g_d}{\partial \theta_k} \\ \dfrac{\partial g_\varphi}{\partial x_k} & \dfrac{\partial g_\varphi}{\partial y_k} & \dfrac{\partial g_\varphi}{\partial \theta_k} \end{bmatrix}\Bigg|_{z_k = \hat{z}_k}$$

$$\tag{10.12}$$

$$= \begin{bmatrix} -\dfrac{x_M - x_k}{\sqrt{(x_M - x_k)^2 + (y_M - y_k)^2}} & -\dfrac{y_M - y_k}{\sqrt{(x_M - x_k)^2 + (y_M - y_k)^2}} & 0 \\ \dfrac{y_M - y_k}{(x_M - x_k)^2 + (y_M - y_k)^2} & -\dfrac{x_M - x_k}{(x_M - x_k)^2 + (y_M - y_k)^2} & 1 \end{bmatrix}\Bigg|_{z_k = \hat{z}_k}$$

and $\xi_k = g(\hat{z}_k) - C_k\hat{z}_k + \sigma_m$. Here, σ_m represents the higher order terms of the Taylor series expansions of the nonlinear function $g(z_k)$. Similar to the discussion in Remark 10.1, $\xi_k \in l_2[0, \infty)$ can be utilized to represent both the linearization error and the external noise.

In mobile robotics applications, where the sensor signal is transmitted through network cables or wireless networks, it is often the case that the measurement outputs are *quantized* before being transmitted to other nodes. Let us denote the quantizer as $h(\cdot) = [h_1(\cdot) \quad h_2(\cdot)]^T$, which is symmetric; that is, $h_j(-v) = -h_j(v)$ $(j = 1, 2)$. The map of the quantization process is

$$h(m_k) = \left[h_1(m_k^{(1)}) \quad h_2(m_k^{(2)}) \right]^T.$$

In this chapter, we are interested in the logarithmic quantization process. For each $h_j(\cdot)$ $(1 \le j \le 2)$, the set of quantization levels is described by

$$\mathcal{U}_j = \{\pm\hat{\mu}_i^{(j)}, \hat{\mu}_i^{(j)} = \chi_j^i\hat{\mu}_0^{(j)}, i = 0, \pm1; \pm2, \ldots\} \cup \{0\}, \quad 0 < \chi_j < 1, \quad \hat{\mu}_0^{(j)} > 0,$$

where χ_j $(j = 1, 2)$ is called the quantization density. Each of the quantization levels corresponds to a segment such that the quantizer maps the whole segment to this quantization level. In addition, the logarithmic quantizer is defined as

$$h_j(m_k^{(j)}) = \begin{cases} \hat{\mu}_i^{(j)}, & \dfrac{1}{1 + \delta_j}\hat{\mu}_i^{(j)} \le m_k^{(j)} \le \dfrac{1}{1 - \delta_j}\hat{\mu}_i^{(j)} \\ 0, & m_k^{(j)} = 0 \\ -h_j(-m_k^{(j)}), & m_k^{(j)} < 0 \end{cases}$$

where

$$\delta_j = \frac{1 - \chi_j}{1 + \chi_j}.$$

It can be easily seen from the above definition that $h_j(m_k^{(j)}) = (1 + \Delta_k^{(j)})m_k^{(j)}$ with $|\Delta_k^{(j)}| \le \delta_j$. According to the transformation discussed above, the quantizing effects can be transformed into the sector-bounded uncertainties [58, 69–71].

Defining $\Delta_k = \text{diag}\{\Delta_k^{(1)}, \Delta_k^{(2)}\}$, the measurements with quantization effects can be expressed as

$$\begin{aligned} h(m_k) &= (I + \Delta_k)m_k \\ &= (I + \Delta_k)(C_k z_k + \xi_k). \end{aligned} \tag{10.13}$$

In practical applications, the measurements received by the robot may not be consecutive but contain missing observations due to various reasons such as the maneuverability of the robot, a failure in the measurement, intermittent sensor failures or accidental loss of some collected data. In the present chapter, we use the following equation to model the measurements with missing observations:

$$\bar{m}_k = \Lambda_k h(m_k)$$

$$= \sum_{i=1}^{2} \alpha_{ik} E_i (I + \Delta_k) m_k \tag{10.14}$$

where $\bar{m}_k$ is the actually signal available for the robot and

$$E_1 = \text{diag}\{1, 0\}, \quad E_2 = \text{diag}\{0, 1\}, \quad \Lambda_k = \text{diag}\{\alpha_{1k}, \alpha_{2k}\}$$

with $\alpha_{ik} \in \mathbb{R}$ $(i = 1, 2)$ being mutually independent random variables that describe the missing measurement phenomenon. It is assumed that α_{ik} has the probability density function $q_i(s)$ on the interval $[0, 1]$ with mathematical expectation $\bar{\alpha}_i$ and variance σ_i^2 $(i = 1, 2)$.

Remark 10.2 *Up to now, we have established the kinematics model for the mobile robot and the measurement model with both quantization effects and missing observations. The next step is to estimate the state of system (10.5) by employing the information received from (10.14). A seemingly natural way is to use the extended Kalman filtering approach based on the assumption that both the process noise w_k and the measurement noise ξ_k are Gaussian white noises. Such an assumption, unfortunately, is impractical in the mobile robot localization problem addressed in this chapter. An alternative yet effective approach to dealing with the mobile robot localization problem is the robust extended H_∞ filter design based on the Krein space theory developed in Yang et al. [180]. In fact, the mobile robot localization system established in this chapter is a substantial extension of that in Yang et al. [180] due to the consideration of both the missing measurements and signal quantizations. Later, a novel recursive RDE approach will be developed to offer the possibility for online applications of the proposed localization algorithm.*

10.2   A Stochastic H_∞ Filter Design

In this section, we aim to design a stochastic H_∞ filter to estimate the state of the mobile robot system (10.5) based on the measurement equation (10.14). For this purpose, we construct the following filter:

$$\hat{z}_{k+1} = A_k \hat{z}_k + K_k (\bar{m}_k - \bar{\Lambda}_k C_k \hat{z}_k), \quad \hat{z}_0 = 0, \tag{10.15}$$

where $\bar{\Lambda}_k = \mathbb{E}\{\Lambda_k\} = \text{diag}\{\bar{\alpha}_1, \bar{\alpha}_2\}$ and K_k is the filter parameter to be determined.

Letting the estimation error be $e_k = z_k - \hat{z}_k$, the dynamics of the estimation error can be obtained from (10.5), (10.14), and (10.15) as follows:

$$e_{k+1} = (A_k - K_k \bar{\Lambda}_k C_k) e_k - \sum_{i=1}^{2} (\alpha_{ik} - \bar{\alpha}_i) K_k E_i (I + \Delta_k)(C_k z_k + \xi_k)$$

$$- K_k \bar{\Lambda}_k (I + \Delta_k) \xi_k + w_k - K_k \bar{\Lambda}_k \Delta_k C_k z_k. \tag{10.16}$$

Our objective of this chapter is to find the sequence of filter parameter matrices K_k such that the filtering error e_k satisfies the following H_∞ performance requirement:

$$J = \mathbb{E}\left\{ \sum_{k=0}^{N-1}(\|e_k\|^2 - \gamma^2\|w_k\|^2 - \gamma^2\|\xi_k\|^2) \right\} - \gamma^2 z_0^{\mathrm{T}} S z_0 < 0, \tag{10.17}$$

$$\forall(\{w_k\}, \{\xi_k\}, z_0) \neq 0,$$

for the given disturbance attenuation level $\gamma > 0$ and the positive-definite matrix $S > 0$.

By defining $\bar{\Delta} = \mathrm{diag}\{\delta_1, \delta_2\}$ and $F_k = \Delta_k \bar{\Delta}^{-1}$, we know that F_k is an uncertain real-valued time-varying matrix satisfying $F_k^{\mathrm{T}} F_k \leq I$. Furthermore, by setting $\eta_k = [\, z_k^{\mathrm{T}} \quad e_k^{\mathrm{T}} \,]^{\mathrm{T}}$ and $[\tilde{w}_k = w_k^{\mathrm{T}} \quad \xi_k^{\mathrm{T}} \,]^{\mathrm{T}}$, the combination of (10.5) and (10.16) yields

$$\begin{cases} \eta_{k+1} = (\tilde{A}_k + \Delta\tilde{A}_k)\eta_k + \sum_{i=1}^{2} \tilde{\alpha}_{ik}(\tilde{C}_{ik} + \Delta\tilde{C}_{ik})\eta_k + (\tilde{K}_{1k} + \Delta\tilde{K}_{1k})\tilde{w}_k \\ \qquad\quad + \sum_{i=1}^{2} \tilde{\alpha}_{ik}(\tilde{K}_{2ik} + \Delta\tilde{K}_{2ik})\tilde{w}_k, \\ e_k = \tilde{L}\eta_k, \end{cases} \tag{10.18}$$

where

$$\tilde{A}_k = \mathrm{diag}\{A_k, A_k - K_k\bar{\Lambda}_k C_k\}, \quad \tilde{C}_{ik} = \begin{bmatrix} 0 & 0 \\ -K_k E_i C_k & 0 \end{bmatrix},$$

$$\tilde{K}_{1k} = \begin{bmatrix} I & 0 \\ I & -K_k\bar{\Lambda}_k \end{bmatrix}, \quad \tilde{K}_{2ik} = \begin{bmatrix} 0 & 0 \\ 0 & -K_k E_i \end{bmatrix}, \quad \tilde{L} = \begin{bmatrix} 0 & I \end{bmatrix},$$

$$\Delta\tilde{A}_k = \tilde{H}_k F_k \tilde{E}_{ck}, \quad \Delta\tilde{C}_{ik} = \tilde{H}_{ki} F_k \tilde{E}_{ck}, \quad \Delta\tilde{K}_{1k} = \tilde{H}_k F_k \tilde{E},$$

$$\Delta\tilde{K}_{2ik} = \tilde{H}_{ki} F_k \tilde{E}, \quad \tilde{H}_k = \begin{bmatrix} 0 & -(K_k\bar{\Lambda}_k)^{\mathrm{T}} \end{bmatrix}^{\mathrm{T}}, \quad \tilde{E}_{ck} = \begin{bmatrix} \bar{\Delta} C_k & 0 \end{bmatrix},$$

$$\tilde{H}_{ki} = \begin{bmatrix} 0 & -(K_k E_i)^{\mathrm{T}} \end{bmatrix}^{\mathrm{T}}, \quad \tilde{E} = \begin{bmatrix} 0 & \bar{\Delta} \end{bmatrix}, \quad \tilde{\alpha}_{ik} = \alpha_{ik} - \bar{\alpha}_i$$

and the H_∞ performance requirement (10.17) can be rewritten as

$$J = \mathbb{E}\left\{ \sum_{k=0}^{N-1}(\|e_k\|^2 - \gamma^2\|\tilde{w}_k\|^2) \right\} - \gamma^2 \eta_0^{\mathrm{T}} R \eta_0 < 0, \quad \forall(\{\tilde{w}_k\}, \eta_0) \neq 0, \tag{10.19}$$

with $R = \Pi^{\mathrm{T}}\mathrm{diag}\{S, mI\}\Pi$, where

$$\Pi = \begin{bmatrix} I & 0 \\ -I & I \end{bmatrix}$$

and m is an arbitrary positive scalar.

To deal with the parameter uncertainties that arise from the quantization effects, we rearrange (10.18) as follows:

$$\eta_{k+1} = \tilde{A}_k \eta_k + \sum_{i=1}^{2} \tilde{\alpha}_{ik} \tilde{C}_{ik} \eta_k + \tilde{B}_{1k} \bar{w}_k + \sum_{i=1}^{2} \tilde{\alpha}_{ik} \tilde{B}_{2ik} \bar{w}_k, \tag{10.20}$$
$$e_k = \tilde{L} \eta_k,$$

where

$$\tilde{B}_{1k} = \begin{bmatrix} \tilde{K}_{1k} & \epsilon_k^{-1} \tilde{H}_k & \epsilon_k^{-1} \tilde{H}_k \end{bmatrix}, \quad \tilde{B}_{2ik} = \begin{bmatrix} \tilde{K}_{2ik} & \epsilon_k^{-1} \tilde{H}_{ki} & \epsilon_k^{-1} \tilde{H}_{ki} \end{bmatrix},$$
$$\bar{w}_k = \begin{bmatrix} \tilde{w}_k^{\mathrm{T}} & (\epsilon_k F_k \tilde{E} \tilde{w}_k)^{\mathrm{T}} & (\epsilon_k F_k \tilde{E}_{ck} \eta_k)^{\mathrm{T}} \end{bmatrix}^{\mathrm{T}}$$

and ϵ_k is a positive function representing a scaling of the perturbation that is introduced to provide more flexibility in the solution [181].

Let $\bar{J}$ be the H_∞ performance requirement for the system (10.20) defined by

$$\bar{J} = \mathbb{E}\left\{ \sum_{k=0}^{N-1} (\|e_k\|^2 - \gamma^2 \|\bar{w}_k\|^2) + \gamma^2 (\|\epsilon_k \tilde{E} \tilde{w}_k\|^2 + \|\epsilon_k \tilde{E}_{ck} \eta_k\|^2) \right\} \tag{10.21}$$
$$- \gamma^2 \eta_0^{\mathrm{T}} R \eta_0 < 0, \quad \forall(\{\tilde{w}_k\}, \eta_0) \neq 0.$$

Before proceeding further, we introduce the following lemmas which will be utilized in the subsequent developments.

Lemma 10.2.1 Consider the performance indices defined in (10.19) and (10.21), respectively. We can conclude that $J \leq \bar{J}$.

Proof.   It is easily seen that

$$J - \bar{J} = \mathbb{E}\left\{ \sum_{k=0}^{N-1} \gamma^2 (\|\bar{w}_k\|^2 - \|\tilde{w}_k\|^2 - \|\epsilon_k \tilde{E} \tilde{w}_k\|^2 - \|\epsilon_k \tilde{E}_{ck} \eta_k\|^2) \right\}$$

$$= \mathbb{E}\left\{ \sum_{k=0}^{N-1} \gamma^2 (\|\epsilon_k F_k \tilde{E} \tilde{w}_k\|^2 + \|\epsilon_k F_k \tilde{E}_{ck} \eta_k\|^2 - \|\epsilon_k \tilde{E} \tilde{w}_k\|^2 - \|\epsilon_k \tilde{E}_{ck} \eta_k\|^2) \right\}$$

$$= \mathbb{E}\left\{ \sum_{k=0}^{N-1} \gamma^2 (\|\epsilon_k [F_k^{\mathrm{T}} F_k - I]^{1/2} \tilde{E} \tilde{w}_k\|^2 + \|\epsilon_k [F_k^{\mathrm{T}} F_k - I]^{1/2} \tilde{E}_{ck} \eta_k\|^2) \right\} \leq 0,$$

which ends the proof.   □

Remark 10.3 *Note that the uncertainties in the parameters of system (10.18) are treated as an additive perturbation $\bar{w}_k$. As can be seen from Lemma 10.2.1, $\bar{J}$ is an upper bound of J; that is, $\bar{J} = \max J$. Therefore, the performance index $\bar{J}$ can be used to replace J in our stochastic H_∞ filtering problem.*

Lemma 10.2.2 [182] Let matrices G, M, and Γ be given with appropriate dimensions. Then, the matrix equation

$$GXM = \Gamma \qquad (10.22)$$

has a solution X if and only if $GG^\dagger \Gamma M^\dagger M = \Gamma$. Moreover, any solution to (10.22) is represented by

$$X = G^\dagger \Gamma M^\dagger + Y - G^\dagger GYMM^\dagger,$$

where Y is a matrix with an appropriate size.

Lemma 10.2.3 Consider the mobile robot localization system (10.5) and (10.14) with a given disturbance rejection attenuation level $\gamma > 0$ and a positive-definite matrix $S > 0$. For each $k = 0, 1, \ldots, N - 1$, the filtering error e_k in (10.16) satisfies the H_∞ performance requirement (10.21) if and only if there exists a positive function $\epsilon_k > 0$ such that the discrete RDE

$$
\begin{cases}
Q_k = \tilde{A}_k^{\mathrm{T}} Q_{k+1} \tilde{A}_k + \displaystyle\sum_{i=1}^{2} \sigma_i^2 \tilde{C}_{ik}^{\mathrm{T}} Q_{k+1} \tilde{C}_{ik} + \tilde{L}^{\mathrm{T}} \tilde{L} + \gamma^2 \epsilon_k^2 \tilde{E}_{ck}^{\mathrm{T}} \tilde{E}_{ck} \\[2mm]
\qquad + \left(\tilde{B}_{1k}^{\mathrm{T}} Q_{k+1} \tilde{A}_k + \displaystyle\sum_{i=1}^{2} \sigma_i^2 \tilde{B}_{2ik}^{\mathrm{T}} Q_{k+1} \tilde{C}_{ik} \right)^{\mathrm{T}} \Theta_k^{-1} \\[4mm]
\qquad \times \left(\tilde{B}_{1k}^{\mathrm{T}} Q_{k+1} \tilde{A}_k + \displaystyle\sum_{i=1}^{2} \sigma_i^2 \tilde{B}_{2ik}^{\mathrm{T}} Q_{k+1} \tilde{C}_{ik} \right), \\[4mm]
Q_N = 0
\end{cases}
\qquad (10.23)
$$

has a solution (Q_k, K_k) satisfying

$$
\begin{cases}
\Theta_k = -\tilde{B}_{1k}^{\mathrm{T}} Q_{k+1} \tilde{B}_{1k} - \displaystyle\sum_{i=1}^{2} \sigma_i^2 \tilde{B}_{2ik}^{\mathrm{T}} Q_{k+1} \tilde{B}_{2ik} - \gamma^2 (\epsilon_k^2 E_I^{\mathrm{T}} \tilde{E}^{\mathrm{T}} \tilde{E} E_I - I) > 0 \\[2mm]
Q_0 < \gamma^2 R,
\end{cases}
\qquad (10.24)
$$

where $E_I = [\, I \quad 0 \quad 0\,]$ and R is defined in (10.19).

Proof.

Sufficiency
By defining

$$\tilde{J}_k = \eta_{k+1}^{\mathrm{T}} Q_{k+1} \eta_{k+1} - \eta_k^{\mathrm{T}} Q_k \eta_k \tag{10.25}$$

and noticing (10.20), we have

$$
\begin{aligned}
\mathbb{E}\{\tilde{J}_k\} &= \mathbb{E}\left\{ \left(\tilde{A}_k \eta_k + \sum_{i=1}^{2} \tilde{\alpha}_{ik} \tilde{C}_{ik} \eta_k + \tilde{B}_{1k} \bar{w}_k + \sum_{i=1}^{2} \tilde{\alpha}_{ik} \tilde{B}_{2ik} \bar{w}_k \right)^{\mathrm{T}} Q_{k+1} \right. \\
&\qquad \left. \times \left(\tilde{A}_k \eta_k + \sum_{i=1}^{2} \tilde{\alpha}_{ik} \tilde{C}_{ik} \eta_k + \tilde{B}_{1k} \bar{w}_k + \sum_{i=1}^{2} \tilde{\alpha}_{ik} \tilde{B}_{2ik} \bar{w}_k \right) - \eta_k^{\mathrm{T}} Q_k \eta_k \right\} \\
&= \mathbb{E}\left\{ \eta_k^{\mathrm{T}} \left(\tilde{A}_k^{\mathrm{T}} Q_{k+1} \tilde{A}_k + \sum_{i=1}^{2} \sigma_i^2 \tilde{C}_{ik}^{\mathrm{T}} Q_{k+1} \tilde{C}_{ik} - Q_k \right) \eta_k \right. \tag{10.26} \\
&\quad + 2\eta_k^{\mathrm{T}} \left(\tilde{A}_k^{\mathrm{T}} Q_{k+1} \tilde{B}_{1k} + \sum_{i=1}^{2} \sigma_i^2 \tilde{C}_{ik}^{\mathrm{T}} Q_{k+1} \tilde{B}_{2ik} \right) \bar{w}_k \\
&\quad \left. + \bar{w}_k^{\mathrm{T}} \left(\tilde{B}_{1k}^{\mathrm{T}} Q_{k+1} \tilde{B}_{1k} + \sum_{i=1}^{2} \sigma_i^2 \tilde{B}_{2ik}^{\mathrm{T}} Q_{k+1} \tilde{B}_{2ik} \right) \bar{w}_k \right\}.
\end{aligned}
$$

Adding the zero term

$$
\|e_k\|^2 - \gamma^2 (\|\bar{w}_k\|^2 - \|\epsilon_k \tilde{E} \bar{w}_k\|^2 - \|\epsilon_k \tilde{E}_{ck} \eta_k\|^2)
$$
$$
- (\|e_k\|^2 - \gamma^2 (\|\bar{w}_k\|^2 - \|\epsilon_k \tilde{E} \bar{w}_k\|^2 - \|\epsilon_k \tilde{E}_{ck} \eta_k\|^2))
$$

to both sides of (10.26) and then taking the mathematical expectation results in

$$
\begin{aligned}
\mathbb{E}\{\tilde{J}_k\} &= \mathbb{E}\left\{ \eta_k^{\mathrm{T}} \left(\tilde{A}_k^{\mathrm{T}} Q_{k+1} \tilde{A}_k + \sum_{i=1}^{2} \sigma_i^2 \tilde{C}_{ik}^{\mathrm{T}} Q_{k+1} \tilde{C}_{ik} + \tilde{L}^{\mathrm{T}} \tilde{L} + \gamma^2 \epsilon_k^2 \tilde{E}_{ck}^{\mathrm{T}} \tilde{E}_{ck} - Q_k \right) \eta_k \right. \\
&\quad + 2\eta_k^{\mathrm{T}} \left(\tilde{A}_k^{\mathrm{T}} Q_{k+1} \tilde{B}_{1k} + \sum_{i=1}^{2} \sigma_i^2 \tilde{C}_{ik}^{\mathrm{T}} Q_{k+1} \tilde{B}_{2ik} \right) \bar{w}_k + \bar{w}_k^{\mathrm{T}} \left(\tilde{B}_{1k}^{\mathrm{T}} Q_{k+1} \tilde{B}_{1k} \right. \\
&\quad \left. \left. + \sum_{i=1}^{2} \sigma_i^2 \tilde{B}_{2ik}^{\mathrm{T}} Q_{k+1} \tilde{B}_{2ik} + \gamma^2 (\epsilon_k^2 E_I^{\mathrm{T}} \tilde{E}^{\mathrm{T}} \tilde{E} E_I - I) \right) \bar{w}_k \right\} \\
&\quad - \mathbb{E}\{\|e_k\|^2 - \gamma^2 (\|\bar{w}_k\|^2 - \|\epsilon_k \tilde{E} \bar{w}_k\|^2 - \|\epsilon_k \tilde{E}_{ck} \eta_k\|^2)\}. \tag{10.27}
\end{aligned}
$$

By applying the completing squares method, we have

$$
\begin{aligned}
\mathbb{E}\{\tilde{J}_k\} = \eta_k^\mathrm{T} \Bigg[& \tilde{A}_k^\mathrm{T} Q_{k+1} \tilde{A}_k + \sum_{i=1}^{2} \sigma_i^2 \tilde{C}_{ik}^\mathrm{T} Q_{k+1} \tilde{C}_{ik} + \tilde{L}^\mathrm{T} \tilde{L} + \gamma^2 \epsilon_k^2 \tilde{E}_{ck}^\mathrm{T} \tilde{E}_{ck} - Q_k \\
& + \left(\tilde{B}_{1k}^\mathrm{T} Q_{k+1} \tilde{A}_k + \sum_{i=1}^{2} \sigma_i^2 \tilde{B}_{2ik}^\mathrm{T} Q_{k+1} \tilde{C}_{ik} \right)^\mathrm{T} \Theta_k^{-1} \left(\tilde{B}_{1k}^\mathrm{T} Q_{k+1} \tilde{A}_k \right. \\
& \left. + \sum_{i=1}^{2} \sigma_i^2 \tilde{B}_{2ik}^\mathrm{T} Q_{k+1} \tilde{C}_{ik} \right) \Bigg] \eta_k - (\bar{w}_k - \bar{w}_k^*)^\mathrm{T} \Theta_k (\bar{w}_k - \bar{w}_k^*) \\
& - \mathbb{E}\{\|e_k\|^2 - \gamma^2 (\|\bar{w}_k\|^2 - \|\epsilon_k \tilde{E} \bar{w}_k\|^2 - \|\epsilon_k \tilde{E}_{ck} \eta_k\|^2)\},
\end{aligned}
\tag{10.28}
$$

where

$$
\bar{w}_k^* = \Theta_k^{-1} \left(\tilde{B}_{1k}^\mathrm{T} Q_{k+1} \tilde{A}_k + \sum_{i=1}^{2} \sigma_i^2 \tilde{B}_{2ik}^\mathrm{T} Q_{k+1} \tilde{C}_{ik} \right) \eta_k.
\tag{10.29}
$$

Taking the sum on both sides of (10.25) from 0 to $N - 1$, we obtain

$$
\begin{aligned}
\mathbb{E}\left\{ \sum_{k=0}^{N-1} \tilde{J}_k \right\} &= \mathbb{E}\{\eta_N^\mathrm{T} Q_N \eta_N\} - \eta_0^\mathrm{T} Q_0 \eta_0 \\
&= - \sum_{k=0}^{N-1} (\bar{w}_k - \bar{w}_k^*)^\mathrm{T} \Theta_k (\bar{w}_k - \bar{w}_k^*) - \sum_{k=0}^{N-1} \mathbb{E}\{\|e_k\|^2 \\
& \quad - \gamma^2 (\|\bar{w}_k\|^2 - \|\epsilon_k \tilde{E} \bar{w}_k\|^2 - \|\epsilon_k \tilde{E}_{ck} \eta_k\|^2)\}.
\end{aligned}
\tag{10.30}
$$

Since $\Theta_k > 0$, $Q_0 - \gamma^2 R < 0$, $Q_N = 0$, and $\eta_0 \neq 0$, it follows that

$$
\begin{aligned}
\bar{J} &= \mathbb{E}\left\{ \sum_{k=0}^{N-1} (\|e_k\|^2 - \gamma^2 (\|\bar{w}_k\|^2 - \|\epsilon_k \tilde{E} \bar{w}_k\|^2 - \|\epsilon_k \tilde{E}_{ck} \eta_k\|^2)) \right\} - \gamma^2 \eta_0^\mathrm{T} R \eta_0 \\
&= \eta_0^\mathrm{T} (Q_0 - \gamma^2 R) \eta_0 - \sum_{k=0}^{N-1} \mathbb{E}\{(\bar{w}_k - \bar{w}_k^*)^\mathrm{T} \Theta_k (\bar{w}_k - \bar{w}_k^*)\} < 0,
\end{aligned}
\tag{10.31}
$$

which is equivalent to (10.21). This means that the prespecified H_∞ performance is satisfied, and therefore the proof of sufficiency is complete.

Necessity
We now proceed to prove that if $\bar{J} < 0$, then there exists a solution Q_k ($0 \leq k < N$) to the recursive equations (10.23) such that $\Theta_k > 0$ is satisfied for all nonzero ($\{\bar{w}_k\}, \eta_0$). In fact, the recursion (10.23) can always be solved backward with the known final condition $Q_N = 0$ if and only if $\Theta_k > 0$ for all $k \in [0, N)$. If (10.23) fails to proceed for some $k = k_0 \in [0, N)$,

then Θ_{k_0} has at least one zero or negative eigenvalue. Therefore, the proof of necessity is equivalent to the proof of the following proposition:

$$\bar{J} < 0 \implies \lambda_i(\Theta_k) > 0, \quad \forall k \in [0, N), \quad i = 1, 2, \ldots, 9, \tag{10.32}$$

where $\lambda_i(\Theta_k)$ denotes the eigenvalue of Θ_k at time i. The rest of the proof is carried out by contradiction. That is, assuming that at least one eigenvalue of Θ_k is either equal to 0 or negative at some time point $k = k_0 \in [0, N)$, we intend to prove that $\bar{J} < 0$ is not true. For simplicity, we denote such an eigenvalue of Θ_k at time k_0 as λ_{k_0}; that is, $\lambda_{k_0} \leq 0$. In the following, we shall use such a nonpositive λ_{k_0} to reveal that there exist certain $(\{\tilde{w}_k\}, \eta_0) \neq 0$ such that $\bar{J} \geq 0$. First, we can choose $\eta_0 = 0$ and

$$\bar{w}_k = \begin{cases} \psi_{k_0}, & k = k_0, \\ \bar{w}_k^*, & k_0 < k < N, \\ 0, & 0 \leq k < k_0, \end{cases} \tag{10.33}$$

where ψ_{k_0} is the eigenvector of Θ_{k_0} with respect to λ_{k_0}. Since $\eta_0 = 0$ and $\xi_k = 0$ when $0 \leq k < k_0$, we can obtain from (10.20), (10.28), and (10.30) that

$$
\begin{aligned}
\bar{J} &= \mathbb{E}\left\{\sum_{k=0}^{N-1}(\|e_k\|^2 - \gamma^2(\|\bar{w}_k\|^2 - \|\epsilon_k \tilde{E}\tilde{w}_k\|^2 - \|\epsilon_k \tilde{E}_{ck}\eta_k\|^2))\right\} - \gamma^2\eta_0^\mathrm{T} R\eta_0 \\
&= \sum_{k=0}^{k_0-1} \mathbb{E}\{\|e_k\|^2 - \gamma^2(\|\bar{w}_k\|^2 - \|\epsilon_k \tilde{E}\tilde{w}_k\|^2 - \|\epsilon_k \tilde{E}_{ck}\eta_k\|^2)\} \\
&\quad + \sum_{k=k_0+1}^{N-1} \mathbb{E}\{\|e_k\|^2 - \gamma^2(\|\bar{w}_k\|^2 - \|\epsilon_k \tilde{E}\tilde{w}_k\|^2 - \|\epsilon_k \tilde{E}_{ck}\eta_k\|^2)\} \\
&\quad + \mathbb{E}\{\|e_{k_0}\|^2 - \gamma^2(\|\bar{w}_{k_0}\|^2 - \|\epsilon_{k_0}\tilde{E}\tilde{w}_{k_0}\|^2 - \|\epsilon_{k_0}\tilde{E}_{ck_0}\eta_{k_0}\|^2)\} \\
&= -\sum_{k=k_0+1}^{N-1}(\bar{w}_k - \bar{w}_k^*)^\mathrm{T}\Theta_k(\bar{w}_k - \bar{w}_k^*) - (\bar{w}_{k_0} - \bar{w}_{k_0}^*)^\mathrm{T}\Theta_{k_0}(\bar{w}_{k_0} - \bar{w}_{k_0}^*) \tag{10.34}\\
&\quad - \sum_{k=k_0+1}^{N-1}\mathbb{E}\{\tilde{J}_k\} + \eta_{k_0}^\mathrm{T}\left[\tilde{A}_{k_0}^\mathrm{T}Q_{k_0+1}\tilde{A}_{k_0} + \sum_{i=1}^{2}\sigma_i^2\tilde{C}_{ik0}^\mathrm{T}Q_{k_0+1}\tilde{C}_{ik0} + \tilde{L}^\mathrm{T}\tilde{L}\right. \\
&\quad + \left(\tilde{B}_{1k}^\mathrm{T}Q_{k+1}\tilde{A}_k + \sum_{i=1}^{2}\sigma_i^2\tilde{B}_{2ik}^\mathrm{T}Q_{k+1}\tilde{C}_{ik}\right)^\mathrm{T}\Theta_k^{-1}\left(\tilde{B}_{1k}^\mathrm{T}Q_{k+1}\tilde{A}_k\right.\\
&\quad \left.\left. + \sum_{i=1}^{2}\sigma_i^2\tilde{B}_{2ik}^\mathrm{T}Q_{k+1}\tilde{C}_{ik}\right) + \gamma^2\epsilon_{k_0}^2\tilde{E}_{ck_0}^\mathrm{T}\tilde{E}_{ck_0} - Q_{k_0}\right]\eta_{k_0} - \mathbb{E}\{\tilde{J}_{k_0}\} \\
&= -\bar{w}_{k_0}^\mathrm{T}\Theta_{k_0}\bar{w}_{k_0} = -\psi_{k_0}^\mathrm{T}\Theta_{k_0}\psi_{k_0} = -\lambda_{k_0}\|\psi_{k_0}\|^2 \geq 0,
\end{aligned}
$$

which contradicts with the condition $\bar{J} < 0$. Therefore, we can conclude that $\Theta_k > 0$ and the proof is now complete. $\square$

So far, we have analyzed the system's H_∞ performance in terms of the solvability of a backward Riccati equation in Lemma 10.2.3. In the next stage, we shall proceed to tackle the design problem of the filter (10.15) such that the filtering error e_k in (10.16) satisfies the H_∞ performance requirement (10.21) in the worst case of the disturbance $\bar{w}_k^*$.

Theorem 10.2.4   *Consider the mobile robot localization system (10.5) and (10.14). Given a disturbance rejection attenuation level $\gamma > 0$ and a positive-definite matrix $S > 0$. For each $k = 0, 1, \ldots, N - 1$, assume that there exists a positive function $\epsilon_k > 0$ such that the discrete RDE (10.23) has a solution (Q_k, K_k) satisfying (10.24) and the discrete RDE*

$$\begin{cases} P_k = A_k^{\mathrm{T}} P_{k+1} A_k + \sum_{i=1}^{2} \sigma_i^2 B_{ik}^{\mathrm{T}} P_{k+1} B_{ik} + \tilde{L}^{\mathrm{T}} \tilde{L} - A_k^{\mathrm{T}} P_{k+1} \Omega_k^{-1} P_{k+1} A_k, \\ P_N = 0, \\ \mathcal{N}_k = \mathcal{N}_k \mathcal{M} \dagger_k \mathcal{M}_k \end{cases} \tag{10.35}$$

has a solution (P_k, K_k) satisfying

$$\Omega_k = P_{k+1} + I > 0, \tag{10.36}$$

where

$$\mathcal{N}_k = \tilde{L} \Omega_k^{-1} P_{k+1} A_k \tilde{L}^{\mathrm{T}}, \quad \mathcal{M}_k = \bar{\Lambda}_k C_k. \tag{10.37}$$

Then, it can be concluded that:

(i) The worst-case disturbance $\bar{w}_k^*$ and the filter gain matrix K_k are given by

$$\bar{w}_k^* = \Theta_k^{-1} \left(\tilde{B}_{1k}^{\mathrm{T}} Q_{k+1} \tilde{A}_k + \sum_{i=1}^{2} \sigma_i^2 \tilde{B}_{2ik}^{\mathrm{T}} Q_{k+1} \tilde{C}_{ik} \right) \eta_k, \tag{10.38}$$

$$K_k = \mathcal{N}_k \mathcal{M} \dagger_k + Y_k - Y_k \mathcal{M}_k \mathcal{M} \dagger_k, \quad Y_k \in \mathbb{R}^{3 \times 2}, \quad k = 1, 2, \ldots, N - 1. \tag{10.39}$$

(ii) The filtering error e_k in (10.16) satisfies the H_∞ performance requirement (10.21).
(iii) The costs or performance objectives of $\bar{J}$ and $\bar{J}_{\bar{w}^*}$ are

$$\bar{J} = \eta_0^{\mathrm{T}} (Q_0 - \gamma^2 R) \eta_0, \quad \bar{J}_{\bar{w}^*} = \eta_0^{\mathrm{T}} P_0 \eta_0,$$

where

$$\mathcal{A}_k = \bar{A}_k + \tilde{B}_{1k} \Theta_k^{-1} \tilde{B}_{1k}^{\mathrm{T}} Q_{k+1} \tilde{A}_k + \tilde{B}_{1k} \Theta_k^{-1} \left(\sum_{i=1}^{2} \sigma_i^2 \tilde{B}_{2ik}^{\mathrm{T}} Q_{k+1} \tilde{C}_{ik} \right),$$

$$\mathcal{B}_{ik} = \tilde{C}_{ik} + \tilde{B}_{2ik} \Theta_k^{-1} \tilde{B}_{1k}^{\mathrm{T}} Q_{k+1} \tilde{A}_k + \tilde{B}_{2ik} \Theta_k^{-1} \left(\sum_{i=1}^{2} \sigma_i^2 \tilde{B}_{2ik}^{\mathrm{T}} Q_{k+1} \tilde{C}_{ik} \right),$$

$$\bar{A}_k = \mathrm{diag}\{A_k, A_k\}.$$

Proof. First, it follows from Lemma 10.2.3 that, when a solution Q_k to (10.23) exists such that $\Theta_k > 0$ and $Q_0 < \gamma^2 R$, then the filtering error e_k satisfies the H_∞ performance requirement (10.21). Moreover, according to (10.31), the worst-case disturbance can be expressed as $\bar{w}_k^* = \Theta_k^{-1}(\tilde{B}_{1k}^T Q_{k+1}\tilde{A}_k + \sum_{i=1}^2 \sigma_i^2 \tilde{B}_{2ik}^T Q_{k+1}\tilde{C}_{ik})\eta_k$ and the performance index is $\bar{J} = \eta_0^T(Q_0 - \gamma^2 R)\eta_0$.

In what follows, we aim to determine the filter gain matrix K_k under the situation of worst-case disturbance $\bar{w}_k^*$. To this end, a cost functional is defined as follows:

$$\bar{J}_{\bar{w}^*} = \mathbb{E}\left\{\sum_{k=0}^{N-1}(\|e_k\|^2 + \|\Upsilon_k\|^2)\right\},\tag{10.40}$$

with $\Upsilon_k = -K_k\bar{\Lambda}_k C_k e_k$; and in this case, the original system (10.20) with the worst-case disturbance $\bar{w}_k^*$ can be rewritten as

$$\begin{cases}\eta_{k+1} = \mathcal{A}_k\eta_k + \sum_{i=1}^2 \tilde{\alpha}_{ik}\mathcal{B}_{ik}\eta_k + \tilde{\Upsilon}_k,\\ e_k = \tilde{L}\eta_k,\end{cases}\tag{10.41}$$

where $\tilde{\Upsilon}_k = [0 \quad \Upsilon_k^T]^T$.

In order to obtain the parametrization of K_k, we define

$$\tilde{\mathcal{J}}_k = \eta_{k+1}^T P_{k+1}\eta_{k+1} - \eta_k^T P_k\eta_k.\tag{10.42}$$

Noticing (10.41) and taking mathematical expectation on both sides of (10.42), we have

$$\mathbb{E}\{\tilde{\mathcal{J}}_k\} = \mathbb{E}\left\{\left(\mathcal{A}_k\eta_k + \sum_{i=1}^2 \tilde{\alpha}_{ik}\mathcal{B}_{ik}\eta_k + \tilde{\Upsilon}_k\right)^T P_{k+1}\left(\mathcal{A}_k\eta_k + \sum_{i=1}^2 \tilde{\alpha}_{ik}\mathcal{B}_{ik}\eta_k + \tilde{\Upsilon}_k\right)\right.$$
$$\left. - \eta_k^T P_k\eta_k\right\}$$
$$= \mathbb{E}\left\{\eta_k^T\left(\mathcal{A}_k^T P_{k+1}\mathcal{A}_k + \sum_{i=1}^2 \sigma_i^2\mathcal{B}_{ik}^T P_{k+1}\mathcal{B}_{ik} - P_k\right)\eta_k + 2\eta_k^T\mathcal{A}_k^T P_{k+1}\tilde{\Upsilon}_k\right.$$
$$\left. + \tilde{\Upsilon}_k^T P_{k+1}\tilde{\Upsilon}_k\right\}.\tag{10.43}$$

Then, it follows that

$$
\mathbb{E}\{\tilde{\mathcal{J}}_k\} = \mathbb{E}\left\{ \eta_k^\mathrm{T}\left(\mathcal{A}_k^\mathrm{T} P_{k+1}\mathcal{A}_k + \sum_{i=1}^{2}\sigma_i^2 \mathcal{B}_{ik}^\mathrm{T} P_{k+1}\mathcal{B}_{ik} - P_k\right)\eta_k + 2\eta_k^\mathrm{T}\mathcal{A}_k^\mathrm{T} P_{k+1}\tilde{\Upsilon}_k \right.
$$

$$
\left. + \tilde{\Upsilon}_k^\mathrm{T} P_{k+1}\tilde{\Upsilon}_k \right\} + \mathbb{E}\{\|e_k\|^2 + \|\Upsilon_k\|^2 - \|e_k\|^2 - \|\Upsilon_k\|^2\}
$$

$$
= \eta_k^\mathrm{T}\left(\mathcal{A}_k^\mathrm{T} P_{k+1}\mathcal{A}_k + \sum_{i=1}^{2}\sigma_i^2 \mathcal{B}_{ik}^\mathrm{T} P_{k+1}\mathcal{B}_{ik} + \tilde{L}^\mathrm{T}\tilde{L} - P_k\right)\eta_k
$$

$$
+ 2\eta_k^\mathrm{T}\mathcal{A}_k^\mathrm{T} P_{k+1}\tilde{\Upsilon}_k + \tilde{\Upsilon}_k^\mathrm{T}(P_{k+1} + I)\tilde{\Upsilon}_k - \mathbb{E}\{\|e_k\|^2 + \|\Upsilon_k\|^2\}.
$$

By applying the completing squares method again, we have

$$
\mathbb{E}\{\tilde{\mathcal{J}}_k\} = \eta_k^\mathrm{T}\left(\mathcal{A}_k^\mathrm{T} P_{k+1}\mathcal{A}_k + \sum_{i=1}^{2}\sigma_i^2 \mathcal{B}_{ik}^\mathrm{T} P_{k+1}\mathcal{B}_{ik} + \tilde{L}^\mathrm{T}\tilde{L} - \mathcal{A}_k^\mathrm{T} P_{k+1}\Omega_k^{-1} P_{k+1}\mathcal{A}_k - P_k\right)
$$

$$
\times \eta_k + (\tilde{\Upsilon}_k - \tilde{\Upsilon}_k^*)^\mathrm{T}\Omega_k(\tilde{\Upsilon}_k - \tilde{\Upsilon}_k^*) - \mathbb{E}\{\|e_k\|^2 + \|\Upsilon_k\|^2\}, \tag{10.44}
$$

where

$$
\tilde{\Upsilon}_k^* = -\Omega_k^{-1} P_{k+1}\mathcal{A}_k\eta_k. \tag{10.45}
$$

Therefore, from (10.35), it is true that

$$
\bar{J}_{\tilde{w}^*} = \mathbb{E}\left\{ \sum_{k=0}^{N-1}(\|e_k\|^2 + \|\Upsilon_k\|^2) \right\}
$$

$$
= \sum_{k=0}^{N-1}(\tilde{\Upsilon}_k - \tilde{\Upsilon}_k^*)^\mathrm{T}\Omega_k(\tilde{\Upsilon}_k - \tilde{\Upsilon}_k^*) + \eta_0^\mathrm{T} P_0\eta_0. \tag{10.46}
$$

In order to suppress the cost of $\bar{J}_{\tilde{w}^*}$, the best choice for K_k is to satisfy $\tilde{\Upsilon}_k = \tilde{\Upsilon}_k^*$, from which we can obtain that

$$
K_k\bar{\Lambda}_k C_k = \tilde{L}\Omega_k^{-1} P_{k+1}\mathcal{A}_k\tilde{L}^\mathrm{T}. \tag{10.47}
$$

According to Lemma 10.2.2, it can be observed that the existence of a solution K_k ($k = 0, 1, \ldots, N - 1$) to (10.47) is equivalent to the feasibility of $\mathcal{N}_k\mathcal{M}\dagger_k\mathcal{M}_k = \mathcal{N}_k$, whose general solution is given by $K_k = \mathcal{N}_k\mathcal{M}\dagger_k + Y_k - Y_k\mathcal{M}_k\mathcal{M}\dagger_k$, $Y_k \in \mathbb{R}^{3\times 2}$, $k = 1, 2, \ldots, N - 1$. Furthermore, the performance index is $\bar{J}_{\tilde{w}^*} = \eta_0^\mathrm{T} P_0\eta_0$, and this completes the proof. $\quad\square$

By means of Theorem 10.2.4, we can summarize the finite-horizon H_∞ filter design (FHFD) algorithm as follows.

Algorithm FHFD

Step 1. Given the H_∞ performance index γ and the positive-definite matrix S, set $k = N - 1$ and an initial value ϵ_{N-1} satisfying $-\gamma^2(\epsilon_{N-1}^2 E_I^T \tilde{E}^T \tilde{E} E_I - I) > 0$. Then, $Q_N = P_N = 0$ are obtained.

Step 2. Calculate Θ_k, Ω_k, $\mathcal{N}_k$, and $\mathcal{M}_k$ with known Q_{k+1} and P_{k+1} via the first equation of (10.24) and equations (10.36) and (10.37), respectively. Furthermore, the filter gain matrix K_k can be obtained by equation (10.39).

Step 3. If $\mathcal{N}_k = \mathcal{N}_k \mathcal{M}_{\dagger k} \mathcal{M}_k$, then solve the first equation of (10.23) and (10.35) to get Q_k and P_k, respectively, and go to the next step, else this algorithm is infeasible, stop.

Step 4. If $k \neq 0$, $\Theta_k > 0$, and $\Omega_k > 0$, set $k = k - 1$ and go to Step 2, else go to the next step.

Step 5. If $Q_0 \geq r^2 R$ or $\Theta_k \leq 0$ or $\Omega_k \leq 0$, then this algorithm is infeasible, stop.

Remark 10.4 *In Theorem 10.2.4, a unified framework is established to solve the mobile robot localization problem with both quantization effects and missing measurements. It is worth mentioning that the proposed technique designed is presented by solving certain coupled recursive RDEs. This way, a suboptimal filter is obtained in terms of (10.35) to realize the H_∞-constraint criterion. The advantages of the proposed stochastic H_∞ filter lie in that it can deal with (1) the linearization error and other non-Gaussian disturbances which are not assumed to have statistic characteristics and (2) the negative effects brought by the possibly nonconsecutive measurements containing quantizations and missing observations.*

10.3 Simulation Results

In this section, we present a simulation example for mobile robot localization to illustrate the effectiveness of the proposed filter design scheme. The filter is established to estimate the mobile robot states (position and orientation in planar motion) using the odometry and the information from the marker detection.

Set $N = 300$. Let the sampling period of the robot's odometer be 150 ms, and let the displacement velocity and the angular velocity be 300 mm/s and 5 rad/s, respectively. The initial states are $x_0 = 0.1$ m, $y_0 = 0.1$ m, and $\theta_0 = 0.1$ rad/s. The position of marker M is $x_M = 5$ m, $y_M = 5$ m. The process and measurement errors are chosen as $w_k = 0.02 \sin(100k)$ and $\xi_k = 0.01 \sin(100k)$. Consider the parameters of the logarithmic quantizer as $\hat{\mu}_0^1 = 0.16$, $\hat{\mu}_0^2 = 0.3$, $\chi_1 = 0.6$, and $\chi_1 = 0.35$. The $\mathcal{H}_\infty$ performance level γ, the positive-definite matrix S, and the initial value of ϵ_k are chosen as $\gamma = 1$, $S = \text{diag}\{1, 1, 1\}$, and $\epsilon_k = 0.7$, respectively. Suppose that the probabilistic density functions of α_{1k} and α_{2k} in $[0, 1]$ are described by

$$q_1(s_1) = \begin{cases} 0.2 & s_1 = 0 \\ 0.3 & s_1 = 0.5 \\ 0.5 & s_1 = 1 \end{cases}, \quad q_2(s_2) = \begin{cases} 0.3 & s_2 = 0 \\ 0.3 & s_2 = 0.5 \\ 0.4 & s_2 = 1 \end{cases}, \tag{10.48}$$

from which the expectations and variances can be easily calculated as $\bar{\alpha}_1 = 0.65$, $\bar{\alpha}_2 = 0.45$, $\sigma_1^2 = 0.39$, and $\sigma_2^2 = 0.43$.

Two indices are employed to evaluate the performance of the filter. Let $Z_k = [X_k, Y_k, \Theta_k]$ be the actual position of robot R at moment k. Define

$$E_e := \frac{1}{N} \sum_{k=1}^{N} \sqrt{(\hat{x}_k - X_k)^2 + (\hat{y}_k - Y_k)^2}, \tag{10.49}$$

which stands for the error mean of filtered estimates of R from moment 1 to N, and

$$M_e := \max_{1 \le k \le N} \sqrt{(\hat{x}_k - X_k)^2 + (\hat{y}_k - Y_k)^2}, \tag{10.50}$$

which means the maximum deviation of filtered estimates of R from moment 1 to N.

The simulation results are shown in Figure 10.3 for the robot position and its estimate and in Figure 10.4, for the robot angle and its estimate. The error mean between the actual trajectory and its estimates is $E_e = 0.2732$ m, and the maximum deviation of filtered estimates is $M_e = 0.4863$ m.

Next, in order to illustrate the effectiveness of our results for different measurement missing situations, we consider the case when the probability for multiple missing measurements becomes lower. Take the probabilistic density functions of α_{3k} and α_{4k} in $[0, 1]$ as

$$q_3(s_1) = \begin{cases} 0 & s_1 = 0 \\ 0.1 & s_1 = 0.5 \\ 0.9 & s_1 = 1 \end{cases}, \quad q_4(s_2) = \begin{cases} 0 & s_2 = 0 \\ 0.01 & s_2 = 0.5 \\ 0.99 & s_2 = 1 \end{cases}, \tag{10.51}$$

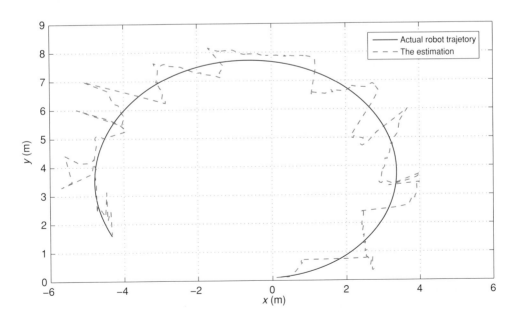

Figure 10.3   Robot trajectory and its estimate

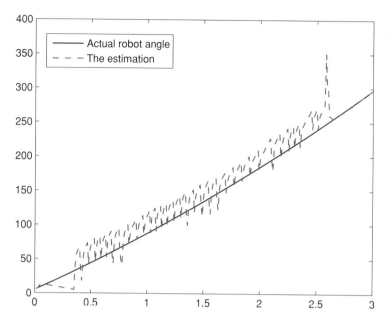

Figure 10.4 Robot angle and its estimate

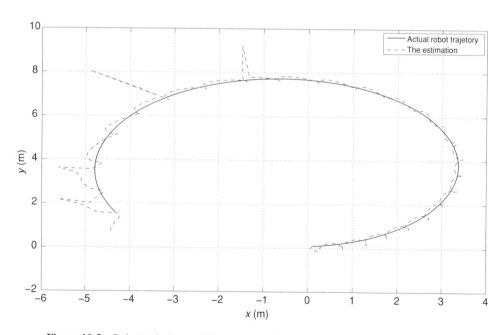

Figure 10.5 Robot trajectory and its estimate when the packet-loss probability is lower

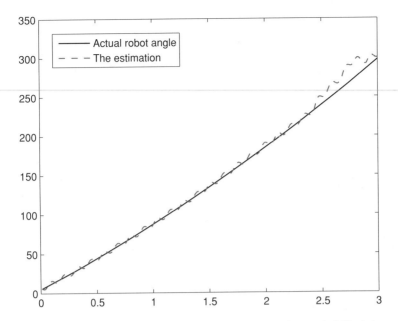

Figure 10.6 Robot angle and its estimate when the packet-loss probability is lower

from which the expectations and variances can be easily calculated as $\bar{\alpha}_3 = 0.95$, $\bar{\alpha}_4 = 0.995$, $\sigma_3^2 = 0.15$, and $\sigma_4^2 = 0.05$. The robot position and its estimate are plotted in Figure 10.5 and the robot angle and its estimate are shown in Figure 10.6. Similarly, the mean error between the actual trajectory and its estimates is $E_e = 0.1647$ m, and the maximum deviation of filtered estimates is $M_e = 0.3711$ m. It can be observed from the simulation results that the lower the multiple packet-loss probabilities are, the better are the accuracy and reliability of the localization performance and, therefore, the more feasible is the stochastic H_∞ filter design problem addressed.

10.4   Summary

In this chapter, the localization problem has been investigated for a mobile robot system with non-Gaussian disturbances, missing measurements, and quantization effects. The missing measurements were modeled by a series of mutually independent random variables satisfying certain probabilistic distributions on the interval [0, 1]. In the framework of logarithmic quantization, by means of stochastic H_∞ filtering technology a new filtering approach has been proposed to ensure the localization of a mobile robot for autonomous navigation equipped with internal sensors and external sensors. Moreover, the filter gains have been explicitly characterized by means of the solvability of certain coupled recursive RDEs. Finally, some simulation results have been provided to show the effectiveness of the proposed localization method.

11

Conclusions and Future Work

This chapter draws conclusions on the book, and points out some possible research directions related to the work done in this book.

11.1 Conclusions

The focus of the book has been placed on filtering, control, and fault detection for some classes of nonlinear systems with randomly occurring incomplete information. Specifically, several research problems have been investigated in detail.

- In Chapter 2, the problem of robust H_∞ finite-horizon filtering and output feedback control problems was discussed for a class of uncertain stochastic nonlinear discrete time-varying systems with error variance constraints, multiple missing measurements, and sensors and actuators subject to saturation. The stochastic nonlinearities under consideration here have been widely used in engineering applications. First, a new robust H_∞ filtering technique was addressed for nonlinear discrete time-varying stochastic systems with norm-bounded uncertainties, multiple missing measurements, and error variance constraints. Sufficient conditions were derived for a finite-horizon filter to satisfy both the estimation error variance constraints and the prescribed H_∞ performance requirement. Subsequently, by using similar analysis techniques, some parallel results were derived for the corresponding robust H_∞ finite-horizon output feedback control problem with both sensor and actuator saturations. Finally, the results of this chapter were demonstrated by some simulation examples.
- In Chapter 3, we studied the robust H_∞ filtering and fuzzy output feedback control problem for nonlinear networked systems with multiple time-varying random communication delays and multiple packet dropouts. The H_∞ filtering problem was first considered for the systems involving parameter uncertainties, state-dependent stochastic disturbances (multiplicative noises or Itô-type noises), multiple stochastic time-varying delays, sector-bounded nonlinearities, and multiple packet dropouts. Sufficient conditions for the robustly exponential stability of the filtering error dynamics were obtained and, at the same time, the prescribed H_∞ disturbance rejection attenuation level was guaranteed. Furthermore, some

Filtering, Control and Fault Detection with Randomly Occurring Incomplete Information, First Edition.
Hongli Dong, Zidong Wang, and Huijun Gao.
© 2013 John Wiley & Sons, Ltd. Published 2013 by John Wiley & Sons, Ltd.

parallel results were also derived for a class of uncertain discrete-time fuzzy systems with both multiple probabilistic delays and multiple missing measurements by using similar analysis techniques. Finally, the results of this chapter were demonstrated by some simulation examples.

- In Chapter 4, the H_∞ filtering and control problems were investigated for systems with repeated scalar nonlinearities and missing measurements. The nonlinear system was described by a discrete-time state equation involving a repeated scalar nonlinearity which typically appears in recurrent neural networks. The missing measurements were modeled by a stochastic variable satisfying the Bernoulli random binary distribution. The quadratic Lyapunov function was used to design both full- and reduced-order H_∞ filters such that, for the admissible random measurement missing and repeated scalar nonlinearities, the filtering error system is stochastically stable and preserves a guaranteed H_∞ performance. Moreover, the multiple missing measurements were included to model the randomly intermittent behaviors of the individual sensors, where the missing probability for each sensor is governed by an individual random variable satisfying a certain probabilistic distribution on the interval [0, 1]. Based on this, an observer-based feedback controller was designed to stochastically stabilize the networked system. Both the stability analysis and controller synthesis problems were investigated in detail. Finally, the results of this chapter were demonstrated by some simulation examples.

- In Chapter 5, the filtering and fault detection problems were investigated for discrete-time MJSs with RVNs and sensor saturation. The issue of RVNs was addressed and the TP matrix considered included the case with polytopic uncertainties and the case with partially unknown TPs. First, the H_∞ filtering problem was investigated, where the randomly occurring nonlinearities were modeled by the Bernoulli-distributed white sequences with known conditional probabilities. Sufficient conditions were derived for the filtering augmented system under consideration to satisfy the H_∞ performance constraint. The corresponding robust H_∞ filters were designed by solving sets of RLMIs. Second, the fault detection problem was investigated for discrete-time MJSs with randomly varying nonlinearities and sensor saturation. Two energy norm indices were used for the fault detection problem, one in order to account for the restraint of disturbance and the other for sensitivity of faults. Based on this, a locally optimized fault detection filter was designed in terms of the solution to certain matrix inequalities. Finally, the results of this chapter were demonstrated by some simulation examples.

- In Chapter 6, the fault detection problems were dealt with for two class of discrete-time nonlinear systems with randomly occurring mixed time-delays, successive packet dropouts, and measurement quantizations. The mixed time-delays involved both multiple time-varying discrete delays and infinite distributed delays. The successive packet dropouts were modeled by a stochastic variable satisfying the Bernoulli random binary distribution. The fault detection problem was first addressed for a class of discrete-time systems with randomly occurring nonlinearities, mixed stochastic time-delays, and measurement quantizations. Sufficient conditions were established via intensive stochastic analysis for the existence of the desired fault detection filters, and then the explicit expression of the desired filter gains was derived by means of the feasibility of certain matrix inequalities. Moreover, the robust fault detection problem was investigated for a class of uncertain discrete-time T–S fuzzy systems comprising randomly occurring mixed time-delays and successive packet dropouts. Some parallel results were also derived by using similar analysis techniques. Two practical examples were provided to show the usefulness and effectiveness of the proposed design methods.

- In Chapter 7, we dealt with the distributed H_∞ filtering problem for a class of nonlinear systems with randomly occurring incomplete information over sensor networks. The incomplete information considered includes both the ROSSs and successive packet dropouts. The issue of ROSSs was addressed to account for the random nature of sensor saturations in a networked environment of sensors, and then the filtering dynamics was analyzed by modeling both the ROSSs and successive packet dropouts in a unified framework. The distributed filters were designed for the filtering dynamics to be exponentially mean-square stable and the filtering errors to satisfy the H_∞ performance constraint. Finally, an illustrative example was provided that highlights the usefulness of the filtering approach developed.

- In Chapter 8, we dealt with the distributed finite-horizon filtering problem for a class of discrete time-varying systems with RVNs over lossy sensor networks involving quantization errors and successive packet dropouts. The filtering dynamics were analyzed by modeling quantization and successive packet dropouts in a unified framework. A new distributed finite-horizon filtering technique by means of a set of recursive linear matrix inequalities was proposed to satisfy the prescribed average filtering performance constraint. Finally, an illustrative example was provided that highlights the usefulness of the developed filtering approach.

- In Chapter 9, we dealt with the distributed H_∞ filtering problem for a class of discrete-time Markovian jump nonlinear time-delay systems with deficient statistics of modes transitions, ROQEs, packet dropouts, and stochastic disturbances in sensor networks. The nonlinearities were described in a Lipschitz-like manner, and the deficient statistics of modes transitions, as a combination of TPs contributing to the practicability of MJSs, was investigated in a unified framework which comprised exactly known TPs, uncertain TPs, and partially unknown TPs. Sufficient conditions were derived for the filtering dynamics system under consideration to satisfy the stochastically stable and the average H_∞ performance constraint. Furthermore, the explicit expression of the desired filter gains was derived. Finally, an illustrative example highlighted the effectiveness of the filtering technology presented in this chapter.

- In Chapter 10, the localization problem was investigated for a mobile robot system with non-Gaussian disturbances, missing measurements, and quantization effects. The missing measurements were modeled by a series of mutually independent random variables satisfying certain probabilistic distributions on the interval [0, 1]. In the framework of logarithmic quantization, by means of stochastic H_∞ filtering technology a new filtering approach was proposed to ensure the localization of a mobile robot for autonomous navigation equipped with internal sensors and external sensors. Moreover, the filter gains were explicitly characterized by means of the solvability of certain coupled recursive RDEs. Finally, some simulation results were provided to show the effectiveness of the proposed localization method.

11.2   Contributions

The main contributions of the book are summarized as follows:

1. The robust filtering problem has been investigated for discrete uncertain nonlinear networked systems with multiple stochastic time-varying communication delays and multiple

missing measurements. A new model has been proposed to describe the multiple network communication delays, each of which satisfies an individual Bernoulli distribution. A combination of important factors contributing to the complexity of NCSs is investigated in a unified framework which comprises partial missing measurements, sector nonlinearities, and parameter uncertainties.

2. The variance-constrained H_∞ filtering problem has been solved for nonlinear time-varying stochastic systems with multiple missing measurements, where a new filtering technique has been developed by employing an algorithm based on the RLMIs which are suitable for online applications.

3. We have addressed and investigated the fault detection problem for discrete-time MJSs with incomplete knowledge of TPs, randomly varying nonlinearities, and sensor saturations. Two energy norm indices have been utilized for the fault detection problem in order to account for the restraint of disturbance and the sensitivity of faults.

4. A new distributed H_∞ filtering problem has been addressed and dealt with for the sensor networks with quantization errors and successive packet dropouts. Also, the RVNs have been proposed to describe the binary switch between two kinds of nonlinear disturbances whose occurrence is governed by a Bernoulli random binary-distributed white sequence with a known conditional probability. We also have proposed a new performance index (i.e., average H_∞ performance), to quantify the overall performance of the distributed filters, which is the extension of the classic H_∞ performance index.

5. We have proposed some new phenomena induced by networks, such as ROMDs and ROQEs. Also, a novel sensor model has been established to account for both the ROMDs and ROQEs in a unified representation which is closer to the real networked environments.

6. A new model of mobile robot localization has been developed that covers general nonlinearities, missing measurements, quantization effects, linearization error, and non-Gaussian noises (in process and measurement), thereby better reflecting the engineering practice. A new stochastic H_∞ filtering approach has been proposed to guarantee the filtering performance of the robot localization process by applying the completing squares method and stochastic analysis technology. The novel coupled recursive RDE approach has been successfully applied to solve the mobile robot localization problem.

11.3   Future Work

Related topics for the future research work are as follows:

- The nonlinearities considered in the book have some condition constraints that bring conservative results. The analysis and synthesis of general nonlinear systems would be one of our future research topics.
- Another future research direction is to investigate multi-objective H_2/H_∞ control and filtering problems for nonlinear systems with randomly occurring incomplete information.
- It would be interesting to investigate the problems of fault detection and fault tolerant control for time-varying systems with randomly occurring incomplete information over a finite time-horizon.

- A trend for future research is to generalize the methods obtained in the book to the control, synchronization, and filtering problems for stochastic nonlinear complex networks with randomly occurring incomplete information.
- Apart from mobile robot navigation, another practical engineering application of the existing theories and methodologies would be fault detection for petroleum well systems.

References

[1] Caballero-Aguila, R., Hermoso-Carazo, A., and Linares-Perez, J. (2012) Least-squares linear estimators using measurements transmitted by different sensors with packet dropouts. *Digital Signal Processing*, **22** (6), 1118–1125.

[2] Caballero-Aguila, R., Hermoso-Carazo, A., and Linares-Perez, J. (2012) Recursive least-squares quadratic smoothing from measurements with packet dropouts. *Signal Processing*, **92** (4), 931–938.

[3] Dong, H., Lam, J., and Gao, H. (2011) Distributed H_∞ filtering for repeated scalar nonlinear systems with random packet losses in sensor networks. *International Journal of Systems Science*, **42** (9), 1507–1519.

[4] Dong, H., Dong, Z., Ho, D.W.C., and Gao, H. (2009) H_∞ fuzzy control for systems with repeated scalar nonlinearities and random packet losses. *IEEE Transactions on Fuzzy Systems*, **17** (2), 440–450.

[5] Shen, B., Wang, Z., Shu, H., and Wei, G. (2011) H_∞ filtering for uncertain time-varying systems with multiple randomly occurred nonlinearities and successive packet dropouts. *International Journal of Robust and Nonlinear Control*, **21** (14), 1693–1709.

[6] Xia, Y., Shang, J., Chen, J., and Liu, G. (2009) Networked data fusion with packet losses and variable delays. *IEEE Transactions on Systems, Man, and Cybernetics, Part B*, **39** (5), 1107–1120.

[7] Kluge, S., Reif, K., and Brokate, M. (2010) Stochastic stability of the extended Kalman filter with intermittent observations. *IEEE Transactions on Automatic Control*, **55** (2), 514–518.

[8] Hu, J., Wang, Z., Gao, H., and Stergioulas, L.K. (2012) Extended Kalman filtering with stochastic nonlinearities and multiple missing measurements. *Automatica*, **48** (9), 2007–2015.

[9] Ishido, Y., Takaba, K., and Quevedo, D.E. (2011) Stability analysis of networked control systems subject to packet-dropouts and finite-level quantization. *Systems & Control Letters*, **60** (5), 325–332.

[10] Ma, W.J. and Gupta, V. (2012) Input-to-state stability of hybrid systems with receding horizon control in the presence of packet dropouts. *Automatica*, **48** (8), 1920–1923.

[11] Moayedi, M., Foo, Y.K., Soh, Y.C., and Nesic, D. (2010) Adaptive Kalman filtering in networked systems with random sensor delays, multiple packet dropouts and missing measurements. *IEEE Transactions on Signal Processing*, **58** (3), 1577–1588.

[12] Nguyen, V.H. and Nguyen, Y.S. (2009) Networked estimation for event-based sampling systems with packet dropouts. *Sensors*, **9** (4), 3078–3089.

[13] Pan, I., Das, S., and Gupta, A. (2011) Handling packet dropouts and random delays for unstable delayed processes in NCS by optimal tuning of (PID mu)-D-lambda controllers with evolutionary algorithms. *ISA Transactions*, **50** (4), 557–572.

[14] Quevedo, D.E. and Nesic, D. (2011) Input-to-state stability of packetized predictive control over unreliable networks affected by packet-dropouts. *IEEE Transactions on Automatic Control*, **56** (2), 370–375.

[15] Tian, Y.C. and Levy, D. (2008) Compensation for control packet dropout in networked control systems. *Information Sciences*, **178** (5), 1263–1278.

[16] Wang, Y. and Yang, G. (2008) Multiple communication channels-based packet dropout compensation for networked control systems. *IET Control Theory & Applications*, **2** (8), 717–727.

Filtering, Control and Fault Detection with Randomly Occurring Incomplete Information, First Edition.
Hongli Dong, Zidong Wang, and Huijun Gao.
© 2013 John Wiley & Sons, Ltd. Published 2013 by John Wiley & Sons, Ltd.

[17] Zhang, L., Yang, L., Guo, L., and Li, J. (2011) Optimal estimation for multiple packet dropouts systems based on measurement predictor. *IEEE Sensors Journal*, **11** (9), 1943–1950.

[18] Wei, G., Wang, Z., and Shu, H. (2009) Robust filtering with stochastic nonlinearities and multiple missing measurements. *Automatica*, **45** (3), 836–841.

[19] Sun, S., Xie, L., Xiao, W., and Soh, Y.C. (2008) Optimal linear estimation for systems with multiple packet dropouts. *Automatica*, **44** (5), 1333–1342.

[20] Sun, S., Xie, L., and Xiao, W. (2008) Optimal full-order and reduced-order estimators for discrete-time systems with multiple packet dropouts. *IEEE Transactions on Signal Processing*, **56** (8), 4031–4038.

[21] Nakamori, S., Caballero-Aguila, R., Hermoso-Carazo, A. *et al.* (2009) Signal estimation with nonlinear uncertain observations using covariance information. *Journal of Statistical Computation and Simulation*, **79** (1), 55–66.

[22] Jimenez-Lopez, J.D., Linares-Perez, J., Nakamori, S. *et al.* (2008) Signal estimation based on covariance information from observations featuring correlated uncertainty and coming from multiple sensors. *Signal Processing*, **88** (12), 2998–3006.

[23] Hounkpevi, F.O. and Yaz, E.E. (2007) Robust minimum variance linear state estimators for multiple sensors with different failure rates. *Automatica*, **43** (7), 1274–1280.

[24] Hounkpevi, F.O. and Yaz, E.E. (2007) Minimum variance generalized state estimators for multiple sensors with different delay rates. *Signal Processing*, **87** (4), 602–613.

[25] Wang, Z., Ho, D.W.C., Liu, Y., and Liu, X. (2009) Robust H_∞ control for a class of nonlinear discrete time-delay stochastic systems with missing measurements. *Automatica*, **45** (3), 684–691.

[26] Gao, H., Zhao, Y., Lam, J., and Chen, K. (2009) H_∞ fuzzy filtering of nonlinear systems with intermittent measurements. *IEEE Transactions on Fuzzy Systems*, **17** (2), 291–300.

[27] Shu, Z., Lam, J., and Xiong, J. (2009) Non-fragile exponential stability assignment of discrete-time linear systems with missing data in actuators. *IEEE Transactions on Automatic Control*, **54**, (3), 625–630.

[28] Xiong, J. and Lam, J. (2007) Stabilization of linear systems over networks with bounded packet loss. *Automatica*, **43** (1), 80–87.

[29] Sahebsara, M., Chen, T., and Shah, S.L. (2007) Optimal H_2 filtering with random sensor delay, multiple packet dropout and uncertain observations. *International Journal of Control*, **80** (2), 292–301.

[30] Zhang, W., Branicky, M., and Phillips, S. (2001) Stability of networked control systems. *IEEE Control Systems Magazine*, **21** (1), 84–99.

[31] Nilsson, J. (1998) Real-time control systems with delays, PhD dissertation, Lund Institute of Technology, Lund, Sweden.

[32] Gökas, F. (2000) Distributed control of systems over communication networks, PhD dissertation, University of Pennsylvania, Philadelphia.

[33] Hua, C., Wang, Q., and Guan, X. (2009) Adaptive fuzzy output-feedback controller design for nonlinear time-delay systems with unknown control direction. *IEEE Transactions on Systems, Man, and Cybernetics, Part B*, **39** (2), 363–374.

[34] Hu, J., Wang, Z., Niu, Y., and Stergioulas, L.K. (2012) H_∞ sliding mode observer design for a class of nonlinear discrete time-delay systems: a delay-fractioning approach. *International Journal of Robust and Nonlinear Control*, **22** (16), 1806–1826.

[35] Gao, H. and Li, X. (2011) H_∞ filtering for discrete-time state-delayed systems with finite frequency specifications. *IEEE Transactions on Automatic Control*, **56** (12), 2935–2941.

[36] Gao, H., Chen, T., and Lam, J. (2008) A new delay system approach to network-based control. *Automatica*, **44** (1), 39–52.

[37] Ma, J. and Sun, S. (2011) Optimal linear estimators for systems with random sensor delays, multiple packet dropouts and uncertain observations. *IEEE Transactions on Signal Processing*, **59** (11), 5181–5192.

[38] Basin, M., Alcorta-Garcia, M.A., and Alanis-Duran, A. (2008) Optimal filtering for linear systems with state and multiple observation delays. *International Journal of Systems Science*, **39** (5), 547–555.

[39] Karimi, H.R., Zapateiro, M., and Luo, N. (2010) A linear matrix inequality approach to robust fault detection filter design of linear systems with mixed time-varying delays and nonlinear perturbations. *Journal of the Franklin Institute*, **347** (6), 957–973.

[40] Shen, B., Wang, Z., Shu, H., and Wei, G. (2009) H_∞ filtering for nonlinear discrete-time stochastic systems with randomly varying sensor delays. *Automatica*, **45** (4), 1032–1037.

[41] Wang, Z. and Ho, D.W.C. (2006) Filtering on nonlinear time-delay stochastic systems. *Automatica*, **39** (1), 101–108.

[42] Chen, B. and Liu, X. (2005) Delay-dependent robust H_∞ control for T–S fuzzy systems with time delays. *IEEE Transactions on Fuzzy Systems*, **13** (4), 544–556.

[43] Liu, M., Ho, D.W.C., and Niu, Y. (2009) Stabilization of Markovian jump linear system over networks with random communication delay. *Automatica*, **45** (2), 416–421.

[44] Zhang, H., Li, M., Yang, J., and Yang, D. (2009) Fuzzy model-based robust networked control for a class of nonlinear systems. *IEEE Transactions on Systems, Man, and Cybernetics, Part A*, **39** (2), 437–447.

[45] Yang, F., Wang, Z., Hung, Y., and Gani, M. (2006) H_∞ control for networked systems with random communication delays. *IEEE Transactions on Automatic Control*, **51** (3), 511–518.

[46] Yue, D., Tian, E., Zhang, Y., and Peng, C. (2009) Delay-distribution-dependent stability and stabilization of T–S fuzzy systems with probabilistic interval delay. *IEEE Transactions on Systems, Man, and Cybernetics, Part B*, **39** (2), 503–516.

[47] Chen, B. and Wornell, G.W. (2001) Quantization index modulation: a class of provably good methods for digital watermarking and information embedding. *IEEE Transactions on Information Theory*, **47** (4), 1423–1443.

[48] Batalin, I.A. and Vilkovisky, G.A. (1983) Quantization of gauge-theories with linearly dependent generators. *Physical Review D*, **28** (10), 2567–2582.

[49] Chang, C.C., Tai, W.L., and Lin, C.C. (2006) A reversible data hiding scheme based on side match vector quantization. *IEEE Transactions on Circuits and Systems for Video Technology*, **16** (10), 1301–1308.

[50] Boche, H. and Moenich, U.J. (2012) Unboundedness of thresholding and quantization for bandlimited signals. *Signal Processing*, **92** (12), 2821–2829.

[51] Ghasemi, N. and Dey, S. (2012) Dynamic quantization and power allocation for multisensor estimation of hidden Markov models. *IEEE Transactions on Automatic Control*, **57** (7), 1641–1656.

[52] Kozica, E. and Kleijn, W.B. (2012) A quantization theoretic perspective on simulcast and layered multicast optimization. *IEEE–ACM Transactions on Networking*, **20** (2), 585–593.

[53] Sun, X. and Coyle, E.J. (2012) Quantization, channel compensation, and optimal energy allocation for estimation in sensor networks. *ACM Transactions on Sensor Networks*, **8** (2), article no. 15.

[54] Shen, B., Wang, Z., Shu, H., and Wei, G. (2010) Robust H_∞ finite-horizon filtering with randomly occurred nonlinearities and quantization effects. *Automatica*, **46** (11), 1743–1751.

[55] Maymon, S. and Oppenheim, A.V. (2012) Quantization and compensation in sampled interleaved multichannel systems. *IEEE Transactions on Signal Processing*, **60** (1), 129–138.

[56] Carli, R., Fagnani, F., Speranzon, A., and Zampieri, S. (2006) Communication constraints in coordinated consensus problems, in *Proceedings of the American Control Conference, Minneapolis, USA*, pp. 4189–4194.

[57] Nakamura, T. (1998) Development of self-learning vision-based mobile robots for acquiring soccer robots behaviors, in *Proceedings of the IEEE International Conference on Robotics and Automation, Leuven, Belgium*, pp. 2592–2598.

[58] Xiao, N., Xie, L., and Fu, M. (2010) Stabilization of Markov jump linear systems using quantized state feedback. *Automatica*, **46** (10), 1696–1702.

[59] Karimi, H.R. (2009) Robust H_∞ filter design for uncertain linear systems over network with network-induced delays and output quantization. *Modeling Identification and Control*, **30** (1), 27–37.

[60] Wang, Z., Shen, B., Shu, H., and Wei, G. (2012) Quantized H_∞ control for nonlinear stochastic time-delay systems with missing measurements. *IEEE Transactions on Automatic Control*, **57** (6), 1431–1444.

[61] Gao, Z. and Ding, S.X. (2007) Actuator fault robust estimation and fault-tolerant control for a class of nonlinear descriptor systems. *Automatica*, **43** (5), 912–920.

[62] Tsai, C.Y. and Song, K.T. (2009) Visual tracking control of a wheeled mobile robot with system model and velocity quantization robustness. *IEEE Transactions on Control Systems Technology*, **17** (3), 520–527.

[63] Dimarogonas, D.V. and Johansson, K.H. (2008) Quantized agreement under time-varying communication topology, in *Proceedings of the American Control Conference, Seattle, USA*, pp. 4376–4381.

[64] Sun, H., Hovakimyan, N., and Başar, T. (2011) L_1 adaptive controller for quantized systems, in *Proceedings of the American Control Conference, San Francisco, USA*, pp. 582–587.

[65] Brockett, R.W. and Liberzon, D. (2000) Quantized feedback stabilization of linear systems. *IEEE Transactions on Automatic Control*, **45** (7), 1279–1289.

[66] Liberzon, D. (2003) On stabilization of linear systems with limited information. *IEEE Transactions on Automatic Control*, **48** (2), 304–307.

[67] Ishii, H. and Francis, B.A. (2002) Stabilizing a linear system by switching control with dwell time. *IEEE Transactions on Automatic Control*, **47** (12), 1962–1973.

[68] Elia, N. and Mitter, S.K. (2001) Stabilization of linear systems with limited information. *IEEE Transactions on Automatic Control*, **46** (9), 1384–1400.

[69] Fu, M. and Xie, L. (2005) The sector bound approach to quantized feedback control. *IEEE Transactions on Automatic Control*, **50** (11), 1698–1711.

[70] Delvenne, J.C. (2006) An optimal quantized feedback strategy for scalar linear systems. *IEEE Transactions on Automatic Control*, **51** (2), 298–303.

[71] Fu, M. and de Souza, C.E. (2009) State estimation for linear discrete-time systems using quantized measurements. *Automatica*, **45** (12), 2937–2945.

[72] Dicarlo, D.A. (2006) Quantitative network model predictions of saturation behind infiltration fronts and comparison with experiments. *Water Resources Research*, **42** (7), W07408, doi:10.1029/2005WR004750.

[73] Gao, W.Z. and Selmic, R.R. (2006) Neural network control of a class of ar systems with actuator saturation. *IEEE Transactions on Neural Networks*, **17** (1), 147–156.

[74] Albertini, F. and Dalessandro, D. (1996) Asymptotic stability of continuous-time systems with saturation nonlinearities. *Systems & Control Letters*, **29** (3), 175–180.

[75] Tsimpanogiannis, I.N. and Yortsos, Y.C. (2004) The critical gas saturation in a porous medium in the presence of gravity. *Journal of Colloid and Interface Science*, **270** (2), 388–395.

[76] Hou, L. and Michel, A.N. (1998) Asymptotic stability of systems with saturation constraints. *IEEE Transactions on Automatic Control*, **43** (8), 1148–1154.

[77] Hu, T. and Lin, Z. (2001) A complete stability analysis of planar discrete-time linear systems under saturation. *IEEE Transactions on Circuits and Systems I: Fundamental Theory and Applications*, **48** (6), 710–725.

[78] Zou, A.M. and Kumar, K.D. (2012) Neural network-based distributed attitude coordination control for spacecraft formation flying with input saturation. *IEEE Transactions on Neural Networks and Learning Systems*, **23** (7), 1155–1162.

[79] Fang, H., Lin, Z., and Hu, T. (2004) Analysis of linear systems in the presence of actuator saturation and L_2 disturbances. *Automatica*, **40** (7), 1229–1238.

[80] Hu, T. and Lin, Z. (2001) *Control Systems with Actuator Saturation: Analysis and Design*, Birkäuser, Boston, MA.

[81] Lv, L. and Lin, Z. (2008) Analysis and design of singular linear systems under actuator saturation and L_2/L_∞ disturbances. *Systems & Control Letters*, **57** (11), 904–912.

[82] Wu, L. and Ho, D.W.C. (2009) Fuzzy filter design for Itô stochastic systems with application to sensor fault detection. *IEEE Transactions on Fuzzy Systems*, **17** (1), 233–242.

[83] Cao, Y., Lin, Z., and Chen, B.M. (2003) An output feedback H_∞ controller design for linear systems subject to sensor nonlinearities. *IEEE Transactions on Circuits and Systems I: Fundamental Theory and Applications*, **50** (7), 914–921.

[84] Wang, Z., Shen, B., and Liu, X. (2012) H_∞ filtering with randomly occurring sensor saturations and missing measurements. *Automatica*, **48** (3), 556–562.

[85] Hu, J., Wang, Z., Gao, H., and Stergioulas, L.K. (2012) Probability-guaranteed H_∞ finite-horizon filtering for a class of nonlinear time-varying systems with sensor saturation. *System & Control Letters*, **61** (4), 477–484.

[86] Ding, D., Wang, Z., Shen, B., and Shu, H. (2012) H_∞ state estimation for discrete-time complex networks with randomly occurring sensor saturations and randomly varying sensor delays. *IEEE Transactions on Neural Networks and Learning Systems*, **23** (5), 725–736.

[87] Zuo, Z., Wang, J., and Huang, L. (2005) Output feedback H_∞ controller design for linear discrete-time systems with sensor nonlinearities. *IEE Proceedings – Control Theory and Applications*, **152** (1), 19–26.

[88] Xiao, Y., Cao, Y., and Lin, Z. (2004) Robust filtering for discrete-time systems with saturation and its application to transmultiplexers. *IEEE Transactions on Signal Processing*, **52** (5), 1266–1277.

[89] Garcia, G., Tarbouriech, S., and Gomes da Silva Jr., J.M. (2007) Dynamic output controller design for linear systems with actuator and sensor saturation, in *Proceedings of the 2007 American Control Conference, New York City, USA*, pp. 5834–5839.

[90] Hu, J., Wang, Z., Gao, H., and Stergioulas, L.K. (2012) Robust sliding mode control for discrete stochastic systems with mixed time-delays, randomly occurring uncertainties and randomly occurring nonlinearities. *IEEE Transactions on Industrial Electronics*, **59** (7), 3008–3015.

[91] Hu, J., Wang, Z., and Gao, H. (2011) A delay fractioning approach to robust sliding mode control for discrete-time stochastic systems with randomly occurring nonlinearities. *IMA Journal of Mathematical Control and Information*, **28** (3), 345–363.

[92] Wang, Z., Wang, Y., and Liu, Y. (2010) Global synchronization for discrete-time stochastic complex networks with randomly occurred nonlinearities and mixed time-delays. *IEEE Transactions on Neural Networks*, **21** (1), 11–25.

[93] Yaz, E.E. and Yaz, Y.I. (2001) State estimation of uncertain nonlinear stochastic systems with general criteria. *Applied Mathematics Letters*, **14** (5), 605–610.

[94] Liu, M. (2009) Network based stability analysis and synthesis for stochastic system with Markovian switching, PhD dissertation, City University of Hong Kong, Hong Kong, China.

[95] Xiong, J., Lam, J., Gao, H., and Ho, D.W.C. (2005) On robust stabilization of Markovian jump systems with uncertain switching probabilities. *Automatica*, **41** (5), 897–903.

[96] Fei, Z., Gao, H., and Shi, P. (2009) New results on stabilization of Markovian jump systems with time delay. *Automatica*, **45** (10), 2300–2306.

[97] Gao, H., Lam, J., Xie, L., and Wang, C. (2005) New approach to mixed H_2/H_∞ filtering for polytopic discrete-time systems. *IEEE Transactions on Signal Processing*, **53** (8,), 3183–3192.

[98] Xiong, J. and Lam, J. (2006) Fixed-order robust H_∞ filter design for Markovian jump systems with uncertain switching probabilities. *IEEE Transactions on Signal Processing*, **54** (4), 1421–1430.

[99] Zhang, X. and Zheng, Y. (2006) Nonlinear H_∞ filtering for interconnected Markovian jump systems. *Journal of Systems Engineering and Electronics*, **17** (1), 138–146.

[100] Zhang, L. (2009) H_∞ estimation for discrete-time piecewise homogeneous Markov jump linear systems. *Automatica*, **45** (11), 2570–2576.

[101] Wu, L., Shi, P., Gao, H., and Wang, C. (2008) H_∞ filtering for 2D Markovian jump systems. *Automatica*, **44** (7), 1849–1858.

[102] Terra, M.H., Ishihara, J.Y., and Jesus, G. (2009) Information filtering and array algorithms for discrete-time Markovian jump linear systems. *IEEE Transactions on Automatic Control*, **54** (1), 158–162.

[103] Liu, H., Sun, F., He, K., and Sun, Z. (2006) Design of reduced-order H_∞ filter for Markovian jumping systems with time delay. *IEEE Transactions on Circuits and Systems II*, **51** (11), 1837–1841.

[104] Ma, L., Da, F., and Zhang, K. (2011) Exponential H_∞ filter design for discrete time-delay stochastic systems with Markovian jump parameters and missing measurements. *IEEE Transactions on Circuits and Systems – Part I: Regular Papers*, **58** (5), 994–1007.

[105] Shi, P., Xia, Y., Liu, G., and Rees, D. (2006) On designing of sliding-mode control for stochastic jump systems. *IEEE Transactions on Automatic Control*, **51** (1), 97–103.

[106] Zhang, J., Xia, Y., and Boukas, E. (2010) New approach to H_∞ control for Markovian jump singular systems. *IET Control Theory & Applications*, **4** (11), 2273–2284.

[107] He, X., Wang, Z., and Zhou, D. (2009) Robust fault detection for networked systems with communication delay and data missing. *Automatica*, **45**, (11), 2634–2639.

[108] Zhong, M., Ye, H., Shi, P., and Wang, G. (2005) Fault detection for Markovian jump systems. *IEE Proceedings – Control Theory and Applications*, **152** (4), 397–402.

[109] Zhang, L., Boukas, E.K., and Lam, J. (2008) Analysis and synthesis of Markov jump linear systems with time-varying delays and partially known transition probabilities. *IEEE Transactions on Automatic Control*, **53** (10), 2458–2464.

[110] Zhao, Y., Zhang, L., Shen, S., and Gao, H. (2011) Robust stability criterion for discrete-time uncertain Markovian jumping neural networks with defective statistics of modes transitions. *IEEE Transactions on Neural Networks*, **22** (1), 164–170.

[111] Shen, B., Wang, Z., and Liu, X. (2011) Bounded H_∞ synchronization and state estimation for discrete time-varying stochastic complex networks over a finite-horizon. *IEEE Transactions on Neural Networks*, **22** (1), 145–157.

[112] Ding, S.X., Jeinsch, T., Frank, P.M. *et al.* (2000) A unified approach to the optimization of fault detection systems. *International Journal of Adaptive Control and Signal Processing*, **14** (7), 725–745.

[113] Zheng, Y., Fang, H., and Wang, H. (2006) Takagi–sugeno fuzzy-model-based fault detection for networked control systems with Markov delays. *IEEE Transactions on Systems, Man, Cybernetics, Part B*, **36** (4), 924–929.

[114] Zhong, M., Ding, S.X., and Ding, E.L. (2010) Optimal fault detection for linear discrete time-varying systems. *Automatica*, **46** (8), 1395–1400.

[115] Wang, Z., Liu, Y., and Liu, X. (2008) H_∞ filtering for uncertain stochastic time-delay systems with sector-bounded nonlinearities. *Automatica*, **44** (5), 1268–1277.

[116] Wei, G., Wang, Z., Shu, H., and Fang, J. (2007) A delay-dependent approach to H_∞ filtering for stochastic delayed jumping systems with sensor non-linearities. *International Journal of Control*, **80** (6), 885–897.

[117] Jiang, X. and Han, Q. (2008) On designing fuzzy controllers for a class of nonlinear networked control systems. *IEEE Transactions on Fuzzy Systems*, **16** (4), 1050–1058.

[118] Wang, Z., Ho, D.W.C., and Liu, X. (2004) A note on the robust stability of uncertain stochastic fuzzy systems with time-delays. *IEEE Transactions on Systems, Man, and Cybernetics, Part A*, **34** (4), 570–576.

[119] Zhang, W., Chen, B., and Tseng, C. (2005) Robust H_∞ filtering for nonlinear stochastic systems. *IEEE Transactions on Signal Processing*, **53** (2), 589–598.

[120] Olfati-Saber, R. (2006) Flocking for multi-agent dynamic systems: algorithms and theory. *IEEE Transactions on Automatic Control*, **51** (3), 401–420.

[121] Yu, W., Chen, G., Wang, Z., and Yang, W. (2009) Distributed consensus filtering in sensor networks. *IEEE Transactions on Systems, Man, and Cybernetics, Part B*, **39** (6), 1568–1577.

[122] Cattivelli, F.S. and Sayed, A.H. (2010) Diffusion strategies for distributed Kalman filtering and smoothing. *IEEE Transactions on Automatic Control*, **55** (9), 1520–1533.

[123] Kamgarpour, M. and Tomlin, C. (2008) Convergence properties of a decentralized Kalman filter, in *Proceedings of the 47th IEEE Conference on Decision and Control, Cancun, Mexico*, pp. 3205–3210.

[124] Olfati-Saber, R. (2009) Kalman-consensus filter optimality stability and performance, in *Proceedings of the 48th IEEE Conference on Decision and Control, Shanghai, China*, pp. 7036–7042.

[125] Shen, B., Wang, Z., and Liu, X. (2011) A stochastic sampled-data approach to distributed H_∞ filtering in sensor networks. *IEEE Transactions on Circuits and Systems I: Regular Papers*, **58** (9), 2237–2246.

[126] Shen, B., Wang, Z., Hung, Y.S., and Chesi, G. (2011) Distributed H_∞ filtering for polynomial nonlinear stochastic systems in sensor networks. *IEEE Transactions on Industrial Electronics*, **58** (5), 1971–1979.

[127] Farina, M., Ferrari-Trecate, G., and Scattolini, R. (2009) Distributed moving horizon estimation for sensor networks, in *Proceedings of 1st IFAC Workshop on Estimation and Control of Networked Systems, Venice, Italy*, pp. 126–131.

[128] Shi, P., Mahmoud, M., Nguang, S.K., and Ismail, A. (2006) Robust filtering for jumping systems with mode-dependent delays. *Signal Processing*, **86** (1), 140–152.

[129] Zhang, H., Feng, G., Duan, G., and Lu, X. (2006) H_∞ filtering for multiple-time-delay measurements. *IEEE Transactions on Signal Processing*, **54** (5), 1681–1688.

[130] Ribeiro, A., Giannakis, G.B., and Roumeliotis, S.I. (2006) SOI-KF: distributed Kalman filtering with low-cost communications using the sign of innovations. *IEEE Transactions on Signal Processing*, **54** (12), 4782–4795.

[131] Hong, L. (1994) Multiresolutional distributed filtering. *IEEE Transactions on Automatic Control*, **39** (4), 853–856.

[132] Tanaka, N., Snyder, S.D., and Hansen, C.H. (1996) Distributed parameter model filtering using smart sensors. *Journal of Vibration and Acoustics: Transactions of the ASME*, **118** (4), 630–640.

[133] Cortes, J. (2009) Distributed kriged Kalman filter for spatial estimation. *IEEE Transactions on Automatic Control*, **54** (12), 2816–2827.

[134] Cheng, P.T. and Lee, T.L. (2006) Distributed active filter systems (DAFSs): a new approach to power system harmonics. *IEEE Transactions on Industry Applications*, **42** (5), 1301–1309.

[135] Oka, A. and Lampe, L. (2008) Energy efficient distributed filtering with wireless sensor networks. *IEEE Transactions on Signal Processing*, **56** (5), 2062–2075.

[136] Ugrinovskii, V. (2011) Distributed robust filtering with H_∞ consensus of estimates. *Automatica*, **47** (1), 1–13.

[137] Assimakis, N.D. (2001) Optimal distributed Kalman filter. *Nonlinear Analysis: Theory Methods & Applications*, **47** (8), 5367–5378.

[138] Theilmann, W. and Rothermel, K. (2000) Optimizing the dissemination of mobile agents for distributed information filtering. *IEEE Concurrency*, **8** (2), 53–61.

[139] Carli, R., Chiuso, A., Schenato, L., and Zampieri, S. (2008) Distributed Kalman filtering based on consensus strategies. *IEEE Journal on Selected Areas in Communications*, **26** (4), 622–633.

[140] Cattivelli, F.S. and Sayed, A.H. (2008) Diffusion strategies for distributed Kalman filtering: formulation and performance analysis, in *Proceedings, Cognitive Information Processing, Santorini, Greece*, pp. 36–41.

[141] Olfati-Saber, R. (2007) Distributed Kalman filtering for sensor networks, in *Proceedings of the 46th IEEE Conference on Decision and Control, New Orleans, LA*, pp. 5492–5498.

[142] Olfati-Saber, R. and Shamma, J.S. (2005) Consensus filters for sensor networks and distributed sensor fusion, in *Proceedings of the 44th IEEE Conference on Decision and Control, and the European Control Conference, Seville, Spain*, pp. 6698–6703.

[143] Shen, B., Wang, Z., and Hung, Y.S. (2010) Distributed consensus H_∞ filtering in sensor networks with multiple missing measurements: the finite-horizon case. *Automatica*, **46** (10), 1682–1688.

[144] Ugrinovskii, V. (2010) Distributed robust filtering with H_∞ consensus of estimations, in *Proceedings of the 2010 American Control Conference, Maryland, USA*, pp. 1374–1379.

[145] Nanacara, W. and Yaz, E.E. (1997) Recursive estimator for linear and nonlinear systems with uncertain observations. *Signal Processing*, **62** (2), 215–228.

[146] Yaz, Y.I. and Yaz, E.E. (1999) On LMI formulations of some problems arising in nonlinear stochastic system analysis. *IEEE Transactions on Automatic Control*, **44** (4), 813–816.

[147] Mao, W. and Chu, J. (2009) Stability and stabilization of linear discrete time-delay systems with polytopic uncertainties. *Automatica*, **45** (3), 842–846.

[148] Yang, F. and Li, Y. (2009) Set-membership filtering for systems with sensor saturation. *Automatica*, **45** (8), 1896–1902.

[149] Boyd, S., Ghaoui, L.E., Feron, E., and Balakrishnan, V. (1994) *Linear Matrix Inequalities in System and Control Theory*, SIAM, Philadelphia, PA.

[150] Khalil, H.K. (1996) *Nonlinear Systems*, Prentice-Hall, Upper Saddle River, NJ.

[151] Han, Q.L. (2005) Absolute stability of time-delay systems with sector-bounded nonlinearity. *Automatica*, **41** (12), 2171–2176.

[152] Lam, J., Gao, H., Xu, S., and Wang, C. (2005) H_∞ and L_2/L_∞ model reduction for system input with sector nonlinearities. *Journal of Optimization Theory and Applications*, **125** (1), 137–155.

[153] He, Y., Wang, Q.G., Lin, C., and Wu, M. (2007) Delay-range-dependent stability for systems with time-varying delay. *Automatica*, **43** (2), 371–376.

[154] Liu, Y., Wang, Z., Liang, J., and Liu, X. (2009) Stability and synchronization of discrete-time Markovian jumping neural networks with mixed mode-dependent time-delays. *IEEE Transactions on Neural Networks*, **20** (7), 1102–1116.

[155] Guan, X. and Chen, C. (2004) Delay-dependent guaranteed cost control for T–S fuzzy systems with time delays. *IEEE Transactions on Fuzzy Systems*, **12** (2), 236–249.

[156] El Ghaoui, L., Oustry, F., and Rami, M.A. (1997) A cone complementarity linearization algorithm for static output-feedback and related problems. *IEEE Transactions on Automatic Control*, **42** (8), 1171–1176.

[157] Tanaka, K., Ikeda, T., and Wang, H.O. (1996) Robust stabilization of a class of uncertain nonlinear systems via fuzzy control: quadratic stabilizability, H_∞ control theory, and linear matrix inequalities. *IEEE Transactions on Fuzzy Systems*, **4** (1), 1–13.

[158] Gao, H., Lam, J., and Wang, C. (2005) Induced l_2 and generalized H_2 filtering for systems with repeated scalar nonlinearities. *IEEE Transactions on Signal Processing*, **53** (11), 4215–4226.

[159] Chu, Y. and Glover, K. (1999) Bounds of the induced norm and model reduction errors for systems with repeated scalar nonlinearities. *IEEE Transactions on Automatic Control*, **44** (3), 4215–4226.

[160] Beck, C., Andrea, R.D., Paganini, F. *et al.* (1996) A statespace theory of uncertain systems, in *Proceedings of the 13th Triennial IFAC World Congress, San Francisco*, pp. 291–296.

[161] Chu, Y. (2001) Further results for systems with repeated scalar nonlinearities. *IEEE Transactions on Automatic Control*, **44** (12), 2031–2035.

[162] Chen, B.S. and Wang, S.S. (1986) The design of feedback controller with nonlinear saturating actuator: time domain approach, in *Proceedings of the 25th Conference on Decision and Control, Athens, Greece*, pp. 2048–2053.

[163] Krikelis, N.J. and Barkas, S.K. (1984) Design of tracking systems subject to actuator saturation and integrator wind-up. *International Journal of Control*, **39** (4), 667–682.

[164] Cao, J. and Wang, J. (2003) Global asymptotic stability of a general class of recurrent neural networks with time-varying delays. *IEEE Transactions on Circuits and Systems I: Fundamental Theory and Applications*, **50** (1), 34–44.

[165] Michel, A.N., Si, J., and Yen, G. (1991) Analysis and synthesis of a class of discrete-time neural networks described on hypercubes. *IEEE Transactions on Neural Networks*, **2** (1), 32–46.

[166] Mahmoud, M.S. (2000) *Robust Control and Filtering for Time-Delay Systems*, Marcel-Dekker, New York, NY.

[167] Liu, D. and Michel, A.N. (1992) Asymptotic stability of discrete-time systems with saturation nonlinearities with applications to digital filters. *IEEE Transactions on Circuits and Systems I*, **39** (10), 789–807.

[168] Tamura, H. (1975) Decentralized optimization for distributed-lag models of discrete systems. *Automatica*, **11** (6), 593–602.

[169] Wang, Z., Yang, F., Ho, D.W.C., and Liu, X. (2006) Robust H_∞ filtering for stochastic time-delay systems with missing measurements. *IEEE Transactions on Signal Processing*, **54** (7), 2579–2587.

[170] Shen, B., Wang, Z., Shu, H., and Wei, G. (2008) On nonlinear H_∞ filtering for discrete-time stochastic systems with missing measurements. *IEEE Transactions on Automatic Control*, **53** (9), 2170–2180.

[171] El Ghaoui, L., Oustry, F., and Rami, M.A. (1997) A cone complementarity linearization algorithm for static output-feedback and related problems. *IEEE Transactions on Automatic Control*, **42** (8), 1171–1176.

[172] Wang, Z., Yang, F., Ho, D.W.C., and Liu, X. (2007) Robust H_∞ control for networked systems with random packet losses. *IEEE Transactions on Systems, Man, and Cybernetics, Part B*, **37** (4), 916–924.

[173] Palm, R. and Driankov, D. (1998) Fuzzy switched hybrid systems-modeling and identification, in *Proceedings of the IEEE ISIC/CIRA/ISAS Joint Conference, Gaithersburg, MD*, pp. 130–135.

[174] Wu, H. and Cai, K. (2006) Mode-independent robust stabilization for uncertain Markovian jump nonlinear systems via fuzzy control. *IEEE Transactions on Systems, Man, and Cybernetics, Part B*, **36** (3), 509–519.

[175] Feng, G. (2009) Nonsynchronized state estimation of discrete time piecewise linear systems. *IEEE Transactions on Signal Processing*, **54** (1), 295–303.

[176] Zuo, Z., Ho, D.W.C., and Wang, Y. (2010) Fault tolerant control for singular systems with actuator saturation and nonlinear perturbation. *Automatica*, **46** (3), 569–576.

[177] Hung, Y.S. and Yang, F. (2003) Robust H_∞ filtering with error variance constraints for discrete time-varying systems with uncertainty. *Automatica*, **39** (7), 1185–1194.

[178] Tarn, T.J. and Rasis, Y. (1976) Observers for nonlinear stochastic systems. *IEEE Transactions on Automatic Control*, **21** (4), 441–448.

[179] Jetto, L., Longhi, S., and Venturini, G. (1999) Development and experimental validation of an adaptive extended Kalman filter for the localization of mobile robots. *IEEE Transactions on Robotics and Automation*, **15** (2), 219–229.

[180] Yang, F., Wang, Z., Lauria, S., and Liu, X. (2009) Mobile robot localization using robust extended H_∞ filtering. *Proceedings of IMechE, Part I: Journal of Systems and Control Engineering*, **223** (8), 1067–1080.

[181] Hung, Y.S. and Yang, F. (2002) Robust H_∞ filtering for discrete time-varying uncertain systems with a known deterministic input. *International Journal of Control*, **75** (15), 1159–1169.

[182] Rami, M.A., Chen, X., and Zhou, X. (2002) Discrete-time indefinite LQ control with state and control dependent noises. *Journal of Global Optimization*, **23** (3–4), 245–265.

Index

Filtering, Control and Fault Detection with Randomly Occurring Incomplete Information, First Edition.
Hongli Dong, Zidong Wang, and Huijun Gao.
© 2013 John Wiley & Sons, Ltd. Published 2013 by John Wiley & Sons, Ltd.